Emerging Technologies and Solutions for the Sustainable Climate Change Challenges

Emerging Technologies and Solutions for the Sustainable Climate Change Challenges

Editor

Ji Whan Ahn

MDPI • Basel • Beijing • Wuhan • Barcelona • Belgrade • Manchester • Tokyo • Cluj • Tianjin

Editor
Ji Whan Ahn
Korea Institute of Geoscience and Mineral Resources (KIGAM)
Korea

Editorial Office
MDPI
St. Alban-Anlage 66
4052 Basel, Switzerland

This is a reprint of articles from the Special Issue published online in the open access journal *Sustainability* (ISSN 2071-1050) (available at: https://www.mdpi.com/journal/sustainability/special_issues/Climate_Change_Challenges).

For citation purposes, cite each article independently as indicated on the article page online and as indicated below:

LastName, A.A.; LastName, B.B.; LastName, C.C. Article Title. *Journal Name* **Year**, *Volume Number*, Page Range.

ISBN 978-3-03928-995-0 (Hbk)
ISBN 978-3-03928-996-7 (PDF)

Cover image courtesy of Ji Whan Ahn.

Contents

About the Editor

Ji Whan Ahn holds B.S., M.S., and Ph.D. degrees in mining and mineral engineering from Inha University and another master's degree in economics from Yonsei University. She also completed CDM research at Seoul National University's Graduate School of Public Administration. Dr. Ahn is the Head of the Center for Carbon Mineralization of the Korea Institute of Geosciences and Mineral Resources (KIGAM), the Chairman of the Korea Institute of Resources Recycling, the Vice-President of the Korea Society of Mineral and Energy Resources Engineering and the Vice-President of the Korea Society for Energy. She is working as a professor in the Resources Recycling Department of UST (University of Science & Technology). She has 30 years of research experience in the field of Carbon Mineralization in KIGAM and she was a representative of Korea of ISO 82 and ISO 102. She has received many remarkable awards like the following.

In 2004, she was awarded the "Scientist of the Month Award" from the Ministry of Science and ICT (MSIT) for the investigation of the mechanism of the global Aragonite Precipitated Calcium Carbonate (PCC) size and shape control technology. By developing mechanisms and core technologies related to the stabilization of environmentally hazardous heavy metals and material treatment using carbon dioxide and inorganic mineral by-products generated from industry and society, she won "The 6th Female Scientist Award" (Engineering Field) from the Ministry of Science and ICT(MSIT) in 2006.

In addition, in 2010, she won the "Best Basic Research Achievement and Technology Award" for research on recycling inorganic waste through carbonation of greenhouse gases, and in 2011, she was awarded "The 21st Science and Technology Best Paper Award" by the Ministry of Science and ICT (MSIT) for research on improving the physical properties of recycled pulp utilizing carbon dioxide using an eco-friendly in-situ process. She received the "Presidential citation for merit" from the Ministry of Science and ICT (MSIT) in 2013 in recognition of her achievement in world-class research on limestone and carbon dioxide which is the 'establishment of the world's first proof of mineral processing of inorganic waste mineral based on the core technology of accelerated carbonation'. In 2018, in recognition of the contribution of climate technology to overseas cooperation, she was awarded the "Commendation for merit" from the minister of Ministry of Science and ICT (MSIT). Also, in 2019, she was awarded the "Commendation for merit" for contributing to environmental science technology by contributing to the success of the Environmental Olympics through pilot production as a representative technology for the application of eco-friendly paper, a global core technology for carbon mineralization.

Preface to "Emerging Technologies and Solutions for the Sustainable Climate Change Challenges"

Carbon neutrality has emerged as a "global new paradigm" accepted by major economic countries such as the European Union, China, and Japan. Considering the seriousness of ever-growing climate change consequences, countries this year collectively declare "carbon neutrality" as an urgent and prominent step against greenhouse gas emissions. The Republic of Korea is also following a preemptive approach, changing a greenhouse gas reduction structure to a new and proactive economic and social structure through the "2050 Carbon Neutral Promotion Strategy." Indeed, these achievements are in dire need of a scientific approach.

Korea being a country with energy-intensive industries needs to develop new and effective proposals such as carbon border tax. However, post-Paris Agreement, the Ministry of Science and ICT has been assigned as the National Designated Entity (NDE) in charge of greenhouse gases in Korea. Accordingly, the National Flagship project (Mineral carbonation) has being carried out in accordance of the Ministry of Trade, Industry and Energy and the Ministry of Environment. In particular, UNFCCC announced a technology that can reduce greenhouse gases through the production of CSA cement. The UNFCCC Methodologies Panel 82th Meeting recommended approval, and CDM Executive Board 107th Meeting announced final approval of the new methodology AM0121. These new technology studies were categorized into greenhouse gas reduction and adaptation. This emerging technology will be used as a fundamental study of 17 sustainable goals as well as reducing carbon dioxide through a lot of cooperation among Asian and African countries.

This special issue focuses on technology reducing the greenhouse gases by carbon mineralization using by-products and CO_2 generated (from a coal power plant, cement, steel industry, etc.) in the multi-energy/multi-resource consumption industry. In addition, topics such as producing valuable products by utilizing associated technologies, technology introduction, and strategy establishment to secure carbon dioxide reductions overseas such as in Vietnam, Philippines, etc. were introduced. Carbon mineralization technology is a technology that utilizes calcium (Ca) and magnesium (Mg) dominant industrial by-products (e.g., power generation ash, slag, etc.) which form stable carbonates with carbon dioxide through a direct carbonation reaction. Potentially, the carbon mineralization technology has the advantage of storing a large amount of CO_2 with high stability. With the development of this technology, we will demonstrate the potential applications of the aforementioned by-products in the Republic of Korea, along with utilization as fillers in abandoned mines. Furthermore, we intended to reduce CO_2 in multi-energy industries and develop product technology applicable to the industries, while securing overseas reductions through technology transfer in developing countries such as Vietnam and the Philippines.

This Special Issue consists of 20 papers with contents such as carbon mineralization technology including the production of complex carbonates using inorganic by-products, high value-added technology utilizing low-concentration CO_2, demonstration of technology for manufacturing calcium sulfo-aluminate (CSA) cement and filling materials for abandoned mines, environmental monitoring of CO_2 and abandoned mine filler material construction technology, CDM registration and CTCN-TA project for overseas transfer of carbon mineralization technology.

Post-publication of this e-book, we intend to update the world regarding the latest research trends in the carbon mineralization technology developed by the Carbon Mineralization Flagship Project with few more continuous publications. Future publications will showcase a new vision for the development of recycling technologies in response to climate change and the fourth industrial revolution through carbon mineralization technology. Through this, we plan to build a new paradigm by taking the lead in securing resources and responding to climate change.

Ongoing research, achievements and accomplishments include:

– Research and demonstration of carbonates production and high value-added/appropriate/package/engineering technology utilizing low-concentration CO_2.

– An empirical study on the application of high value-added carbonates derived from inorganic by-products such as household combustion ash.

– Construction and verification of cover cement production and filler material for abandoned mine.

– Environment/ground improvement/CO_2 monitoring study of backfill construction at abandoned mine.

– Study on applicability of carbon mineralization technology through international mechanism.

– Research on new CDM methodology for carbon mineralization.

– A study of carbon mineralization technology achievement, economic feasibility, LCA analysis and policy strategy in domestic and overseas.

Ji Whan Ahn
Editor

 *sustainability*

Article

Semi-Dry Carbonation Process Using Fly Ash from Solid Refused Fuel Power Plant

Jung Hyun Kim and Woo Teck Kwon *

Energy & Environmental Division, Korea Institute of Ceramic Engineering & Technology (KICET), 101 Soho-ro, Jinju-si, Gyeongsangnam-do 52581, Korea; junghyunkim@kicet.re.kr
* Correspondence: wtkwon@kicet.re.kr

Received: 22 January 2019; Accepted: 8 February 2019; Published: 11 February 2019

Abstract: The increasing CO_2 concentration in the Earth's atmosphere, mainly caused by fossil fuel combustion, has led to concerns about global warming. Carbonation is a technique that can be used as a carbon capture and storage (CCS) technology for CO2 sequestration. In this study, the utilization of the fly ash from a solid refused fuel (SRF) power plant as a solid sorbent material for CO_2 capture via semi-dry carbonation reaction was evaluated as a simple process to reduce CO_2. The fly ash was exposed to accelerated carbonation conditions at a relative humidity of 25, 50, 75, and 100%, to investigate the effects of humidity on the carbonation kinetics of the fly ash. The reaction conditions such as moisture, concentration of CO_2, and reaction time can affect CO_2 capture performance of fly ash. Due to a short diffusion length of H_2CO_3 in water, the semi-dry process exhibits faster carbonation reaction than the wet process. Especially, the semi-dry process does not require a wastewater treatment plant because it uses a small amount of water. This study may have important implications, illustrating the possibility of replacing the wet process with the semi-dry process.

Keywords: SRF; fly ash; carbon dioxide; carbonation; CO_2 capture

---

1. Introduction

CO_2 is a main greenhouse gas and undoubtedly a major contributor to global warming. Capturing CO_2 from the atmosphere is an essential parameter of the carbon management for sequestrating CO_2 from our environment. The concentration of CO_2 in our atmosphere is promoted by the combustion of fossil fuels for generating electricity [1,2]. In addition, the amount of the fly ash produced from power plants is expected to increase continuously as the demand for electricity increases. Thus, the interest in fly ash utilization has increased [3–6].

Among many CO_2 reduction and sequestration techniques, solid-looping using calcium oxide (CaO) is a promising CO_2 capture process, known as carbonate looping [7–13]. This process is based on the reversible reaction between CaO and CO_2 to form calcium carbonate. The carbonation reaction can be used to remove CO_2 from the atmosphere. Derevschikov et al. prepared a CaO/Y_2O_3 sorbent to capture CO_2 at high temperatures [14]. J. Shi et al. synthesized a CaO/sepiolite sorbent by the hydration reaction [15]. However, these manufacturing processes are complex and uneconomical for industrial application.

Meanwhile, solid refused fuel (SRF), a highly heterogeneous mixture of high calorific fraction of non-hazardous waste materials, has been recognized as a viable alternative to fossil fuels, and is already being used as a fuel in various industrial sectors, including power plants [16]. The fly ash from SRF plants contains about 20 wt% of lime (CaO) which can be used to sequester CO_2 by aqueous carbonation. Thus, studies about utilizing fly ash as a solid sorbent material for CO_2 capture have increased. In the carbonation reaction between CaO and CO_2, water plays an essential role in hydrating the calcium-based materials to form calcium hydroxide and most studies were conducted in environments with a high amount of water. Loo et al. conducted accelerated wet carbonation experiments on circulating fluidized bed combustion (CFBC) ash (the prepared solution possessed an ash-to-water ratio of 1:10) [17]. Dananjayan et al. and Ebrahimi et al. conducted aqueous carbonation with the water-to-ash ratio of 15:1 with carbonation capacities of 50.3 g and 32 g of CO_2/kg of fly ash, respectively [18,19].

However, although a $Ca(OH)_2$ aqueous solution can be effectively used as an absorbent to capture CO_2, commercialization of the CO_2 capture process is still hindered because of the high amount of energy required to evaporate water. Moreover, the process needs a wastewater treatment plant which can be a source of greenhouse gasses.

In this study, we investigated the utilization of fly ash as a solid sorbent material for CO_2 capture via semi-dry carbonation reaction. The effects of the amount of moisture, CO_2 concentration, and reaction time on the performance of fly ash was investigated. The rest of the paper is structured as follows: Section 2 describes the experimental framework used to evaluate the semi-dry carbonation. Section 3 contains the composition changes after the carbonation of SRF ash. Sections 4 and 5 present our discussions and conclusions.

2. Materials and Methods

2.1. Materials

Fly ash obtained from Korea District Heating Corporation in Korea was used for this research. The fly ash was dried in an oven at 100 °C for 10 h to remove moisture.

2.2. Semi-Dry Carbonation Reactor

CO_2 capture was conducted in a round-bottom flask with a single neck as shown in Figure 1. The mixture of fly ash (200 g) and water was delivered into the flask. The rotation speed of the flask was fixed at 40 RPM to shake the mixture. The CO_2 stream (99.99%) and N_2 stream (99.99%) were flowed into the reactor using the regulator and the flow meter to control the CO_2 concentration. The total flow rate of the mixture gas was 10 L min^{-1}. The temperature in the reactor was maintained at 25 °C. The flow chart of the experimental method of semi-dry carbonation is shown in Figure 2.

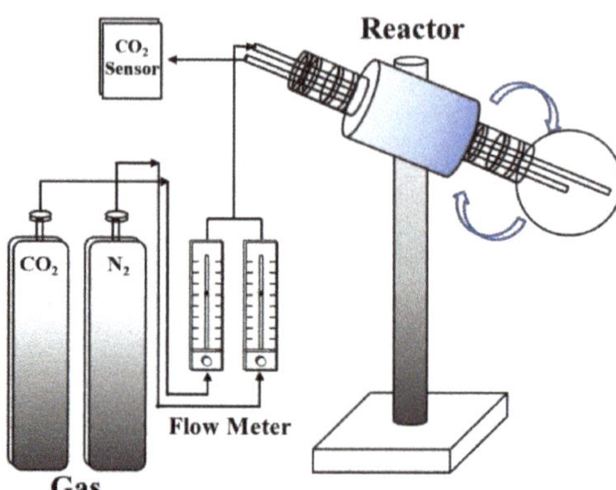

Figure 1. Schematic diagram of the semi-dry carbonation system for the CO_2 capture.

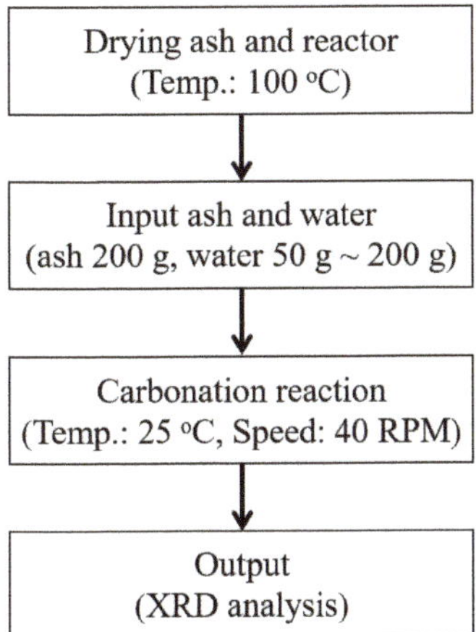

Figure 2. Experimental method of semi-dry carbonation.

2.3. Characterizations

The morphologies and structures of the fly ash were analyzed by SEM (JSM-6700F, JEOL, Tokyo, Japan). The crystal structures were investigated by XRD (X'pert^3 Powder, Malvern PANalytical) with Cu Ka radiation (λ = 1.5418 Å) at the Ssangyong Cement Co. (Daejeon). The particle size was measured using a laser diffraction particle size analyzer (PSA) (Mastersizer S Ver. 2.15, Malvern Instruments, Malvern, UK). The elemental compositions of the fly ash were investigated using X-ray fluorescence analysis (XRF, Primus2, Rigaku, Japan).

3. Results

When SRF is burned in a bottom boiler, most of the unburned material is caught in the flue gas and captured as fly ash. Bottom ash is an incombustible byproduct that is collected from the bottom of the furnaces that burn SRF for generating steam. Therefore, fly ash and bottom ash are quite different physically and chemically. The fly ash and the bottom ash used in the study are shown in Figure 3.

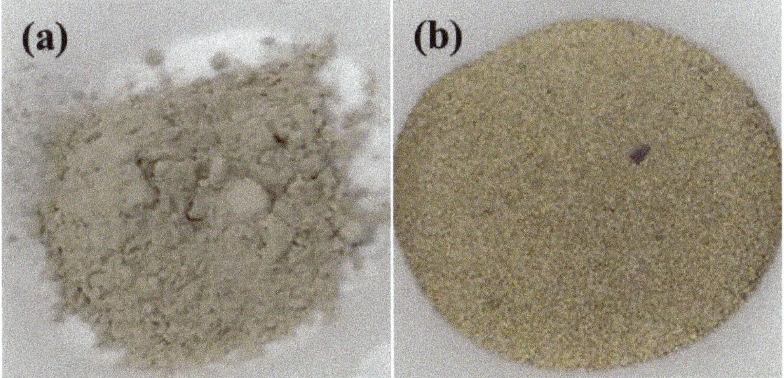

Figure 3. Photographs of (**a**) fly ash and (**b**) bottom ash.

The fly ash was composed of fine particles, while the bottom ash was coarse and granular. The color of the bottom ash was dark, dull brown while the color of the fly ash was light brown. Figure 4 shows the morphologies of the fly ash and the bottom ash. The microstructure of the bottom ash (Figure 4b) had a much larger particle size compared to fly ash, which was about the size of sand but was more porous. As shown in Figure 5, the particle size of the fly ash was between 2 and 130 μm, with D50 (medium diameter) of 25 μm, D10 of 7 μm and D90 of 160 μm.

Figure 4. Morphologies of the (**a**) fly ash and (**b**) bottom ash.

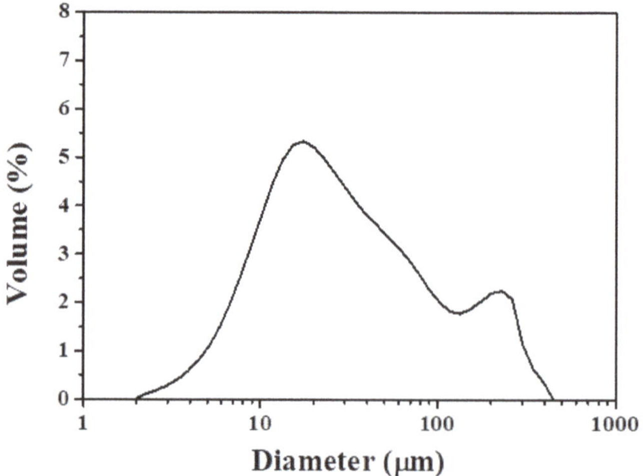

Figure 5. Particle size distribution of fly ash.

Table 1 shows the results of X-ray fluorescence analysis of the fly ash and bottom ash. Concentration of the major elements in the fly ash was given in the form of oxides by XRF. The chemical compositions of the fly ash showed that SiO_2, Al_2O_3, Fe_2O_3, CaO, and Na_2O were the main oxides. On the other hand, the major chemical composition of the bottom ash was SiO_2. Lead and copper were the most common heavy metals in both ashes (Table 2).

Table 1. Oxide compositions in fly ash and bottom ash (wt%).

	SiO_2	Al_2O_3	Fe_2O_3	CaO	MgO	N_2O	K_2O	SO_3	P_2O_5	Cl
Fly Ash	24.9	13.2	2.56	17.1	1.82	13.1	2.36	1.29	2.96	12.8
Bottom Ash	85.0	2.90	0.87	7.27	0.50	1.01	0.35	0.18	0.94	0.2

Table 2. Heavy metals in fly ash and bottom ash (mg kg^{-1}).

	Pb	Cd	As	Hg	Cu
Fly Ash	785	33	N.D.	N.D.	5620
Bottom Ash	74	N.D.	N.D.	N.D.	2240

The content of CaO (17.1%) in the fly ash was much higher than that in the bottom ash (7.27%). Therefore, the fly ash, which has high calcium-based contents, was selected for CO_2 capture.

Figure 6 shows the crystal structure of the fly ash. The XRD pattern in Figure 6 confirms the formation of composite powders with mixed crystal structures consisting of periclase (MgO), lime (CaO), calcite ($CaCO_3$), anhydrite ($CaSO_4$), quartz (SiO_2), mullite ($3Al_2O_3$-$2SiO_2$), hematite (Fe_2O_3), gehlenite ($Ca_2Al[AlSiO_7]$), halite (NaCl), and sylvite (KCl). The calcium-based materials such as calcite, anhydrite, and gehlenite could be prepared by a high temperature combustion. By using the relative intensity ratio (RIR) technique from the XRD we found that the weight percentage of the periclase, lime, calcite, anhydrite, quartz, mullite, hematite, gehlenite, halite, and sylvite were 0.89, 0.40, 3.02, 7.57, 12.78, 0.54, 2.27, 24.58, 42.46, and 5.48%, respectively (Table 3).

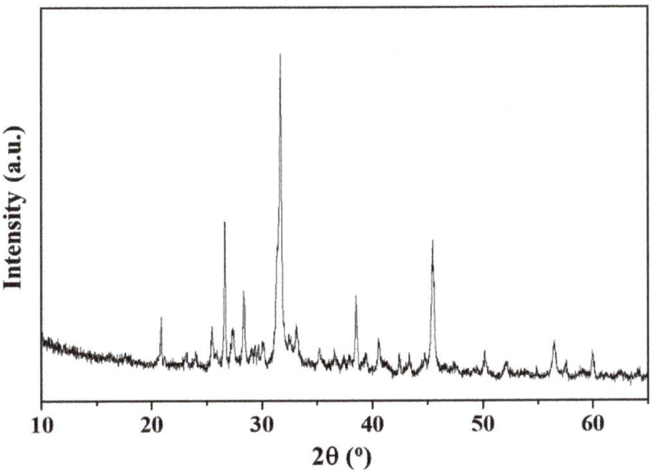

Figure 6. X-ray diffraction pattern of fly ash.

Table 3. Quantitative analytical results obtained by relative intensity ratio (RIR) method (wt%) of the fly ash.

Periclase	Lime	Calcite	Anhydrite	Quartz	Mullite	Hematite	Gehlenite	Halite	Sylvite
0.89	0.40	3.02	7.57	12.78	0.54	2.27	24.58	42.46	5.48

Among the above materials, calcium-based materials (lime, anhydrite, and gehlenite) can react with CO_2:

Lime [20]:

$$CaO + H_2O \rightarrow Ca(OH)_2 \tag{1}$$

$$Ca(OH)_2 + CO_2 \rightarrow CaCO_3 + H_2O \tag{2}$$

Gehlenite [21]:

$$4Ca_2Al(AlSiO_7) + 18H_2O \rightarrow CaAl_2Si_4O_{12}\cdot2H_2O + 7Ca(OH)_2 + 6Al(OH)_3 \tag{3}$$

To investigate the effect of the amount of moisture on the performance of fly ash for CO_2 capture, reaction with CO_2 was conducted with the following amounts of water: 25 (25-W/A), 50 (50-W/A), 75 (75-W/A), and 100% (100-W/A). The CO_2 concentration and reaction time were fixed at 100% and 60 min, respectively.

Figure 7 and Table 4 show the crystal structure of the fly ash after CO_2 capture using different amounts of moisture and materials by using the relative intensity ratio (RIR) technique from the XRD. The $CaCO_3$ components of the 25-W/A, 50-W/A, 75-W/A, and 100-W/A were 14.13, 21.31, 26.16, and 21.98%, respectively. The amount of $CaCO_3$ increased with increasing the amount of water. This suggests that the presence of water played an essential role in hydrating the calcium-based materials to form calcium hydroxide, which then sequestrated carbon by forming calcium carbonate. Notably, the amount of $CaCO_3$ increased from 3.02 to 26.16% when the ratio of the ash-to-water was 1:0.75. The results showed that fly ash of 100 g from a SRF fired power plant captured CO_2 of 10.17 g. However, when the ratio of the ash-to-water was 1:1, the amount of $CaCO_3$ decreased. CO_2 in gaseous phase does not react with calcium-based materials; it has to dissolve in the water first to form carbonic acid (H_2CO_3) [22]. As shown in Figure 8, the diffusion length of the H_2CO_3 from the outside of the water to the calcium-based materials increased with the increasing amount of water. Therefore, the fly ash with too much water reacted slowly with CO_2.

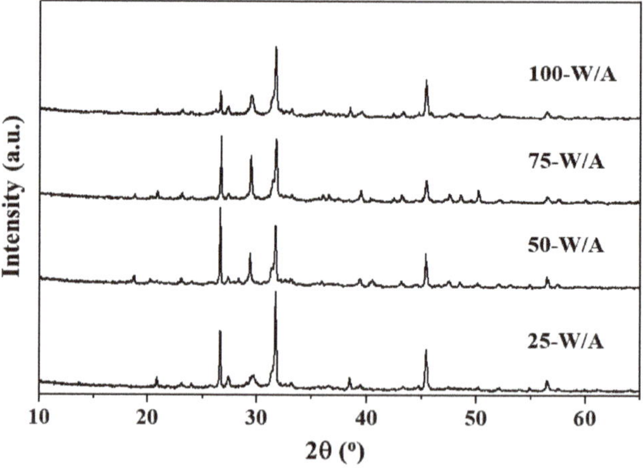

Figure 7. X-ray diffraction patterns of fly ashes after carbonation with various amount of water.

Table 4. Quantitative analytical results obtained by RIR method (wt%) of the fly ash after carbonation with various amounts of water.

Samples	Periclase	Lime	Calcite	Anhydrite	Quartz	Mullite	Hematite	Gehlenite	Halite	Sylvite
25-W/A	0.35	0.00	14.13	1.68	18.07	0.28	2.14	22.50	40.79	0.06
50-W/A	0.88	0.14	21.31	1.92	18.91	0.65	3.66	23.03	26.03	3.21
75-W/A	0.58	0.06	26.16	2.12	17.34	0.72	3.31	23.25	25.15	1.30
100-W/A	0.95	0.24	21.98	2.25	9.32	2.47	0.82	24.01	37.96	0.00

To investigate the properties of the fly ash with a small amount of water in the environments with low CO_2 concentrations, CO_2 captures were conducted at CO_2 concentrations of 10 (10-C/A), 20 (20-C/A), 50 (50-C/A), and 100% (100-C/A). The moisture content and reaction time were fixed at 20% and 10 min, respectively.

Figure 9 and Table 5 show the crystal structure of the fly ash after CO_2 capture using different concentrations of CO_2 and the amount of materials by using the relative intensity ratio (RIR) technique from the XRD. The $CaCO_3$ components of the 10-C/A, 20-C/A, 50-C/A, and 100-C/A were 15.21,

19.46, 19.86, and 21.98%, respectively. As shown in Table 5, the $CaCO_3$ component stabilized at CO_2 concentration of 20%. This result indicates that the semi-dry process can be applied to the power plant without a CO_2 concentration process.

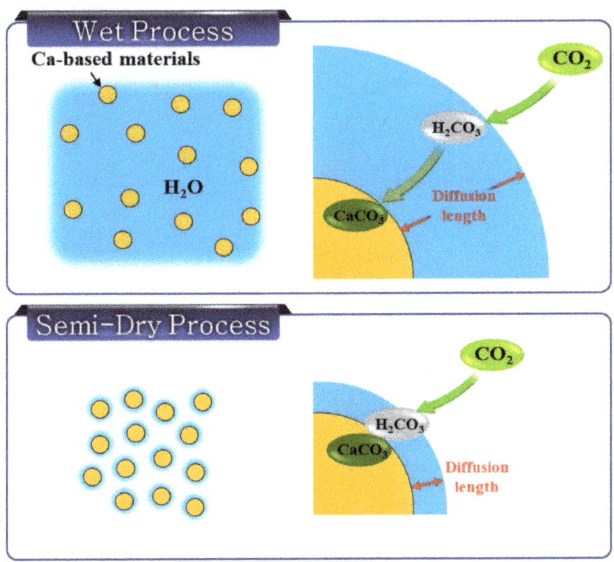

Figure 8. Schematic diagram of the semi-dry process and wet process for the CO_2 capture.

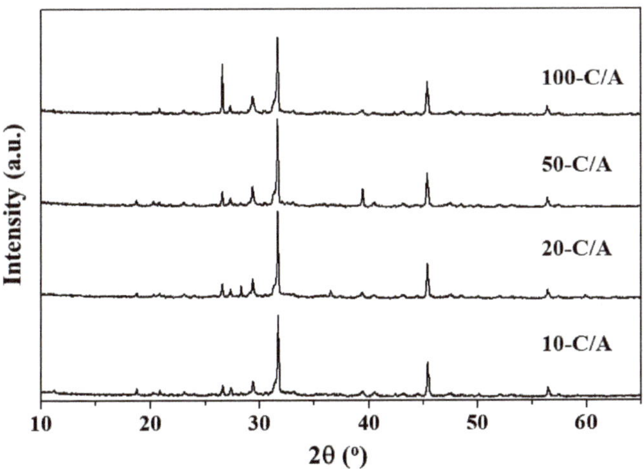

Figure 9. X-ray diffraction patterns of fly ashes after carbonation with various CO_2 concentrations.

Table 5. Quantitative analytical results obtained by RIR method (wt%) of the fly ash after carbonation with various CO_2 concentrations.

Samples	Periclase	Lime	Calcite	Anhydrite	Quartz	Mullite	Hematite	Gehlenite	Halite	Sylvite
10-C/A	0.34	0.00	15.21	1.56	14.32	1.09	1.22	25.11	40.79	0.06
20-C/A	0.34	0.02	19.46	1.33	11.35	0.84	1.39	24.39	38.34	2.54
50-C/A	0.38	0.01	19.86	1.12	14.84	0.33	0.91	22.55	38.72	1.17
100-C/A	0.95	0.24	21.98	2.25	13.17	2.47	0.82	24.01	37.96	0.00

Figure 10 and Table 6 show the crystal structure of the fly ash after CO_2 capture in a reaction time of 1 (1-T/A), 5 (5-T/A), 10 (10-T/A), and 30 min (30-T/A), and the amount of materials. The $CaCO_3$ component of the 1-T/A, 5-T/A, 10-T/A, and 30-T/A was 16.26, 15.16, 15.21, and 16.72%, respectively. As shown in Table 6, the carbonation performed well despite the short reaction time of 1 minute. This result indicates that the semi-dry process can be designed as a continuous process to capture large scales of CO_2.

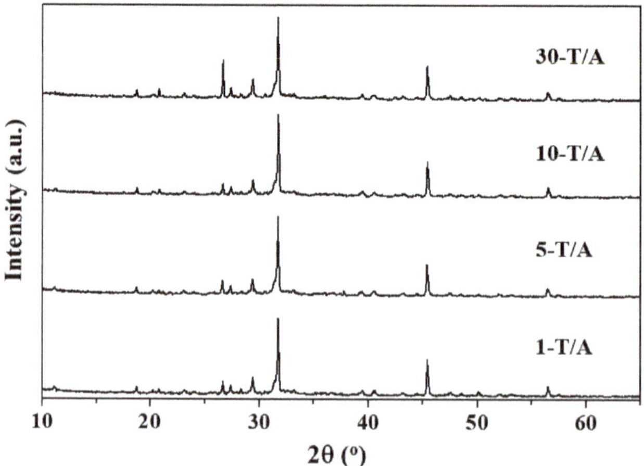

Figure 10. X-ray diffraction patterns of fly ashes after carbonation with various reaction times.

Table 6. Quantitative analytical results obtained by RIR method (wt%) of the fly ash after carbonation with various reaction times.

Samples	Periclase	Lime	Calcite	Anhydrite	Quartz	Mullite	Hematite	Gehlenite	Halite	Sylvite
10-C/A	0.34	0.00	15.21	1.56	14.32	1.09	1.22	25.11	40.79	0.06
20-C/A	0.34	0.02	19.46	1.33	11.35	0.84	1.39	24.39	38.34	2.54
50-C/A	0.38	0.01	19.86	1.12	14.84	0.33	0.91	22.55	38.72	1.17
100-C/A	0.95	0.24	21.98	2.25	13.17	2.47	0.82	24.01	37.96	0.00

4. Discussions

The utilization of SRF ash has emerged as one of the most important interventions in SRF power plants. Due to the calcium-based materials in the SRF ash, the SRF ash can be reused as a material for CO_2 capture. According to Loo et al., who studied the utilization of circulating fluidized bed combustion (CFBC) ash, high humidity conditions during the carbonation process were a critical factor that significantly enhanced the CO_2 uptake in the CFBC ash [17]. However, in this study, the results show that the high amount of water led to a low efficiency of carbonation reaction. This may be due to the fact that the diffusion length of the H_2CO_3 increased correlatively with the amount of water. These results demonstrate that the semi-dry carbonation process is more effective in the aspect of cost and time compared to the wet process. The carbonated SRF ash can be used in various construction industries such as a mineral admixture for concrete [23]. For the utilization of the carbonated SRF ash prepared via semi-dry carbonation process, further research should be conducted on controlling the conditions of the process such as temperature and pH value.

5. Conclusions

In this study, the utilization of the fly ash from SRF power plant as a solid sorbent material for CO_2 capture via semi-dry carbonation reaction was evaluated as a simple CO_2 reduction technique which does not require a high cost wastewater treatment plant and an evaporating process. Quantitative analysis was conducted by using the relative intensity ratio technique from the XRD to study the potential for fly ash carbonation under semi-dry carbonation process. The amount of CO_2 capture increased as the amount of water increased. However, when the ratio of the ash-to-water was 1:1, the amount of CO_2 capture decreased because the diffusion length of the H_2CO_3, from the outside of the water to the calcium-based materials, increased with the increasing amount of water. The results indicated that H_2O plays an important role in the carbonation of calcium-based materials. The fly ash of 100 g can capture 10.17 g of CO_2 by semi-dry carbonation reaction without any treatment. The moisture content can affect the CO_2 capture capacity but CO_2 concentration and reaction time do not significantly affect the carbonation reaction. In the semi-dry carbonation process, the carbonation was well performed in the short reaction time and with low CO_2 concentration. These results show that the semi-dry process can be designed as a continuous process and applied to a power plant directly without CO_2 concentration processes.

Author Contributions: Conceived and wrote the manuscripts, J.H.K.; advice and revised the manuscript, W.T.K.; both authors agreed with the final version of manuscript.

Funding: National Research Foundation of Korea: 2017M3D8A2086035.

Acknowledgments: This research was supported by the National Strategic Project-Carbon Upcycling of the National Research Foundation of Korea (NRF) funded by the Ministry of Science and ICT (MSIT), the Ministry of Environment (ME) and the Ministry of Trade, Industry and Energy (MOTIE) (2017M3D8A2086035).

Conflicts of Interest: The authors declare no conflict of interest.

References

1. Plaza, M.G.; Pevida, C.; Arenillas, A.; Rubiera, F.; Pis, J.J. CO_2 capture by adsorption with nitrogen enriched carbons. *Fuel* **2007**, *14*, 2204–2212. [CrossRef]
2. Yang, H.; Xu, Z.; Fan, M.; Gupta, R.; Slimane, R.B.; Bland, A.E.; Wright, I. Progress in carbon dioxide separation and capture: A review. *J. Environ. Sci.* **2008**, *20*, 14–27. [CrossRef]
3. Al-Shawabkeh, A.; Matsuda, H.; Hasatani, M. Comparative reactivity of treated FBC- and PCC-fly ash for SO_2 removal. *Can. J. Chem. Eng.* **1995**, *15*, 193–201.
4. Davini, P. Flue gas treatment by activated carbon obtained from oil-fired fly ash. *Carbon* **2002**, *40*, 1973–1979. [CrossRef]
5. Lu, G.Q.; Do, D.D. Adsorption properties of fly ash particles for NO_x removal from flue gases. *Fuel Process Technol.* **1991**, *27*, 95–107. [CrossRef]
6. Rubel, A.; Andrews, R.; Gonzalez, R.; Groppo, J.; Robl, T. Adsorption of Hg and NOx on coal by-products. *Fuel* **2005**, *84*, 911–916. [CrossRef]
7. Florin, N.; Fennell, P. Synthetic CaO-based sorbent for CO_2 capture. *Energy Procedia* **2011**, *4*, 830–838. [CrossRef]
8. Han, S.-J.; Yoo, M.; Kim, D.-W.; Wee, J.-H. Carbon dioxide capture using calcium hydroxide aqueous solution as the absorbent. *Energy Fuels* **2011**, *25*, 3825–3834. [CrossRef]
9. Manovic, V.; Anthony, E.J. Lime-based sorbents for high-temperature CO_2 capture-a review of sorbent modification methods. *Int. J. Environ. Res. Public Health* **2010**, *7*, 3129–3140. [CrossRef] [PubMed]
10. Zhang, H.; Liu, R.; Ning, T.; Lal, R. Higher CO_2 absorption using a new class of calcium hydroxide ($Ca(OH)_2$) nanoparticles. *Environ. Chem. Lett.* **2018**, *16*, 1095–1100. [CrossRef]
11. Skoufa, Z.; Antzara, A.; Heracleous, E.; Lemonidou, A.A. Evaluating the activity and stability of CaO-based sorbents for post-combustion CO_2 capture in fixed-bed reactor experiments. *Energy Procedia* **2016**, *86*, 171–180. [CrossRef]
12. Ibrahim, A.-R.; Vuningoma, J.B.; Huang, Y.; Wang, H.; Li, J. Rapid carbonation for calcite from a solid-liquid-gas system with an imidazolium-based ionic liquid. *Int. J. Mol. Sci.* **2014**, *15*, 11350–11363. [CrossRef] [PubMed]
13. Solieman, A.A.A.; Dijkstra, J.W.; Haije, W.G.; Cobden, P.D.; van den Brink, R.W. Calcium oxide for CO_2 capture: Operational window and efficiency penalty in sorption-enhanced steam methane reforming. *Int. J. Greenh. Gas Control* **2009**, *3*, 393–400. [CrossRef]

14. Derevschikov, V.S.; Lysikov, A.L.; Okunev, A.G. CaO/Y_2O_3 pellets for reversible CO_2 capture in sorption enhanced reforming process. *Catal. Sustain. Energy* **2012**, *1*, 53–59. [CrossRef]
15. Shi, J.; Li, Y.; Zhang, Q.; Ma, X.; Duan, L.; Zhou, X. CO_2 capture performance of a novel synthetic CaO/sepiolite sorbent at calcium looping conditions. *Appl. Energy* **2017**, *203*, 412–421. [CrossRef]
16. Garg, A.; Smith, R.; Hill, D.; Simms, N.; Pollard, S. Wastes as co-fuels: The policy framework for solid recovered fuel (SRF) in Europe, with UK implications. *Environ. Sci. Technol.* **2007**, *41*, 4868–4874. [CrossRef] [PubMed]
17. Loo, L.; Maaten, B.; Konist, A.; Siirde, A.; Neshumayev, D.; Pihu, T. Carbon dioxide emission factors for oxy-fuel CFBC and aqueous carbonation of the Ca-rich oil shale ash. *Energy Procedia* **2017**, *128*, 144–149. [CrossRef]
18. Dananjayan, R.R.T.; Kandasamy, P.; Andimuthu, R. Direct mineral carbonation of coal fly ash for CO_2 sequestration. *J. Clean. Prod.* **2016**, *112*, 4173–4182. [CrossRef]
19. Ebrahimi, A.; Saffari, M.; Milani, D.; Montoya, A.; Valix, M.; Abbas, A. Sustainable transformation of fly ash industrial waste into a construction cement blend via CO_2 carbonation. *J. Clean. Prod.* **2017**, *156*, 660–669. [CrossRef]
20. Wang, C.; Jia, L.; Tan, Y.; Anthony, E.J. Carbonation of fly ash in oxy-fuel CFB combustion. *Fuel* **2008**, *87*, 1108–1114. [CrossRef]
21. Hoschek, G. Gehlenite stability in the system CaO-Al_2O_3-SiO_2-H_2O-CO_2. *Contrib. Mineral. Petrol.* **1974**, *47*, 245–254. [CrossRef]
22. Chun, Y.M.; Naik, T.R.; Kraus, R.N. Carbon dioxide sequestration in concrete in different curing environments. In Proceedings of the Conference on Sustainable Construction Materials and Technologies, Coventry, UK, 11–13 June 2007; pp. 18–24.
23. Siriruang, C.; Toochinda, P.; Julnipitawong, P. CO_2 capture using fly ash from coal fired power plant and applications of CO_2-captured fly ash as a mineral admixture for concrete. *J. Environ. Manag.* **2016**, *170*, 70–78. [CrossRef] [PubMed]

Review

Opportunities for Mineral Carbonation in Australia's Mining Industry

Mehdi Azadi [1],*, Mansour Edraki [1], Faezeh Farhang [2] and Jiwhan Ahn [3]

[1] Centre for Mined Land Rehabilitation, Sustainable Minerals Institute, The University of Queensland, St Lucia, QLD 4072, Australia; m.edraki@cmlr.uq.edu.au
[2] School of Chemical Engineering, University of Newcastle, Callaghan, NSW 2308, Australia; faezeh.farhang@newcastle.edu.au
[3] Centre for Carbon Mineralisation, Climate Change Mitigation and Sustainability Division, Korea Institute of Geoscience and Mineral Resources, Daejeon 34132, Korea; ahnjw@kigam.re.kr
* Correspondence: mehdi.azadi@uq.edu.au; Tel.: +61-7-334-64061

Received: 30 January 2019; Accepted: 22 February 2019; Published: 27 February 2019

Abstract: Carbon capture, utilisation and storage (CCUS) via mineral carbonation is an effective method for long-term storage of carbon dioxide and combating climate change. Implemented at a large-scale, it provides a viable solution to harvesting and storing the modern crisis of GHGs emissions. To date, technological and economic barriers have inhibited broad-scale utilisation of mineral carbonation at industrial scales. This paper outlines the mineral carbonation process; discusses drivers and barriers of mineral carbonation deployment in Australian mining; and, finally, proposes a unique approach to commercially viable CCUS within the Australian mining industry by integrating mine waste management with mine site rehabilitation, and leveraging relationships with local coal-fired power station. This paper discusses using alkaline mine and coal-fired power station waste (fly ash, red mud, and ultramafic mine tailings, i.e., nickel, diamond, PGE (platinum group elements), and legacy asbestos mine tailings) as the feedstock for CCUS to produce environmentally benign materials, which can be used in mine reclamation. Geographical proximity of mining operations, mining waste storage facilities and coal-fired power stations in Australia are identified; and possible synergies between them are discussed. This paper demonstrates that large-scale alkaline waste production and mine site reclamation can become integrated to mechanise CCUS. Furthermore, financial liabilities associated with such waste management and site reclamation could overcome many of the current economic setbacks of retrofitting CCUS in the mining industry. An improved approach to commercially viable climate change mitigation strategies available to the mining industry is reviewed in this paper.

Keywords: mineral carbonation; greenhouse gas (GHG) emission; climate change; Australian mining; mining waste; mine reclamation

---

1. Introduction

Global concerns about climate change have been growing over the last few decades. Greenhouse gas (GHG) emissions with anthropogenic origins are recognised to be the major driver of these climate change effects [1]. The increased industrial developments and human activities which have been traditionally reliant on burning fossil fuels have resulted in a critical accumulation of CO_2 in the atmosphere. Largely anthropogenic GHG emissions have raised CO_2 concentrations significantly during the last century, with up to 408 ppm in November 2018 [2] in comparison to under 300 ppm in the 19th century. During the last three decades, the surface temperature on earth has increased progressively [3]. The increasing trend of global warming is attributed to the increased rate of CO_2 emissions in the last few decades. All these data indicate that disastrous climate change effects can occur by the end of the 21st century if the current trends continue. It is important that countries with major emissions take a leading role to address GHG emissions. Australia contributes about 0.33 percent of the world population, but is accountable for nearly 1.3 percent of the global emissions [4]. As a country with access to advanced technologies and developing industries, Australian industrial sectors have the opportunity to improve toward long-term decarbonisation.

The Australian mining industry, as a major industrial sector, has been one of the contributors to the GHG emissions and climate change effects. This has been through energy consumption and fugitive GHG emissions associated with mining operations and processing of minerals and metals, as well as burning coal, one of Australia's most important commodities, in power stations. Australia's GHG emissions for the year to June 2018 were 547.0 Mt CO_2-e. Emissions from the electricity sector account for 35% of Australia's greenhouse gas emission [5], and mining accounts for approximately 10% of Australia's energy consumption, and it has increased 6% per year during the last decade [6]. The demand for electricity production has been growing with the development of industries. In fact, Australia's expanding resource sector relies heavily on electricity consumption to operate mines, mineral processing plants, refineries, chemical manufacturing facilities, and its infrastructure. Under The Paris Agreement [7] Australia made the commitment to reduce its GHG emissions by 26 to 28% of 2005 levels by 2030 [4,8]. In order to be on track to meet this target, it is crucial that all sectors in Australia improve their emission mitigation strategies. Therefore, it is important for the mining industry to adopt appropriate technologies, which support offsetting GHG emissions and decarbonisation of the entire industry in the long-term.

Many attempts have been made to address the effect of climate change and reduce GHG emissions from Australian industries. It is widely accepted that a portfolio of strategies needs to be applied to reduce CO_2 emission in the long-term. Considering the energy sector as the source of the largest GHG emissions, replacement of fossil fuels with renewable energies has been suggested repeatedly as a major climate change mitigation strategy. However, there is a general consensus that in the short-term it is not possible to stop using fossil fuels and switch to renewable energies without major economic disruption [9]. Therefore, transition to a portfolio of energy supply chains with a larger proportion of renewable energies, and hence reducing GHG emissions, also requires the development of cleaner fossil fuel energy production. Due to the increasing demand for minerals and metals production, resource mining worldwide should develop such decarbonisation strategies, and particularly in countries such as Australia, with globally significant mining activities.

Carbon capture, utilisation, and storage (CCUS) is an emerging strategy, which has proven to be a promising tool as a long-term solution to address CO_2 emissions [10]. Estimates by the International Energy Agency (IEA) suggest that carbon capture and storage (CCS) deployed to coal-fired power stations globally might remove 10% of the CO_2 emissions from the energy sector by 2050 [11,12]. However, recent estimates by IEA, at the 24th Conference of the Parties to the United Nations Framework Convention on Climate Change (COP24), [10] suggest that CCUS remains well off-track to meet the IEA's Sustainable Development Scenario (SDS) [13], which is aligned with the energy-related concerns of the Sustainable Development Goals (SDGs) [14].

All of these recent trends show the importance of the application of CCUS in various industrial sectors, including the mining industry. Successful application of CCUS in the mining industry has the potential for substantial carbon offsetting of mining operations and minimising CO_2 emissions from coal-fired power stations. Global acceptance and successful integration of CCUS in large-scale industries, such as mining, requires successful trials. Australia as a country with an advanced resources industry is a perfect place for trial and development of mineral carbonation for climate change mitigation in the mining industry, which can then be utilised globally.

Economic drawbacks have been the major barrier to the deployment of CCUS technologies at a large-scale [15]. However, opportunities exist which can assist in reducing the deployment cost of mineral carbonation. Magnesium and calcium silicate mineral sources, which are generally considered for mineral carbonation, can be expensive to pretreat and energy-intensive to activate. However, industrial wastes are usually more reactive and require minimum pretreatment and activation. Furthermore, industrial wastes are liabilities which are available at no or minimum cost and are often costly to dispose [16]. Successful utilisation of alkaline industrial wastes, including furnace slag, cement kiln dust, red mud, coal combustion by-products, and mine tailings as the feedstock of mineral carbonation has been reported [15–21].

Recent investment and interest by major mining companies [22–24], governments (i.e., UK council for CCUS [25]), and research institutions for deploying CCUS at industrial scale, indicate the significance of these technologies. In particular, investigating the potential role of mining waste in mineral carbon sequestration, for example, investigation of mineral carbonation in diamond tailings at Venetia Mine in South Africa and the Gahcho Kué Mine in Canada [24] and process optimisation toward more economically viable technologies with reduced energy requirements [26], more efficient process routes [27,28], and recycled industrial waste feedstock for mineral carbonation [29,30] will contribute to the low carbon future. Rehabilitation of mined land often includes various factors related to the risks associated with residue during and after mining operations, for example, management of acid mine drainage (AMD). Even where high standards of remediation have been applied, there is a requirement for ongoing monitoring and management of residual risk. Nevertheless, remediation poses significant expenses to the mining industry as shown, for example, in the reports by Rio Tinto [31], BHP [32], and Glencore [33].

This paper discusses decarbonisation in Australian mining by recycling waste as a viable climate change mitigation strategy. The potentials that exist for the utilisation of mineral carbon sequestration in the Australian mining industry are reviewed. The drivers of and barriers to the deployment of the mineral carbonation, and an integrated mineral carbonation model for Australia's mining industry are discussed. Some of the many potential and nearer-term opportunities and synergies that exist in using mining waste for climate change adaptation are identified, in this paper.

2. Mineral Carbonation as a Route for Reducing CO$_2$ Emission

Geological sequestration [34–37], ocean disposal [38–41], and biological fixation [42–44] have been reported extensively in the literature, while mineral carbonation, also called CO$_2$ mineralisation, a less explored method of sequestering CO$_2$, in a disposal option that has great potential due to the availability of large quantities of suitable mineral materials [45]. Mineral carbonation involves the chemical conversion of CO$_2$ to solid inorganic carbonates. It commonly refers to the fixation of CO$_2$ using alkaline and alkaline-earth oxides, such as calcium oxide (CaO) and magnesium oxide (MgO) [46], for which the respective basic reactions are [47]

$$CaO + CO_2 \rightarrow CaCO_3 + 179 \text{ kJ/mol} \tag{1}$$

$$MgO + CO_2 \rightarrow MgCO_3 + 118 \text{ kJ/mol} \tag{2}$$

Pure calcium and magnesium oxide minerals are rare in nature, but abundant quantities of minerals containing chemical combinations of these oxides occur naturally as silicate minerals [45]. For example, ultramafic igneous rocks (primarily peridotites and serpentinites) contain large amounts of MgO bound into a silicate structure.

The process of mineral carbonation occurs naturally, where it is known as one of the weathering phenomena [48,49]. Silicate minerals form approximately 92 wt.% of the earth's crust which makes the weathering of silicate minerals a quantitatively important process [50]. Weathering of serpentine is known as a primary source of natural magnesium carbonates and is responsible for the majority of all large-scale sedimentary magnesite (MgCO$_3$) deposits [51]. However, the kinetics of chemical weathering reactions under natural conditions are not sufficient to absorb all current anthropogenic CO$_2$ emissions [52]. Therefore, the employment of ultramafic rocks for industrial-scale mineral carbonation processes was proposed for the first time in 1995 by Lackner et al. [47]. As mentioned above, the precursor material for mineral carbonation must be rich in alkaline earth metals, either

Mg or, more preferably but less abundant, Ca. The alkaline earth metals are present as oxides in silicate minerals such as wollastonite $CaSiO_3$, olivine Mg_2SiO_4, or the more abundant ultramafic rock serpentine $Mg_3Si_2O_5(OH)_4$, which is hydrolysed olivine [47,53]. Dunite is a rock made up entirely of olivine. Serpentinites and dunites are attractive feedstock for the mineral carbonation process because of their great abundance and the high Mg content by weight (35 to 49 wt-% MgO) [47,54,55]. There are large deposits of these minerals available worldwide and, in particular, in Australia, USA, Canada and parts of Europe [45]. Serpentinised peridotite reserves are estimated to be in the hundreds of thousands of gigatonnes [56]. In an industrial application between 1.6 to 3.7 tonnes of ore (depending on the magnesium content of the ore) are typically required to fix one tonne of CO_2 [57]. In comparison, Brent and Petrie [58] showed the mining practice of serpentinite required for mineral carbonation is of a similar scale of coal-based mining operations for coal-fired power generation.

Mg/Ca-rich precursors can be converted into carbonates using two methods: directly via the gas-solid reaction and in aqueous media. The aqueous route has clear advantages in terms of the speed of reaction and process flexibility. Therefore, the majority of research on CO_2 mineralisation focuses on the aqueous route. Ex situ aqueous mineral carbonation, which is the accelerated form of natural weathering of silicate rocks through above-ground processes, involves three steps: (1) the dissolution of CO_2 in the aqueous phase, (2) the extraction of the alkaline earth metal from the mineral structure while leaving a SiO_2 rich layer, and (3) the precipitation of stable metal carbonate. The sequence of the reactions is illustrated through Equations (3)–(5).

CO_2 dissolution:

$$\begin{aligned} CO_{2\,(aq)} + H_2O_{(l)} \;&\rightleftharpoons\; H_2CO_{3\,(aq)} \\ &\rightleftharpoons\; H^+_{(aq)} + HCO^-_{3\,(aq)} \end{aligned} \tag{3}$$

Serpentine dissolution:

$$Mg_3Si_2O_5(OH)_{4(s)} + 6H^+_{(aq)} \rightleftharpoons 3Mg^{2+}_{(aq)} + 2SiO_{2(s)} + 5H_2O \tag{4}$$

Carbonate formation:

$$HCO^-_{3(aq)} + Mg^{2+}_{(aq)} \rightleftharpoons MgCO_{3(s)} + 2H^+ \tag{5}$$

CO_2 is first dissolved in water to form carbonic acid (H_2CO_3), which is then dissociated to protons (H^+ or H_3O^+) and bicarbonate ions ($HCO_3{}^-$). The protons hydrolyse the mineral, resulting in the liberation of magnesium cations and the formation of free silica and water. The magnesium cations in the solution react with the bicarbonate ions to form solid magnesium carbonate ($MgCO_3$) [56]. The Albany Research Centre (ARC) conducted the best-studied case of the aqueous mineral carbonation including over 700 kinetic tests [57]. They investigated carbon mineralisation in a single step mode, where the steps in Equations (3) to (5) take place in the same reactor simultaneously. The overall reaction sequence in Equations 3 to 5 is exothermic, the enthalpy of reaction of the given example being -64 kJ mol^{-1} of CO_2. However, the kinetics of the reactions involved are very slow [48]. Therefore, different strategies to increase the speed and efficiency of the carbonation reaction by, among others, Los Alamos National Laboratory (LANL) [47,59], Åbo Akademi University [60,61], and ARC [57], have proved at least partially effective. Table 1 summarises the main methods used by the researchers to date.

Table 1. Strategies and methods used to increase the speed and efficiency of the carbonation reaction.

Strategy	Reference	Method	Benefit
Mechanical pretreatment of serpentine rock	[62–66]	High-energy crushing and grinding	Increases the overall specific surface area of the minerals, i.e., the surface per unit mass available for Mg/Ca-extraction
Thermal activation of the hydrated Mg-silicate serpentine	[67–73]	Heating the mineral to 630 °C or above	Destabilises the crystal lattice, thus increasing the reactivity of the mineral Heat activation also creates an even higher specific surface area [46]

Table 1. *Cont.*

Strategy	Reference	Method	Benefit
Increasing reaction temperatures	[74,75]	Applying operating temperatures above 100 °C	Accelerate the kinetics of the reaction
Increasing reaction pressure	[76,77]	Applying operating CO_2 pressure above 150 bar	Increases the activity of protons needed for metal extraction. Also counteracts the low solubility of CO_2 at high temperature
Using organic and inorganic chemicals	[61,78–80]	Adding different substances as catalysts and additives to the reactants	Enhances the kinetics of the carbonation process and precipitation of magnesium carbonate
Applying New double-step and multistep process designs	[81]	The operating conditions are changed between the different stages	Allows to control and promote extraction and precipitation separately
Direct capture of CO_2 from flue gas stream	[81,82]	Capturing CO_2 directly from a flue gas stream by mineral carbonation	Provides a simple and straightforward process route

In addition to the minerals mentioned above, MgO and CaO can also be found in alkaline industrial residues, i.e., in steel slags, fly ashes, or cement wastes, as well as in waste brines and slurries such as alumina production residue (red mud). The sequestration of carbon using industrial and commercial waste is an attractive option and offers the potential to utilise residues. Steel manufacturers can benefit significantly from use of a carbonation process to improve the environmental and mechanical properties of the large amounts of alkaline solid residues that are generated during steel manufacturing such as basic oxygen furnace (BOF) slag, a by-product of the conversion of iron into steel [83–86]. This research has recently been tested on a pilot scale and resulted in a successful outcome [87]. In other studies, coal combustion fly-ash, an industrial waste that contains about 4.1 wt.% of lime (CaO), was used to sequester CO_2 by aqueous carbonation [88–90]. All in all, while in many industries these waste materials are available only in small quantities [48] which can be a barrier to commercial viable CO_2 fixation at the large-scale, they can be found in large quantities as by-products of many mining-related operations in Australia. This can be an opportunity that may help simultaneously to mature the technology towards commercial application and solve the problem of massive waste management in mining and other industries. Mines with mineral carbonation capabilities could use their waste rock as a value-added product. This could effectively lower the grade of material that is economical to mine. This additional revenue stream could also lower the strip ratio of a surface mining operation by virtue of a greater proportion of the material being mined, having value [91]. As a result, there may be the potential to enhance a marginal project to a level in which that mining project could become economically feasible and environmentally attractive [92].

3. Mineral Carbonation in Australia

One of first applications of mineral carbonation at an industrial scale in Australia was in Alcoa's Kwinana aluminium refinery, Western Australia [93]. A carbonation plant was developed by Alcoa for bauxite residue treatment in order to reduce long-term storage risk and adverse environmental effects. Deployment of the mineral carbonation plant assisted neutralising red mud by reducing its alkalinity from pH 13 to 10.5 and by sequestrating CO_2 from the nearby CSBP ammonia plant [93].

Another case for mineral carbonation in Australia was outlined in 2008 by Brent and Petrie [58]. They focused on carbonation of serpentine as feedstock for CCS via mineral carbonation, based on the fact that large outcropping deposits of serpentinite are located in close proximity to existing large emission sources in Australia [94]. This study showed the feasibility of achieving 80% net sequestration of CO_2 using Aspen (Aspen Technology Inc., Bedford, MA, USA, 2007) to model a standalone mineral carbonation process based on the ARC/NETL aqueous process [57]. Following this research, Mineral Carbonation International (MCi), an Australian-based company, has been developing technology for carbon capture and utilisation (CCU) since 2013. Their research and technical facilities are centred at the University of Newcastle, Australia. MCi's suite of projects includes fundamental theoretical and laboratory research [66,95–100] as well as one of the world's first mineral carbonation pilot plant facilities [101,102]. MCi's project has primarily been studying the carbonation of serpentine, which as mentioned above is an abundant mineral in New South Wales (NSW), Australia and globally [102]. It is believed that the future feedstock supply needs will be available on the gigatonne scale. For example, the size of only one resource in the north of the town of Barraba, NSW, Australia, is sufficient to sequester all NSW stationary emissions for over 300 years [94]. The MCi's project would be the first to encompass the full CCUS chain from mineral mapping, characterisation, processing, and pretreatment through to integration with power generation and CO_2 capture processes, carbonation, value-adding and final product storage. Another aim of the MCi project is determining the potential for utilising the carbonated product in next-generation "green" building products [102].

Other research groups at the University of Queensland and Monash University, Australia have been working on different aspects of mineral carbonation, field studies and the use of waste materials [103–110]. Recent research investigated ultramafic mine tailings as feedstock because they contain abundant Mg-rich silicate and hydroxide minerals and have a smaller grain size and increased reactive surface area due to ore processing [106,107]. Table 2 presents a summary of research on mineral carbonation in Australia.

Table 2. Summary of the mineral carbonation research undertaken in Australia institutes and industries.

Institute/Industry	Research Focus	Goal
Alcoa's Kwinana Aluminium Refinery	Bauxite residue treatment	To reduce long-term storage risk and adverse environmental effects
Mineral Carbonation International (MCi)	Carbonation of serpentine as an abundant mineral in New South Wales (NSW), Australia and globally	integration with power generation and CO_2 capture processes, carbonation, value-adding and final product storage
The University of Queensland	field studies	use of waste materials and ultramafic mine tailings as feedstock
Monash University	field studies	use of waste materials and ultramafic mine tailings as feedstock

4. Drivers of Utilisation of Mineral Carbonation in the Australian Mining Industry

Mining is a primary industry that forms one of the pillars of Australia's economy [111]. Due to Australia's role as a world leader in resource mining and production, it is crucial to integrate climate change adaptation technologies in mining to address the ever-growing environmental awareness and the increasing demand for minerals and metals in the future for Australia simultaneously. Integration of mineral carbonation in the Australian mining industry is one of the climate change adaptation strategies which will benefit sustainable development initiatives of the mining sector by addressing GHG mitigation, improved mining waste management and mine reclamation. Australia has a large and diverse portfolio of mineral commodity production and is one of the world's biggest producers of minerals and metals.

Table 3 lists Australian mineral resources and the commodities, indicating the potential of tailings and by-products for mineral carbonation, according to the mineralogy of the waste. It also shows how critical some commodities are by presenting Australia's production levels and rankings in the world. Table 3 also highlights the potential of tailings and waste by-products which can be further investigated to evaluate their potential for mineral carbonation, according to the mineralogy of tailings and available research in the literature. This is in accordance with the fact that mineral carbonation can occur as the result of the reaction between CO_2 with elements such as calcium, magnesium, and iron [112]. The rankings and production data were extracted from the Australian Atlas of mineral resources, mines, & processing centres, Australian Government as of December 2016 [111].

Table 3. Australia's mineral resources and potential of its mining waste for carbon capture and storage (CCS) by mineral carbonation.

Australia's Mineral Resources (Alphabetic Order)	Can Be Found in Mafic-Ultramafic Ores	Mineral Carbonation Potential Elements in Waste Products	Examples of Mineral Phases Prone to Carbonation	Australia's World Ranking for Resources	% of World Resources	Australia's World Ranking for Production	% of World Production	Australia's Production (Mt)	World's Production (Mt)
Antimony				4	9	4	4	5.5 (kt)	130 (kt)
Bauxite		Ca, Mg, Fe, Na	Magnetite (Fe_3O_4), Sodium aluminosilicate ($Na(AlSiO_4)$)	2	22	1	31	82.152	271.5
Black Coal		Ca, Mg	Hydrocalumite ($Ca_2Al(OH)_6Cl \cdot 2H_2O$), Ettringite ($Ca_6Al_2(SO_4)_3(OH)_{12} \cdot 26H_2O$), Portlandite ($Ca(OH)_2$)	4	10	4	7	566.3	7795
Brown Coal		Ca, Mg	Hydrocalumite ($Ca_2Al(OH)_6Cl \cdot 2H_2O$), Ettringite ($Ca_6Al_2(SO_4)_3(OH)_{12} \cdot 26H_2O$), Portlandite ($Ca(OH)_2$)	2	24	5	6	63.3	783.3
Chromium	X	Mg		n.a.	n.a.	n.a.	n.a.	n.a.	30.4
Cobalt	X	Ca, Fe		2	14	5	4	5.47 (kt)	123 (kt)
Copper	X	Ca, Mg, Fe		2	12	5	5	0.948	19.4
Diamond	X	Ca, Mg, Fe	Serpentine-Group minerals ($Mg_3(Si_2O_5)(OH)_4$), Forsterite (Mg_2SiO_4)	3	18	2	24	13.958 (Mc)	127 (Mc)
Flourine				n.a.	n.a.	n.a.	n.a.	n.a.	3.1
Gold	X	Ca, Mg, Fe		1	17	2	9	288 (t)	3255 (t)
Ilmenite				2	19	3	13	1.4	11.6
Iron Ore	X	Fe, Ca, Mg	Magnetite (Fe_3O_4), goethite ($FeO(OH)$), biotite ($K(Mg,Fe)_3AlSi_3O_{10}(F,OH)_2$)	1	29	1	38	858	2230
Lead				1	40	2	9	0.45	4.82
Lithium				3	18	1	41	14 (kt)	34.7 (kt)
Magnesite		Ca, Mg, Fe		5	4	8	2	0.554	27.7

Table 3. *Cont.*

Australia's Mineral Resources (Alphabetic Order)	Can Be Found in Mafic-Ultramafic Ores	Mineral Carbonation Potential Elements in Waste Products	Examples of Mineral Phases Prone to Carbonation	Australia's World Ranking for Resources	% of World Resources	Australia's World Ranking for Production	% of World Production	Australia's Production (Mt)	World's Production (Mt)
Manganese Ore				4	13	4	9	3.2	44
Molybdenum	X			7	1	n.a.	n.a.	n.a.	227 (kt)
Nickel	X	Mg, Ca,	Enstatite ($MgSiO_3$), Diopside ($MgCaSi_2O_6$), Talc ($Mg_3Si_4O_{10}(OH)_2$), Serpentine-Group minerals ($Mg_3(Si_2O_5)(OH)_4$)	1	24	5	9	0.204	2.25
Niobium				2	6	minor	minor	minor	64 (kt)
Phosphate				10	2	minor	minor	1.54	264
PGEs (Platinum-group elements)	X	Mg, Ca, Fe, Na	Enstatite ($MgSiO_3$), Talc ($Mg_3Si_4O_{10}(OH)_2$), Bytownite [(Ca, Na)[Al(Al, Si)Si_2O_8], Diopside ($MgCaSi_2O_6$)	n.a.	n.a.	n.a.	n.a.	678 (kg)	380 (t)
Potash				minor	2	minor	minor	minor	39
Rare Earths (REO & Y_2O_3)				6	3	2	11	0.014	0.126
Rutile				1	50	1	42	0.3	0.7
Silver				2	16	5	5	1.418 (kt)	27 (kt)
Tantalum				1	n.a.	n.a.	n.a.	183 (t)	1.1 (kt)
Tin				4	10	7	2	6.635	278
Tungsten				2	12	n.a.	n.a.	0.11	86.5
Uranium				1	29	3	10	6.314	62 (kt)
Vanadium	X	Ca, Mg, Fe		4	11	minor	minor	minor	76 (kt)
Zinc				1	28	3	7	0.884	11.9
Zircon				1	67	1	31	0.6	2.4

Notes: t: tonne, kt: kilotonnes, Mt: million tonnes, Mc: million carrat, n.a.: not available.

4.1. Integrated Mineral Carbonation Model for Australia's Mining Industry

A major barrier of deploying mineral carbonation, at an industrial scale, is the financial requirements necessary for obtaining appropriate feedstock for the process [15]. Integration of mine waste recycling with mineral carbonation provides a distinctive opportunity for the mining industry to overcome some of the economic barriers in adopting an effective climate change adaptation technology.

Calcium- and magnesium-rich mining wastes, particularly tailings, are capable of reacting with CO_2 to form stable carbonate minerals. Mineral carbonation of alkaline mining waste under controlled conditions not only accelerates this process for CO_2 sequestration, but it can also neutralise the alkalinity and potentially immobilise and fix heavy metals and other mine waste contaminants in stable mineral carbonate forms. The research on mineral carbonation has been focused mainly on magnesium and calcium silicate minerals, while iron-containing minerals have not been investigated as much, to be used for the feedstock of mineral carbonation, due largely to their value for steel production. However, as shown in Table 3, Australia holds the world's largest resources and is the biggest producer of iron ore. That also includes large deposits of low-grade magnetite that can have the potential to be used for mineral carbon sequestration. Magnetite can react with carbon dioxide to form siderite according to the basic reaction below [112].

$$Fe_3O_4 + CO_2 \rightarrow FeCO_3 + Fe_2O_3 \tag{6}$$

However, during magnetite carbonation, formation of hematite as a passive layer can limit the carbonation kinetic, resulting in a slow reaction rate [113].

In addition, many of the Australian minerals are found in ultramafic rocks, which are rich in magnesium, iron, and calcium, as shown in Table 3. Alkaline mining wastes provide several economic advantages for CCUS by mineral carbonation. Mineral carbonation can reduce the financial burden associated with the disposal of mine waste materials. The availability of these wastes, such as red mud and fly ash, reduces the cost of sorbents which are currently used as the feedstock for mineral carbonation. Furthermore, mining wastes, tailings, and by-products are often fine and comminution energy for pretreatment of minerals is not required. In addition, neutralising the mining waste, which can be potentially hazardous to the environment, by mineral carbonation makes it safe for recycling and disposal which reduces the cost of waste management, and carbonated minerals can be used as locally available, environmentally benign materials for mine reclamation.

Figure 1 shows the schematic model of climate change adaptation technology for the Australian mining industry. Integrated mineral carbonation of alkaline mining waste in Australia brings forth the chance for waste management of some of the critical commodities' by-products, such as red mud and fly ash, at the same time reducing Australia's GHG emission and mine reclamation. This is particularly significant for countries such as Australia with many active and legacy mines. There is an estimate of over 1 million legacy mines worldwide [114], and more than 60,000 in Australia, many of which require remediation to solve and mitigate environmental and social issues such as AMD associated with past mining activities in those regions [115].

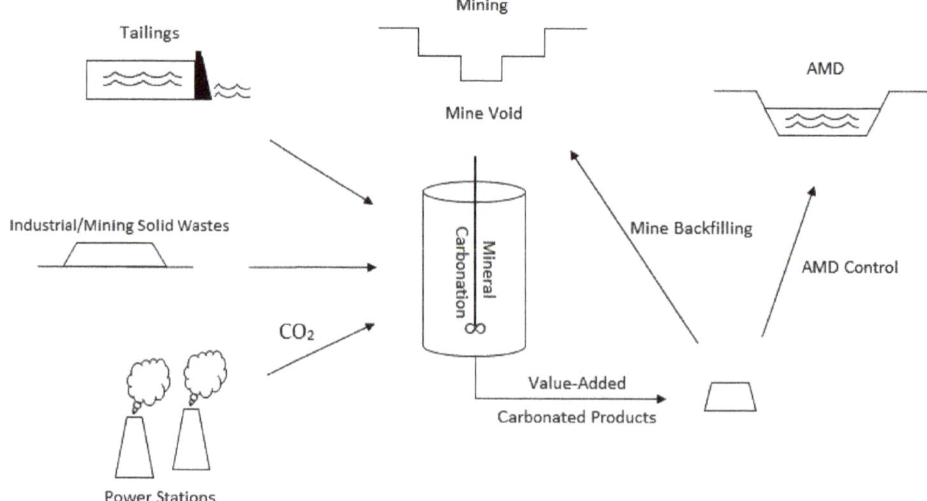

Figure 1. Integrated mineral carbonation of mining waste for waste management and mine reclamation.

The integrated mineral carbonation model will contribute to the development of an innovative climate change adaptation technology by addressing three critical issues at the same time including GHG reduction (CO_2 sequestration), mine waste management (neutralisation and reuse) and mine reclamation (i.e., mine backfilling and AMD control using the environmentally benign products of mineral carbonation).

According to Table 3, the magnitude of Australia's mineral production is very significant considering Australia's ranking for production in many critical commodities. Hence, there are accordingly large amount of mining waste, by-products and GHG production. Integration of mineral carbonation provides the opportunity to offset some of the GHG emission of the mineral's life cycle in a sustainable way. The outcome of this approach will have the potential to provide national economic, environmental and social benefits for Australia by contributing to Australia's adaptation to climate change.

4.2. Synergies between Coal Mines and Coal-Fired Power Stations in Close Proximity

One-third of Australia's GHG emissions are from the electricity generating sector, 80% of which is based on coal-fired power generation, which has been critical for the operation of many Australian industries [116]. Furthermore, it has created many jobs and is central to the Australian economy. For instance, one out of eight employment in the whole of Queensland and one out of four in Central Queensland, are dependent on the resource industry [116]. Therefore, because of the effect of coal industry on Australia's economy it is crucial that adaptation technologies such as CCUS be integrated into the industry to support coal use while offsetting its emissions, especially due to the fact that fossil fuels will still remain a significant portion of Australia and world's energy portfolio in the short-term.

Coal has historically been the primary source of electricity generation in Australia, and as the result, numerous coal-fired power stations have been built to support this demand, mainly in proximity of some of the major coal mines. The shorter distance between coal mines and power stations reduces the cost of coal transport between the source and end use. However, it also provides a unique synergy between the two for coal mine reclamation via the integrated mineral carbonation model. Figure 2 shows the locations of coal-fired power stations in proximity to coal mines in Australia.

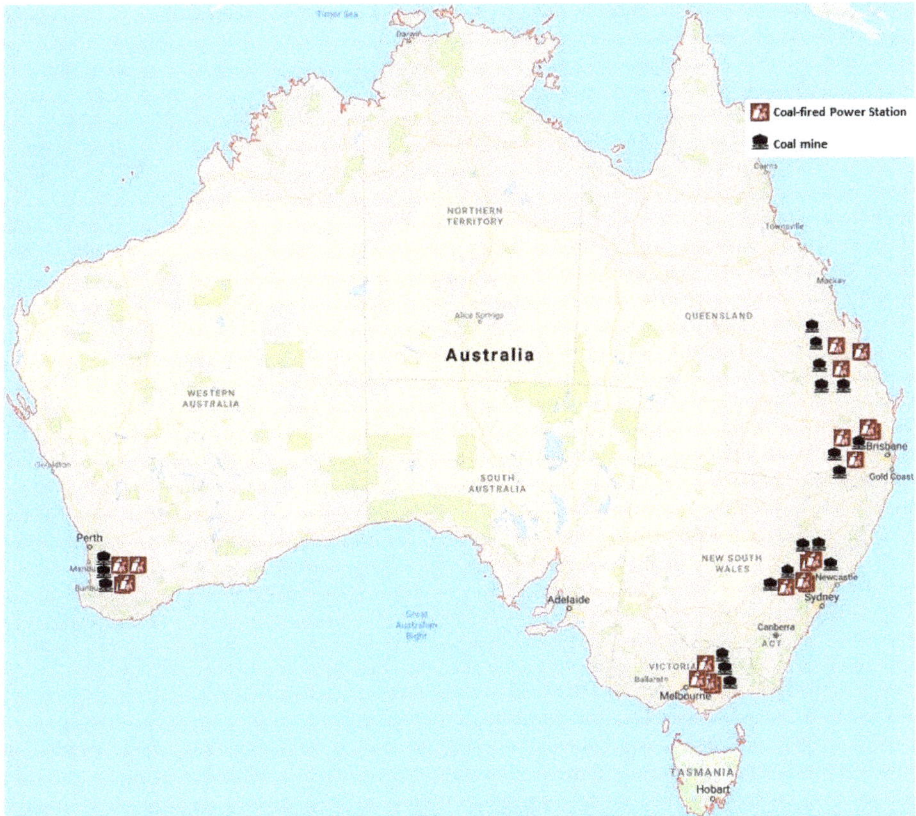

Figure 2. Coal-fired power stations in proximity of coal mines in Australia.

Fly ash is the primary by-product of coal combustion in power stations. Approximately 777.1 million tonnes (Mt) of coal combustion by-products (CCBs) are produced globally every year as the result of burning coal in power stations [117]. The global average of fly ash reuse and utilisation is 53.5% [117]. In the National waste report 2018, the Australian Government reported, for the financial year of 2016–2017, the generation of 12.3 Mt of ash by coal-fired power stations, 43% of which was recycled [118]. It was emphasised in this report that there are opportunities for fly ash utilisation if the contamination issues are addressed. Comparison of Australia's CCBs recycling ratio with other regions such as Japan (96.4 %) and Europe-EU15 (90.9 %) [117] points out the existing opportunity of fly ash utilisation in Australia. However, fly ash can be a hazardous material due to the presence of potentially toxic trace elements in the flue gas [119]. This is a major liability in Australia due to the scale of coal-fired electricity production, as it has the potential to cause serious environmental issues if it is not managed properly.

The scale of which fly ash is produced in Australia provides the opportunity for its recycling as the feedstock for mineral carbonation and, consequently, mine reclamation. Mineral carbonation has the potential to neutralise the hazardous effect of fly ash, turning it into environmentally benign minerals. This can be achieved by neutralising the alkalinity of fly ash due to the reaction with CO_2, and potentially immobilisation of trace elements due to carbonate formation. The carbonated minerals can then to be used for reclamation of the coal mines, such as backfilling and acid mine drainage treatment.

Short distances between mines and coal-fired power stations not only reduces the cost of coal transport between the coal mines and power stations, but it also provides the opportunity to transport the carbonated fly ash back to the coal mine, at low cost. For instance, Stanwell Corporation, which is owned by the Queensland Government and is the state's largest electricity generator, operates coal-fired power stations at Central and South East Queensland. The 1843 MW Tarong Power Stations are located in the South Burnet in South East Queensland [120], only a few kilometres away from Meandu Mine that provides their coal. Several other coal-fired power stations in Australia are similarly located in close proximity with coal mines (Figure 2) providing a unique synergy and the opportunity to overcome the economic barrier of deploying mineral carbonation technology in Australia's mining industry, as an economically viable climate change adaptation technology.

4.3. Red Mud

Australia produced 1.49 Mt of aluminium, as the world's 6th largest producer, in 2017 [121]. Consequently, large quantities of red mud, a by-product of the Bayer process in the alumina industry, are generated every year in Australia. Red mud is regarded as highly hazardous, with significantly high alkalinity and pH values up to 13. It often contains considerable amounts of trace elements and poses critical hazards to the environment. For instance, the dam wall of Ajka Alumina plant's red mud reservoir in Hungary, collapsed on October 4th 2010. As the result, 10 people were killed, over 100 were injured, and the lakes, rivers and surrounding ecosystems were significantly polluted [122–124]. In Australia, disposal of 26.4 Mt of red mud was reported in 2018, at sites in Queensland and Western Australia. Over the last 50 years, 812 Mt of red mud has been deposited in Australia [118]. While many methods have been trialled, such as red mud neutralisation with sea water, associated waste management concerns still exist, and millions of dollars are spent every year for rehabilitation.

Mineral carbonation can be used to reduce and neutralise the alkalinity and contaminations in red mud. The reaction pathways of red mud carbonation are reported in the literature [125,126]. It reduces the environmental risk of long-term red mud storage and utilisation, as well as offsetting CO_2 emission associated with aluminium production [127]. Treatment of bauxite residue with mineral carbonation has the advantage of not introducing impurities to the refining process [93]. It also reduces the impact of residue leachate on clay seal material and its long-term deterioration at storage facilities. It also alleviates the effect of red mud residue leachate on groundwater contamination. Furthermore, carbonated red mud requires reduced drying area due to higher shear strength and potentially faster drying time. In addition, the surface of carbonated red mud is less susceptible to aerosol emissions, besides from the fact that reduced drying area limits the exposure of red mud to dust emissions [128–130].

4.4. Ultramafic Mine Tailings

Many of the mineral commodities in Australia are mined from ultramafic ores. The growing demand for the development of infrastructure for low carbon technologies has resulted in increased ultramafic metals' mining operations such as nickel and cobalt. For example, BHP has recently shown interest in increased cobalt and nickel production driven by the market call for rechargeable batteries. BHP's plan for increasing the production of nickel sulphate to 200,000 metric tons per year is reported, which will make Nickel West in Western Australia the largest plant in the world [131]. Furthermore, a number of other commodities are mined in Australia from ultramafic rocks such as PGEs (platinum-group elements), diamond and legacy asbestos mines.

Many concerns exist for the climate change impacts of mining wastes and by-products of ultramafic mine tailings. They can be hazardous due to containing many toxic elements and mineral groups, and in the case of asbestos tailings, pose serious human health risks if not managed properly [132].

Ultramafic mine tailings are often very rich in magnesium, which makes them a suitable feedstock for CO_2 sequestration by mineral carbonation. In addition, they are often ground and do not require high activation energy and grinding, which reduces some of the related costs. More importantly, the carbonation process can fix and immobilise hazardous elements and mineral groups in carbonate forms, and in some cases, produce environmentally benign minerals which can be used for mine site rehabilitation. In case of legacy asbestos mine tailings, mineral carbonation can potentially reduce contamination and dust emissions, as well as providing the benefit of capturing CO_2 permanently in a stable form. Table 4 lists the potential mining wastes in Australia to be used for the integrated mineral carbonation.

Table 4. Applicability of Australian mining waste for integrated mineral carbonation for mine waste management and mine reclamation.

Mine Waste	Carbonation Advantages	Product Applicability	Carbonation References
Red mud	• Neutralising the high alkalinity for safe waste management • Immobilisation and fixation of heavy metals and contaminations • Reduced the risk of groundwater contamination • Reduced drying area • Reduced hazardous dust emissions • Large-scale availability of red mud due to Australia's extent of aluminium production	• Mine backfilling • Acid mine drainage (AMD) control	Cooling et al. 2002 [130] Tran 2016 [125] Revathy et al. 2017 [133] Liang et al. 2018 [134] Sahu et al. 2010 [135]
Fly ash	• Produced at power stations as a major source of CO_2 emission • Reduced aerosol emissions • Large-scale Availability due to Australia's extent of coal-fired power generation • No requirement for grinding	• Mine backfilling • Acid mine drainage (AMD) control • Soil amendment	Liu et al. 2018 [136] Jaschik et al. 2016 [137] Tamilselvi Dananjayan et al. 2016 [138] Ukwattage et al. 2015 [139]
Ultramafic mine tailings (nickel, diamond, PGE and asbestos tailings)	• Neutralising the alkalinity • Immobilisation of environmentally hazardous metals in tailings • No requirement for grinding • Reducing hazardous/fatal dust emissions	• Acid mine drainage (AMD) control	Nickel tailings: Teir et al. 2007 [140] Teir et al. 2009 [141] Diamond tailings: Mervine et al. 2018 [142] PGE tailings: Meyer et al. 2014 [143] Vogeli et al. 2011 [144] Asbestos tailings: McCutcheon et al. 2017 [145] Oskierski et al. 2016 [146]

5. Barriers to Utilisation of Integrated Mineral Carbonation

The initial capital investment is one of the primary barriers for utilisation of mineral carbonation in Australia's mining industry. The range of the cost estimates of implementing mineral carbonation technology reported in the literature varies in a wide range of values [57,147–149]. Although some of these initial cost estimations (i.e., \$69/ton CO_2 [57]) are too high to be economically feasible, the fact that this method has the potential for permanent storage over geologic time makes it desirable [45].

Despite the fact that many strategies used to improve the mineral carbonation process, they have been successful in terms of increasing the overall yields of carbonation process; each method has its own disadvantages. The energy cost associated with fine grinding of the feedstock directly contributes to decreasing the process viability [149–152]. Thermal activation of serpentine, for instance, requires a considerable amount of energy that results in a not economically feasible process unless methods of thermal activation can be optimised or reconsidered [69,153]. Obviously, high operating temperatures and pressures also add significantly to the cost of operation. Therefore, studies have been investigating the extraction of Mg from minerals under low temperature and pressure conditions using mineral acids [61,64,154]. Also, any chemical that is added to the reactants must be fully recoverable in a real process to make it commercially viable. The rate-limiting step for serpentine carbonation is the dissolution step [96]. Many previous investigations have reported the potential passivating quality of silica layers during the dissolution of serpentinite minerals [74,155]. Other research has shown reprecipitation of amorphous silica in the solution that causes a precipitate layer to form on the surface of the mineral particles and significantly limits further extraction of magnesium [96]. The removal of silica layers is possible through a concurrent mechanical exfoliation process could enhance carbonation yields [66,156].

Obtaining a constant source of high-concentration CO_2 can also be a major issue to deploy mineral carbonation plant in Australian mining industries. There might not always be a reliable source of CO_2 in the vicinity of mining waste storage facilities to support mineral carbonation plants. Therefore, the transportation of CO_2 to the carbonation plant can potentially add to the cost of the operation. However, this is not always the case and for example Alcoa used the CO_2 from the waste stream of an ammonia plant near to the Alcoa Kwinana Refinery for the operation of their mineral carbonation plant [93].

Mineral carbonation can reduce the alkalinity of the mining waste, by the reactions between CO_2 and alkali and alkaline earth metals. However, in some cases there is risk of changing the alkalinity back to higher pH again after carbonation. Alkali elements such as sodium, which can be found in waste streams like red mud, can react with CO_2 and generate bicarbonates that are less stable for long-term storage of CO_2 above the ground due to their solubility in water [16,47,157]. The carbonation reaction of CO_2 with the sodium hydroxide, which is found in red mud, is relatively rapid, while the reaction with the alkaline earth metals can be slower. In addition, the presence of tricalcium aluminate (TCA6) in red mud is reported to cause technical issues by increasing the pH in the residue over time [130].

6. Conclusions

The effects of climate change have been observed on a global scale. CCUS via mineral carbonation appears to be a promising emerging technology to be included in the portfolio of climate change mitigation options, as an effective long-term strategy. Barriers mainly due to the cost of deployment and the lack of a reliable source of alkaline waste have hindered its application at an industrial scale. Utilisation of mining waste provides the opportunity to overcome this challenge. The abundance of mining wastes and processing by-products such as fly ash, red mud and ultramafic mine tailings in Australia creates a unique opportunity for retrofitting mineral carbonation in its mining sector. The high environmental and financial liabilities of mining wastes in Australia justify investments in detailed investigation into mineral carbonation opportunities. Integrated mineral carbonation of mining wastes contributes to mining waste management, mine reclamation, and GHG emissions. Development of climate change adaptation technologies within the mining industry in Australia, as a country with an advanced resource industry, can also initiate global models for utilisation in developing countries.

This review discussed the opportunities and barriers that exist for unlocking the value of mining waste by mineral carbonation. Changing CO_2 policies, further research for improved technology through better understanding of the mineral carbonation process, the development of pilot-scale operations and controlling mining waste contamination could collectively promote the progression of ex situ mineral carbonation to make it applicable at industrial scales.

Author Contributions: Conceptualisation, M.A. and M.E.; Investigation, M.A., M.E. and F.F.; Resources, M.E. and J.A.; Writing—Original Draft Preparation, M.A. and F.F.; Writing—Review and Editing, M.A., F.F., M.E. and J.A.; Visualisation, M.A.; Supervision, M.E. and J.A.

Funding: This research was supported by the National Strategic Project-Carbon Upcycling of the National Research Foundation of Korea (NRF) funded by the Ministry of Science and ICT (MSIT), the Ministry of Environment (ME) and the Ministry of Trade, Industry and Energy (MOTIE) (2017M3D8A2084752).

Acknowledgments: M.A. thanks David Cooling from Alcoa for technical discussion and A/David Doley for reviewing an early draft of the manuscript.

References

1. Scoping of the IPCC Sixth Assessment Report (AR6). Background, Cross Cutting Issues and the AR6 Synthesis Report. Available online: https://www.ipcc.ch/site/assets/uploads/2018/04/040820171122-Doc.-6-SYR_Scoping.pdf (accessed on 20 January 2019).
2. Trends in Atomospheric Carbon Dioxide. 2018. Available online: https://www.esrl.noaa.gov/gmd/ccgg/trends/data.html (accessed on 20 January 2019).
3. *Climate Change 2014: Synthesis Report. Summary for Policymaker*; Working Groups of the Intergovernmental Panel on Climate Change, IPCC: Geneva, Switzerland, 2014.
4. Australian Government, Department of the Environment and Energy. Australia's 2030 Climate Change Target. 2015. Available online: http://www.environment.gov.au/system/files/resources/c42c11a8-4df7-4d4f-bf92-4f14735c9baa/files/factsheet-australias-2030-climate-change-target.pdf (accessed on 20 January 2019).
5. Australian Government, Department of the Environment and Energy. Quarterly Update of Australia's National Greenhouse Gas Inventory for June 2018. Available online: http://www.environment.gov.au/system/files/resources/e2b0a880-74b9-436b-9ddd-941a74d81fad/files/nggi-quarterly-update-june-2018.pdf (accessed on 20 January 2019).
6. SunSHIFT Pty Ltd. *Renewable Energy in the Australian Mining Sector*; Australian Government White Paper; Australian Renewable Energy Agency: Sydney, Australia, 2017.
7. United Nations. *The Paris Agreement*; United Nations: Paris, France, 2015.
8. The Parliament of the Commonwealth of Australia. Retirement of Coal Fired Power Stations. Interim Report. 2016. Available online: https://www.aph.gov.au/Parliamentary_Business/Committees/Senate/Environment_and_Communications/Coal_fired_power_stations/Interim_Report (accessed on 20 January 2019).
9. Jewell, J.; Cherp, A.; Riahi, K. Energy security under de-carbonization scenarios: An assessment framework and evaluation under different technology and policy choices. *Energy Policy* **2014**, *65*, 743–760. [CrossRef]
10. IEA COP24 Brochure. International Energy Agency at COP24. 2018. Available online: https://www.iea.org/media/topics/climatechange/IEA-COP24-brochure.pdf?utm_campaign=IEA%20newsletters&utm_source=SendGrid&utm_medium=Email (accessed on 20 January 2019).
11. International Energy Agency (IEA). *Energy Technology Perspectives*; OECD/IEA: Paris, France, 2010.
12. International Energy Agency (IEA). *CCS Retrofit, Analysis of the Global Installed Power Plant Fleet*; OECD/IEA: Paris, France, 2012.
13. International Energy Agency (IEA). Sustainable Development Scenario. 2015. Available online: https://www.iea.org/weo/weomodel/sds/ (accessed on 20 January 2019).
14. United Nations. The Sustainable Development Goals. 2015. Available online: https://www.un.org/sustainabledevelopment/sustainable-development-goals/ (accessed on 20 January 2019).
15. Sanna, A.; Uibu, M.; Caramanna, G.; Kuusik, R.; Maroto-Valer, M.M. A review of mineral carbonation technologies to sequester CO_2. *Chem. Soc. Rev.* **2014**, *43*, 8049–8080. [CrossRef] [PubMed]
16. Bobicki, E.R.; Liu, Q.; Xu, Z.; Zeng, H. Carbon capture and storage using alkaline industrial wastes. *Prog. Energy Combust. Sci.* **2012**, *38*, 302–320. [CrossRef]
17. Bobicki, E.R.; Liu, Q.; Xu, Z. Mineral carbon storage in pre-treated ultramafic ores. *Miner. Eng.* **2015**, *70*, 43–54. [CrossRef]
18. Sarvaramini, A.; Assima, G.P.; Beaudoin, G.; Larachi, F. Biomass torrefaction and CO_2 capture using mining wastes—A new approach for reducing greenhouse gas emissions of co-firing plants. *Fuel* **2014**, *115*, 749–757. [CrossRef]
19. Sanna, A.; Dri, M.; Hall, M.R.; Maroto-Valer, M. Waste materials for carbon capture and storage by mineralisation (CCSM)—A UK perspective. *Appl. Energy* **2012**, *99*, 545–554. [CrossRef]
20. Li, J.; Hitch, M.; Power, I.M.; Pan, Y. Integrated mineral carbonation of ultramafic mine deposits—A review. *Minerals* **2018**, *8*, 147. [CrossRef]
21. Araizi, P.-K. Accelerated Carbonation of Wastes and Minerals. Ph.D. Thesis, University of Greenwich, London, UK, 2015.
22. *BHP Sustainability Report*; BHP plc: Melbourne, Australia, 2017.
23. *Glencore Sustainability Report*; Glencore plc: Baar, Switzerland, 2017.

24. *AngloAmerican Sustainability Report*; Anglo American plc: London, UK, 2017.

25. CCUS Council. Available online: https://www.gov.uk/government/groups/ccus-council (accessed on 20 January 2019).

26. Xu, X.; Liu, W.; Chu, G.; Zhang, G.; Luo, D.; Yue, H.; Liang, B.; Li, C. Energy-efficient mineral carbonation of $CaSO_4$ derived from wollastonite via a roasting-leaching route. *Hydrometallurgy* **2019**, *184*, 151–161. [CrossRef]

27. Deng, C.; Liu, W.; Chu, G.; Luo, D.; Zhang, G.; Wang, L.; Yue, H.; Liang, B.; Li, C. Aqueous carbonation of $MgSO_4$ with $(NH_4)_2CO_3$ for CO_2 sequestration. *Greenh. Gases Sci. Technol.* **2019**. [CrossRef]

28. Wang, F.; Dreisinger, D.; Jarvis, M.; Hitchins, T. Kinetics and mechanism of mineral carbonation of olivine for CO_2 sequestration. *Miner. Eng.* **2019**, *131*, 185–197. [CrossRef]

29. Liu, W.; Yin, S.; Luo, D.; Zhang, G.; Yue, H.; Liang, B.; Wang, L.; Li, C. Optimising the recovery of high-value-added ammonium alum during mineral carbonation of blast furnace slag. *J. Alloys Compd.* **2019**, *774*, 1151–1159. [CrossRef]

30. Ragipani, R.; Bhattacharya, S.; Suresh, A.K. Towards efficient calcium extraction from steel slag and carbon dioxide utilisation via pressure-swing mineral carbonation. *React. Chem. Eng.* **2019**, *4*, 52–66. [CrossRef]

31. *Rio Tinto Sustainable Development Report*; Rio Tinto plc: London, UK, 2017.

32. *BHP Annual Report*; BHP plc: Melbourne, Australia, 2017.

33. *Glencore Annual Report*; Glencore plc: Baar, Switzerland, 2017.

34. Nguyen, D. *Carbon Dioxide Geological Sequestration: Technical and Economic Reviews*; Society of Petroleum Engineers: San Antonio, TX, USA, 2003. [CrossRef]

35. Ennis-King, J.; Paterson, L. Engineering aspects of geological sequestration of carbon dioxide. In Proceedings of the SPE Asia Pacific Oil and Gas Conference and Exhibition, Melbourne, Australia, 8–10 October 2002.

36. Holloway, S. Underground sequestration of carbon dioxide—A viable greenhouse gas mitigation option. *Energy* **2005**, *30*, 2318–2333. [CrossRef]

37. Marini, L. *Geological Sequestration of Carbon Dioxide: Thermodynamics, Kinetics, and Reaction Path Modeling*; Elsevier: Amsterdam, The Netherlands, 2006; Volume 11.

38. Wilson, T. The deep ocean disposal of carbon dioxide. *Energy Convers. Manag.* **1992**, *33*, 627–633. [CrossRef]

39. Handa, N.; Ohsumi, T. *Direct ocean Disposal of Carbon Dioxide*; Terrapub: Tokyo, Japan, 1995.

40. Brewer, P.G.; Friederich, G.; Peltzer, E.T.; Orr, F.M. Direct experiments on the ocean disposal of fossil fuel CO_2. *Science* **1999**, *284*, 943–945. [CrossRef] [PubMed]

41. Palmgren, C.R.; Morgan, M.G.; Bruine de Bruin, W.; Keith, D.W. Initial Public Perceptions of Deep Geological and Oceanic Disposal of Carbon Dioxide. *Environ. Sci. Technol.* **2004**, *38*, 6441–6450. [CrossRef] [PubMed]

42. Usui, N.; Ikenouchi, M. The biological CO_2 fixation and utilization project by RITE (1)—Highly-effective photobioreactor system—. *Energy Convers. Manag.* **1997**, *38*, S487–S492. [CrossRef]

43. Yun, Y.S.; Lee, S.B.; Park, J.M.; Lee, C.I.; Yang, J.W. Carbon dioxide fixation by algal cultivation using wastewater nutrients. *J. Chem. Technol. Biotechnol.* **1997**, *69*, 451–455. [CrossRef]

44. Aresta, M. *Carbon Dioxide as Chemical Feedstock*; John Wiley & Sons: Hoboken, NJ, USA, 2010.

45. Park, A.-H.A. Carbon Dioxide Sequestration: Chemical and Physical Activation of Aqueous Carbonation of Mg-Bearing Minerals and pH Swing Process. Ph.D. Thesis, The Ohio State University, Columbus, OH, USA, 2005.

46. Werner, M.S. Combined CO_2 Capture and Storage by Mineralization: Dissolution and Carbonation of Activated Serpentine. Ph.D. Thesis, ETH Zurich, Zürich, Switzerland, 2014.

47. Lackner, K.S.; Wendt, C.H.; Butt, D.P.; Joyce, E.L., Jr.; Sharp, D.H. Carbon dioxide disposal in carbonate minerals. *Energy* **1995**, *20*, 1153–1170. [CrossRef]

48. Metz, B.; Davidson, O.; De Coninck, H. *Carbon Dioxide Capture and Storage: Special Report of the Intergovernmental Panel on Climate Change*; Cambridge University Press: Cambridge, MA, USA, 2005.

49. Loughnan, F.C. *Chemical Weathering of the Silicate Minerals*; Elsevier: New York, NY, USA, 1969; p. 154.

50. Garrels, R.M. The carbonate-silicate geochemical cycle and its effect on atmospheric carbon dioxide over the past 100 million years. *Am. J. Sci.* **1983**, *283*, 641–683.

51. Möller, P. *Magnesite: Geology, Mineralogy, Geochemistry, Formation of Mg-Carbonates*; Gebrüder Borntraeger: Berlin, Germany, 1989.

52. Seifritz, W. CO_2 disposal by means of silicates. *Nature* **1990**, *345*, 486. [CrossRef]

53. Hasan, S.N.M.S.; Kusin, F.M.; Jusop, S.; Yusuff, F.M. Potential of soil, sludge and sediment for mineral carbonation process in selinsing gold mine, Malaysia. *Minerals* **2018**, *8*, 257. [CrossRef]

54. Lackner, K.S. Carbonate chemistry for sequestering fossil carbon. *Annu. Rev. Energy Environ.* **2002**, *27*, 193–232. [CrossRef]

55. Goff, F.; Lackner, K. Carbon dioxide sequestering using ultramafic rocks. *Environ. Geosci.* **1998**, *5*, 89–101. [CrossRef]

56. Zafaranloo, A. Kinetics of Magnesium Extraction from Activated Serpentine by Carbonic Acid. Ph.D. Thesis, University of Sydney, Sydney, Australia, 2015.

57. O'Connor, W.; Dahlin, D.; Rush, G.; Gerdemann, S.; Penner, L.; Nilsen, D. *Aqueous Mineral Carbonation: Mineral Availability, Pretreatment, Reaction Parametrics, and Process Studies*; Office of Process Development, National Energy Technology Laboratory (Formerly Albany Research Center) Office of Fossil Energy, US DOE: Albany, OR, USA, 2005; p. 462.

58. Brent, G.F.; Petrie, J.G. CO_2 sequestration by mineral carbonation in the Australian context. In Proceedings of the Chemeca 2008: Towards a Sustainable Australasia, Newcastle, Australia, 28 September–1 October 2008; p. 1273.

59. Lackner, K.S.; Butt, D.P.; Wendt, C.H. Progress on binding CO_2 in mineral substrates. *Energy Convers. Manag.* **1997**, *38*, S259–S264. [CrossRef]

60. Fagerlund, J. Carbonation of $Mg(OH)_2$ in a Pressurised Fluidised Bed for CO_2 Sequestration. Ph.D Thesis, Åbo Akademi University, Turku, Finland, 2012.

61. Teir, S.; Revitzer, H.; Eloneva, S.; Fogelholm, C.-J.; Zevenhoven, R. Dissolution of natural serpentinite in mineral and organic acids. *Int. J. Miner. Process.* **2007**, *83*, 36–46. [CrossRef]

62. Drief, A.; Nieto, F. The effect of dry grinding on antigorite from Mulhacen, Spain. *Clays Clay Miner.* **1999**, *47*, 417–424. [CrossRef]

63. Kim, D.-J.; Chung, H.-S. Effect of grinding on the structure and chemical extraction of metals from serpentine. *Part. Sci. Technol.* **2002**, *20*, 159–168. [CrossRef]

64. Van Essendelft, D.T.; Schobert, H.H. Kinetics of the Acid Digestion of Serpentine with Concurrent Grinding. 3. Model Validation and Prediction. *Ind. Eng. Chem. Res.* **2010**, *49*, 1588–1590. [CrossRef]

65. Kleiv, R.A.; Thornhill, M. The effect of mechanical activation in the production of olivine surface area. *Miner. Eng.* **2016**, *89*, 19–23. [CrossRef]

66. Rashid, M.I.; Benhelal, E.; Farhang, F.; Oliver, T.K.; Rayson, M.S.; Brent, G.F.; Stockenhuber, M.; Kennedy, E.M. Development of Concurrent grinding for application in aqueous mineral carbonation. *J. Clean. Prod.* **2019**, *212*, 151–161. [CrossRef]

67. Balucan, R.D.; Dlugogorski, B.Z. Thermal Activation of Antigorite for Mineralization of CO_2. *Environ. Sci. Technol.* **2013**, *47*, 182–190. [CrossRef] [PubMed]

68. Balucan, R.D.; Kennedy, E.M.; Mackie, J.F.; Dlugogorski, B.Z. Optimization of antigorite heat pre-treatment via kinetic modeling of the dehydroxylation reaction for CO_2 mineralization. *Greenh. Gases Sci. Technol.* **2011**, *1*, 294–304. [CrossRef]

69. Farhang, F.; Kennedy, E.; Stockenhuber, M.; Brent, G.; Rayson, M. Dehydroxylation of Magnesium Silicate Minerals for Carbonation. WIPO Patent No.: WO/2017/106929, 29 June 2017.

70. Brindley, G.W.; Hayami, R. Kinetics and mechanisms of dehydration and recrystallization of serpentine: II Spectrum of activation energies for recrystallization. In Proceedings of the 12th National Conference on Clays and Clay Minerals, Atlanta, GA, USA, 30 September–2 October 1963; pp. 49–54.

71. Mann, J.P. Serpentine Activation for CO_2 Sequestration. Ph.D. Thesis, University of Sydney, Faculty of Engineering & IT, School of Chemical & Biomolecular Engineering, Sydney, Australia, 2014.

72. Maroto-Valer, M.M.; Fauth, D.J.; Kuchta, M.E.; Zhang, Y.; Andrésen, J.M. Activation of magnesium rich minerals as carbonation feedstock materials for CO_2 sequestration. *Fuel Process. Technol.* **2005**, *86*, 1627–1645. [CrossRef]

73. McKelvy, M.J.; Chizmeshya, A.V.G.; Diefenbacher, J.; Béarat, H.; Wolf, G. Exploration of the Role of Heat Activation in Enhancing Serpentine Carbon Sequestration Reactions. *Environ. Sci. Technol.* **2004**, *38*, 6897–6903. [CrossRef] [PubMed]

74. Daval, D.; Hellmann, R.; Martinez, I.; Gangloff, S.; Guyot, F. Lizardite serpentine dissolution kinetics as a function of pH and temperature, including effects of elevated pCO_2. *Chem. Geol.* **2013**, *351*, 245–256. [CrossRef]

75. Prigiobbe, V.; Hänchen, M.; Costa, G.; Baciocchi, R.; Mazzotti, M. Analysis of the effect of temperature, pH, CO_2 pressure and salinity on the olivine dissolution kinetics. *Energy Procedia* **2009**, *1*, 4881–4884. [CrossRef]

76. Felmy, A.R.; Qafoku, O.; Arey, B.W.; Hu, J.Z.; Hu, M.; Todd Schaef, H.; Ilton, E.S.; Hess, N.J.; Pearce, C.I.; Feng, J.; et al. Reaction of water-saturated supercritical CO_2 with forsterite: Evidence for magnesite formation at low temperatures. *Geochim. Cosmochim. Acta* **2012**, *91*, 271–282. [CrossRef]

77. Hanchen, M.; Prigiobbe, V.; Baciocchi, R.; Mazzotti, M. Precipitation in the Mg-carbonate system—Effects of temperature and CO_2 pressure. *Chem. Eng. Sci.* **2008**, *63*, 1012–1028. [CrossRef]

78. Bonfils, B.; Julcour-Lebigue, C.; Guyot, F.; Bodénan, F.; Chiquet, P.; Bourgeois, F. Comprehensive analysis of direct aqueous mineral carbonation using dissolution enhancing organic additives. *Int. J. Greenh. Gas Control* **2012**, *9*, 334–346. [CrossRef]

79. Swanson, E.J. Catalytic Enhancement of Silicate Mineral Weathering for Direct Carbon Capture and Storage. Ph.D. Thesis, Columbia University, New York City, NY, USA, 2014.

80. Park, A.-H.A.; Fan, L.-S. CO_2 mineral sequestration: Physically activated dissolution of serpentine and pH swing process. *Chem. Eng. Sci.* **2004**, *59*, 5241–5247. [CrossRef]

81. Werner, M.; Hariharan, S.; Mazzotti, M. Flue gas CO_2 mineralization using thermally activated serpentine: From single-to double-step carbonation. *Phys. Chem. Chem. Phys.* **2014**, *16*, 24978–24993. [CrossRef] [PubMed]

82. Lee, J.-Y. *Research and Education of CO_2 Separation from Coal Combustion Flue Gases with Regenerable Magnesium Solutions*; Chemical Engineering Program Department of Biomedical, Chemical, and Environmental Engineering, University of Cincinnati: Cincinnati, OH, USA, 2013.

83. Huijgen, W.J.J.; Comans, R.N.J. Mineral CO_2 Sequestration by Steel Slag Carbonation. *Environ. Sci. Technol.* **2005**, *39*, 9676–9682. [CrossRef] [PubMed]

84. Baciocchi, R.; Costa, G.; Di Gianfilippo, M.; Polettini, A.; Pomi, R.; Stramazzo, A. Thin-film versus slurry-phase carbonation of steel slag: CO_2 uptake and effects on mineralogy. *J. Hazard. Mater.* **2015**, *283*, 302–313. [CrossRef] [PubMed]

85. Huijgen, W.J.J.; Comans, R.N.J. Carbonation of Steel Slag for CO_2 Sequestration: Leaching of Products and Reaction Mechanisms. *Environ. Sci. Technol.* **2006**, *40*, 2790–2796. [CrossRef] [PubMed]

86. Morone, M.; Costa, G.; Polettini, A.; Pomi, R.; Baciocchi, R. Valorization of steel slag by a combined carbonation and granulation treatment. *Miner. Eng.* **2014**, *59*, 82–90. [CrossRef]

87. Librandi, P.; Costa, G.; Souza, A.C.B.D.; Stendardo, S.; Luna, A.S.; Baciocchi, R. Carbonation of Steel Slag: Testing of the Wet Route in a Pilot-scale Reactor. *Energy Procedia* **2017**, *114*, 5381–5392. [CrossRef]

88. Montes-Hernandez, G.; Pérez-López, R.; Renard, F.; Nieto, J.M.; Charlet, L. Mineral sequestration of CO_2 by aqueous carbonation of coal combustion fly-ash. *J. Hazard. Mater.* **2009**, *161*, 1347–1354. [CrossRef] [PubMed]

89. Dindi, A.; Quang, D.V.; Vega, L.F.; Nashef, E.; Abu-Zahra, M.R.M. Applications of fly ash for CO_2 capture, utilization, and storage. *J. CO_2 Util.* **2019**, *29*, 82–102. [CrossRef]

90. Ji, L.; Yu, H.; Yu, B.; Zhang, R.; French, D.; Grigore, M.; Wang, X.; Chen, Z.; Zhao, S. Insights into Carbonation Kinetics of Fly Ash from Victorian Lignite for CO_2 Sequestration. *Energy Fuels* **2018**, *32*, 4569–4578. [CrossRef]

91. Jacobs, A.D. Quantifying the Mineral Carbonation Potential of Mine Waste Material: A New Parameter for Geospatial Estimation. Ph.D. Thesis, University of British Columbia, Vancouver, BC, Canada, 2014.

92. Hitch, M.; Ballantyne, S.M.; Hindle, S.R. Revaluing mine waste rock for carbon capture and storage. *Int. J. Min. Reclam. Environ.* **2010**, *24*, 64–79. [CrossRef]

93. Alcoa. Long Term Residue Management Strategy. Alcoa Australia Aluminium Kwainana 2012 Report. Available online: https://www.alcoa.com/australia/en/pdf/kwinana_refinery_ltrms_report_2012.pdf (accessed on 20 January 2019).

94. Brent, G.F.; Allen, D.J.; Eichler, B.R.; Petrie, J.G.; Mann, J.P.; Haynes, B.S. Mineral Carbonation as the Core of an Industrial Symbiosis for Energy-Intensive Minerals Conversion. *J. Ind. Ecol.* **2012**, *16*, 94–104. [CrossRef]

95. Farhang, F.; Oliver, T.K.; Rayson, M.; Brent, G.; Stockenhuber, M.; Kennedy, E. Experimental study on the precipitation of magnesite from thermally activated serpentine for CO_2 sequestration. *Chem. Eng. J.* **2016**, *303*, 349–449. [CrossRef]

96. Farhang, F.; Rayson, M.; Brent, G.; Hodgins, T.; Stockenhuber, M.; Kennedy, E. Insights into the dissolution kinetics of thermally activated serpentine for CO_2 sequestration. *Chem. Eng. J.* **2017**, *330*, 1174–1186. [CrossRef]

97. Oliver, T.K.; Farhang, F.; Hodgins, T.W.; Rayson, M.S.; Brent, G.F.; Molloy, T.S.; Stockenhuber, M.; Kennedy, E.M. CO_2 Capture Modeling Using Heat-Activated Serpentinite Slurries. *Energy Fuels* **2018**. [CrossRef]

98. Rashid, M.I.; Benhelal, E.; Farhang, F.; Oliver, T.K.; Rayson, M.S.; Brent, G.F.; Stockenhuber, M.; Kennedy, E.M. ACEME: Direct Aqueous Mineral Carbonation of Dunite Rock. *Environ. Prog. Sustain. Energy* **2018**. [CrossRef]

99. Benhelal, E.; Rashid, M.I.; Holt, C.; Rayson, M.S.; Brent, G.; Hook, J.M.; Stockenhuber, M.; Kennedy, E.M. The utilisation of feed and byproducts of mineral carbonation processes as pozzolanic cement replacements. *J. Clean. Prod.* **2018**, *186*, 499–513. [CrossRef]

100. Farhang, F.; Oliver, T.K.; Rayson, M.S.; Brent, G.F.; Molloy, T.S.; Stockenhuber, M.; Kennedy, E.M. Dissolution of heat activated serpentine for CO_2 sequestration: The effect of silica precipitation at different temperature and pH values. *J. CO_2 Util.* **2019**, *30*, 123–129. [CrossRef]

101. Benhelal, E.; Rashid, M.I.; Rayson, M.S.; Prigge, J.-D.; Molloy, S.; Brent, G.F.; Cote, A.; Stockenhuber, M.; Kennedy, E.M. Study on mineral carbonation of heat activated lizardite at pilot and laboratory scale. *J. CO_2 Util.* **2018**, *26*, 230–238. [CrossRef]

102. Brent, G.; Rayson, M.; Kennedy, E.; Stockenhuber, M.; Collins, W.; Prigge, J.; Hynes, R.; Molloy, S.; Zulfiqar, H.; Farhang, F.; et al. Mineral carbonation of serpentinite: From the laboratory to pilot scale—The MCi Project. In Proceedings of the 5th International Conference on Accelerated Carbonation for Environmental and Material Engineering ACEME, New York, NY, USA, 21–24 January 2015.

103. McCutcheon, J.; Wilson, S.A.; Southam, G. Microbially Accelerated Carbonate Mineral Precipitation as a Strategy for in Situ Carbon Sequestration and Rehabilitation of Asbestos Mine Sites. *Environ. Sci. Technol.* **2016**, *50*, 1419–1427. [CrossRef] [PubMed]

104. McCutcheon, J.; Dipple, G.M.; Wilson, S.A.; Southam, G. Production of magnesium-rich solutions by acid leaching of chrysotile: A precursor to field-scale deployment of microbially enabled carbonate mineral precipitation. *Chem. Geol.* **2015**, *413*, 119–131. [CrossRef]

105. Hamilton, J.L.; Wilson, S.A.; Morgan, B.; Turvey, C.C.; Paterson, D.J.; MacRae, C.; McCutcheon, J.; Southam, G. Nesquehonite sequesters transition metals and CO_2 during accelerated carbon mineralisation. *Int. J. Greenh. Gas Control* **2016**, *55*, 73–81. [CrossRef]

106. Turvey, C.C.; Wilson, S.A.; Hamilton, J.L.; Tait, A.W.; McCutcheon, J.; Beinlich, A.; Fallon, S.J.; Dipple, G.M.; Southam, G. Hydrotalcites and hydrated Mg-carbonates as carbon sinks in serpentinite mineral wastes from the Woodsreef chrysotile mine, New South Wales, Australia: Controls on carbonate mineralogy and efficiency of CO_2 air capture in mine tailings. *Int. J. Greenh. Gas Control* **2018**, *79*, 38–60. [CrossRef]

107. Hamilton, J.L.; Wilson, S.A.; Morgan, B.; Turvey, C.C.; Paterson, D.J.; Jowitt, S.M.; McCutcheon, J.; Southam, G. Fate of transition metals during passive carbonation of ultramafic mine tailings via air capture with potential for metal resource recovery. *Int. J. Greenh. Gas Control* **2018**, *71*, 155–167. [CrossRef]

108. Steel, K.M.; Alizadehhesari, K.; Balucan, R.D.; Bašić, B. Conversion of CO_2 into mineral carbonates using a regenerable buffer to control solution pH. *Fuel* **2013**, *111*, 40–47. [CrossRef]

109. Steel, K.; Alizadehhesari, K.; Fox, K.; Balucan, R. Potential for CO_2 sequestration as mineral carbonate within Ni laterite processing. In Proceedings of the ALTA 2013: Nickel-Cobalt-Copper, Uranium-REE and Gold Conference & Exhibition, Perth, Australia, 25 May–1 June 2013; pp. 424–431.

110. Morgan, B.; Wilson, S.A.; Madsen, I.C.; Gozukara, Y.M.; Habsuda, J. Increased thermal stability of nesquehonite ($MgCO_3 \cdot 3H_2O$) in the presence of humidity and CO_2: Implications for low-temperature CO_2 storage. *Int. J. Greenh. Gas Control* **2015**, *39*, 366–376. [CrossRef]

111. Australian Government, Geoscience Australia. Australia's Identified Mineral Resources. 2017. Available online: http://www.ga.gov.au/__data/assets/pdf_file/0005/58874/Australias-Identified-Mineral-Resources-2017.pdf (accessed on 20 January 2019).

112. Burgess, J.; Jefery, L.; Lowe, A.; Schuck, S.; Flentje, W. *Novel CO_2 Capture Task Force Report*; Global CCS Institute: Melbourne, Australia, 2011.

113. Kemache, N.; Pasquier, L.-C.; Cecchi, E.; Mouedhen, I.; Blais, J.-F.; Mercier, G. Aqueous mineral carbonation for CO_2 sequestration: From laboratory to pilot scale. *Fuel Process. Technol.* **2017**, *166*, 209–216. [CrossRef]

114. Coelho, P.C.S.; Teixeira, J.P.F.; Gonçalves, O.N.B.S.M. Mining Activities: Health Impacts. In *Encyclopedia of Environmental Health*; Nriagu, J.O., Ed.; Elsevier: Burlington, NJ, USA, 2011. [CrossRef]

115. Mining Report Finds 60,000 Abandoned Sites, Lack of Rehabilitation and Unreliable Data. 2017. Available online: https://www.abc.net.au/news/2017-02-15/australia-institute-report-raises-concerns-on-mine-rehab/8270558 (accessed on 20 January 2019).

116. The Parliament of the Commonwealth of Australia. Between a Rock and a Hard Place the Science of Geosequestration. 2007. Available online: https://www.aph.gov.au/Parliamentary_Business/Committees/House_of_Representatives_Committees?url=scin/geosequestration/report.htm (accessed on 20 January 2019).

117. Heidrich, C.; Feuerborn, H.-J.; Weir, A. Coal Combustion Products: A Global Perspective. In Proceedings of the 2013 World of Coal Ash (WOCA) Conference, Lexington, KY, USA, 22–25 April 2013.

118. The Australian Government, Department of the Environment and Energy. National Waste Report. 2018. Available online: http://www.environment.gov.au/system/files/resources/7381c1de-31d0-429b-912c-91a6dbc83af7/files/national-waste-report-2018.pdf (accessed on 20 January 2019).

119. Ahmaruzzaman, M. A review on the utilization of fly ash. *Prog. Energy Combust. Sci.* **2010**, *36*, 327–363. [CrossRef]

120. Stanwell Corporation. 2018. Available online: https://www.stanwell.com/energy-assets/our-power-stations/coal/ (accessed on 20 January 2019).

121. USGS. *Mineral Commodity Summaries*; U.S. Department of the Interior U.S. Geological Survey: Reston, VA, USA, 2018.

122. Gelencsér, A.; Kováts, N.; Turóczi, B.; Rostási, A.; Hoffer, A.; Imre, K.; Nyiró-Kósa, I.; Csákberényi-Malasics, D.; Tóth, A.; Czitrovszky, A.; et al. The red mud accident in Ajka (Hungary): Characterization and potential health effects of fugitive dust. *Environ. Sci. Technol.* **2011**, *45*, 1608–1615. [CrossRef] [PubMed]

123. Ruyters, S.; Mertens, J.; Vassilieva, E.; Dehandschutter, B.; Poffijn, A.; Smolders, E. The red mud accident in Ajka (Hungary): Plant toxicity and trace metal bioavailability in red mud contaminated soil. *Environ. Sci. Technol.* **2011**, *45*, 1616–1622. [CrossRef] [PubMed]

124. Winkler, D.; Bidló, A.; Bolodár-Varga, B.; Erdő, Á.; Horváth, A. Long-term ecological effects of the red mud disaster in Hungary: Regeneration of red mud flooded areas in a contaminated industrial region. *Sci. Total Environ.* **2018**, *644*, 1292–1303. [CrossRef] [PubMed]

125. Tran, C.P. Red Mud Minimisation and Management for the Alumina Industry by the Carbonation Method. Ph.D. Thesis, The University of Adelaide, Adelaide, Australia, 2016.

126. Lu, G.; Zhang, T.A.; Zhu, X.; Liu, Y.; Wang, Y.; Guo, F.; Zhao, Q.; Zheng, C. Calcification–carbonation Method for Cleaner Alumina Production and CO_2 Utilization. *JOM* **2014**, *66*, 1616–1621. [CrossRef]

127. Ahn, J.W.; Thriveni, T.; Nam, S.Y. Sustainable recycling technologies for bauxite residue (red mud) utilization. In *Energy Technology 2015: Carbon Dioxide Management and Other Technologies*; Springer International Publishing: Basel, Switzerland, 2016. [CrossRef]

128. Carter, C.M.; Van Der Sloot, H.A.; Cooling, D.; Van Zomeren, A.; Matheson, T. Characterization of untreated and neutralized bauxite residue for improved waste management. *Environ. Eng. Sci.* **2008**, *25*, 475–488. [CrossRef]

129. Nikraz, H.R.; Bodley, A.J.; Cooling, D.J.; Kong, P.Y.; Soomro, M. Comparison of physical properties between treated and untreated bauxite residue mud. *J. Mater. Civ. Eng.* **2007**, *19*, 2–9. [CrossRef]

130. Cooling, D.J.; Hay, P.S.; Guilfoyle, L. Carbonation of Bauxite Residue. In Proceedings of the 6th International Alumina Quality Workshop, Brisbane, Australia, 8–13 September 2002.

131. Stringer, D. World's Top Miner Charges into Cobalt. 2018. Available online: https://www.bloomberg.com/news/articles/2018-08-06/world-s-top-miner-charges-into-cobalt-as-battery-ambitions-grow (accessed on 20 January 2019).

132. Mao, F. Wittenoom: Tourists Urged to Stay Away from Asbestos Town. 2018. Available online: https://www.bbc.com/news/world-australia-44816907 (accessed on 20 January 2019).

133. Revathy, T.D.R.; Palanivelu, K.; Ramachandran, A. Sequestration of Carbon dioxide by red mud through direct mineral carbonation at room temperature. *Int. J. Glob. Warm.* **2017**, *11*, 23–37. [CrossRef]

134. Liang, G.; Chen, W.; Nguyen, A.V.; Nguyen, T.A.H. Red mud carbonation using carbon dioxide: Effects of carbonate and calcium ions on goethite surface properties and settling. *J. Colloid Interface Sci.* **2018**, *517*, 230–238. [CrossRef] [PubMed]

135. Sahu, R.C.; Patel, R.K.; Ray, B.C. Neutralization of red mud using CO_2 sequestration cycle. *J. Hazard. Mater.* **2010**, *179*, 28–34. [CrossRef] [PubMed]

136. Liu, W.; Su, S.; Xu, K.; Chen, Q.; Xu, J.; Sun, Z.; Wang, Y.; Hu, S.; Wang, X.; Xue, Y.; et al. CO_2 sequestration by direct gas–solid carbonation of fly ash with steam addition. *J. Clean. Prod.* **2018**, *178*, 98–107. [CrossRef]

137. Jaschik, J.; Jaschik, M.; Warmuzinski, K. The utilisation of fly ash in CO_2 mineral carbonation. *Chem. Process Eng. Inzynieria Chemiczna i Procesowa* **2016**, *37*, 29–39. [CrossRef]

138. Tamilselvi Dananjayan, R.R.; Kandasamy, P.; Andimuthu, R. Direct mineral carbonation of coal fly ash for CO_2 sequestration. *J. Clean. Prod.* **2016**, *112*, 4173–4182. [CrossRef]

139. Ukwattage, N.L.; Ranjith, P.G.; Yellishetty, M.; Bui, H.H.; Xu, T. A laboratory-scale study of the aqueous mineral carbonation of coal fly ash for CO_2 sequestration. *J. Clean. Prod.* **2015**, *103*, 665–674. [CrossRef]

140. Teir, S.; Kuusik, R.; Fogelholm, C.J.; Zevenhoven, R. Production of magnesium carbonates from serpentinite for long-term storage of CO_2. *Int. J. Miner. Process.* **2007**, *85*, 1–15. [CrossRef]

141. Teir, S.; Eloneva, S.; Fogelholm, C.J.; Zevenhoven, R. Fixation of carbon dioxide by producing hydromagnesite from serpentinite. *Appl. Energy* **2009**, *86*, 214–218. [CrossRef]

142. Mervine, E.M.; Wilson, S.A.; Power, I.M.; Dipple, G.M.; Turvey, C.C.; Hamilton, J.L.; Vanderzee, S.; Raudsepp, M.; Southam, C.; Matter, J.M.; et al. Potential for offsetting diamond mine carbon emissions through mineral carbonation of processed kimberlite: An assessment of De Beers mine sites in South Africa and Canada. *Mineral. Petrol.* **2018**, *112*, 755–765. [CrossRef]

143. Meyer, N.A.; Vögeli, J.U.; Becker, M.; Broadhurst, J.L.; Reid, D.L.; Franzidis, J.P. Mineral carbonation of PGM mine tailings for CO_2 storage in South Africa: A case study. *Miner. Eng.* **2014**, *59*, 45–51. [CrossRef]

144. Vogeli, J.; Reid, D.L.; Becker, M.; Broadhurst, J.; Franzidis, J.P. Investigation of the potential for mineral carbonation of PGM tailings in South Africa. *Miner. Eng.* **2011**, *24*, 1348–1356. [CrossRef]

145. McCutcheon, J.; Turvey, C.C.; Wilson, S.A.; Hamilton, J.L.; Southam, G. Experimental deployment of microbial mineral carbonation at an asbestos mine: Potential applications to carbon storage and tailings stabilization. *Minerals* **2017**, *7*, 191. [CrossRef]

146. Oskierski, H.C.; Dlugogorski, B.Z.; Oliver, T.K.; Jacobsen, G. Chemical and isotopic signatures of waters associated with the carbonation of ultramafic mine tailings, Woodsreef Asbestos Mine, Australia. *Chem. Geol.* **2016**, *436*, 11–23. [CrossRef]

147. Newall, P. *CO_2 Storage as Carbonate Minerals*; IEA Greenhouse R&D Programme Report; Report Number PH3/17; CSMA Consultants: Cheltenham, UK, 2000.

148. Penner, L.; O'Connor, W.K.; Dahlin, D.C.; Gerdemann, S.; Rush, G.E. Mineral carbonation: Energy costs of pretreatment options and insights gained from flow loop reaction studies. In Proceedings of the Third Annual Conference on Carbon Capture & Sequestration, Alexandria, VA, USA, 3–6 May 2004.

149. Gerdemann, S.J.; O'Connor, W.K.; Dahlin, D.C.; Penner, L.R.; Rush, H. Ex Situ Aqueous Mineral Carbonation. *Environ. Sci. Technol.* **2007**, *41*, 2587–2593. [CrossRef] [PubMed]

150. Huijgen, W.J.J.; Comans, R.N.J. *Carbon Dioxide Sequestration by Mineral Carbonation. Literature Review*; ECN-C-03-016; Energy research Centre of the Netherlands ECN: Petten, The Netherlands, 2003.

151. Huijgen, W.J.J.; Comans, R.N.J.; Witkamp, G.-J. Cost evaluation of CO_2 sequestration by aqueous mineral carbonation. *Energy Convers. Manag.* **2007**, *48*, 1923–1935. [CrossRef]

152. Haug, T.A.; Kleiv, R.A.; Munz, I.A. Investigating dissolution of mechanically activated olivine for carbonation purposes. *Appl. Geochem.* **2010**, *25*, 1547–1563. [CrossRef]

153. Geerlings, J.J.C.; Wesker, E. Process for Sequestration of Carbon Dioxide by Mineral Carbonation. U.S. Patent 7722850, 25 May 2010.

154. Park, A.-H.A.; Jadhav, R.; Fan, L.-S. CO_2 mineral sequestration: Chemically enhanced aqueous carbonation of serpentine. *Can. J. Chem. Eng.* **2003**, *81*, 885–890. [CrossRef]

155. Daval, D.; Sissmann, O.; Menguy, N.; Saldi, G.D.; Guyot, F.; Martinez, I.; Corvisier, J.; Garcia, B.; Machouk, I.; Knauss, K.G.; et al. Influence of amorphous silica layer formation on the dissolution rate of olivine at 90 °C and elevated pCO_2. *Chem. Geol.* **2011**, *284*, 193–209. [CrossRef]

156. Julcour, C.; Bourgeois, F.; Bonfils, B.; Benhamed, I.; Guyot, F.; Bodenan, F.; Petiot, C.; Gaucher, E.C. Development of an attrition-leaching hybrid process for direct aqueous mineral carbonation. *Chem. Eng. J.* **2015**, *262*, 716–726. [CrossRef]

157. *Carbon Dioxide Storage by Mineral Carbonation*; IEA Greenhouse R&D Programme Report; Report Number 2005/11; IEA Greenhouse Gas R&D Programme: Cheltenham, UK, 2005.

Article

Utilization of Calcium Carbonate-Coated Wood Flour in Printing Paper and Their Conservational Properties

Yung Bum Seo *, Dong Suk Kang and Jung Soo Han

Department of Bio-based Materials, Chungnam National University, Yousung-Gu, Gung-Dong 220, Daejun 305-764, Korea; creammupin@naver.com (D.S.K.); han1606@naver.com (J.S.H.)
* Correspondence: ybseo@cnu.ac.kr

Received: 2 January 2019; Accepted: 27 March 2019; Published: 28 March 2019

Abstract: Wood flours (WFs) are bulky lignocellulosic materials that can increase the bulk and stiffness of paper. To be used in printing paper for replacing chemical pulp, WFs were first fractionated by a 200-mesh screen to improve smoothness; second, they were coated with calcium carbonate by an *in-situ* $CaCO_3$ formation method (coated wood flours, CWFs) to improve brightness. The performance of CWFs for printing paper was compared to those of bleached wood flours (BWFs) and bleached chemical pulp. Equivalent brightness and much higher smoothness were obtained for the CWFs compared to the BWFs. Furthermore, BWFs caused a significant loss of yield and required wastewater treatment in the bleaching process, while the CWFs increased the yield greatly by attaching $CaCO_3$ to the wood flours, and caused no wastewater burden. An accelerated aging test showed that the CWFs caused lesser brightness and strength loss than the bleached chemical pulp and BWFs. CWFs still had room for improvement to replace chemical pulp, but showed slower aging in optical and close strength properties.

Keywords: wood flours; *in-situ* $CaCO_3$ formation; brightness; smoothness; printing paper; conservational property

---

1. Introduction

Wood flours (WFs) are used in the filler layer of multiple paperboards and papers to achieve high bulk [1–5]. If higher paperboard bulk than necessary is achieved, one can reduce the basis weight of the paperboard down to the customer-acceptable bulk and achieve savings in wood fiber and drying energy without losing essential board properties. The use of WFs may also be extended to printing paper [6–10]. If WFs can be used in printing paper without causing optical and physical problems, the bulk and stiffness will be greatly improved, and again, basis weight reduction may be possible while essential properties are kept at acceptable levels. Therefore, the aim of this study was to find a technology that enables us to use WFs in printing paper more effectively. In reality, printing paper containing WFs always suffers from low brightness, low smoothness, and fast color deterioration caused by aging. To improve the brightness of WFs, Shin et al. [7] applied an alkaline pulping process. After pulping, they mixed WFs and GCC together for further pulverizing to reduce their size and to improve optical properties. Kim et al. [8] applied a bleaching process. They prepared WFs that passed 200mesh screen, and applied two-step bleaching process by using chlorine dioxide and peroxide. Their results showed that the addition of bleached wood flours still gave low brightness and high roughness, but high bulk in printing paper. Later, Kim et al. [9] tried to add an optical brightener in addition to wood flour bleaching, and obtained more improved brightness by addition of optical brightener.

In-situ $CaCO_3$ formation on the surface of lignocellulosic materials [10–12] greatly improved brightness without loss of yield. In contrast, applying pulping and bleaching processes to WFs caused significant yield loss and water contamination [10]. Seo et al. [10] claimed that the *in-situ* $CaCO_3$ formation on WFs increased furnish yield without water contamination because the attached calcium carbonates on wood flours became the inexpensive part of the entire furnish. Low smoothness caused by WFs is generally due to their large size and rigid structure, and can be improved by optimizing wood flour size and size distribution. To overcome low smoothness, we used a 200-mesh hole type screen to control the maximum particle size in this study.

Wood flours contain lignin, and those wood-containing papers become dark when they are exposed to the air, sunshine, and heat for an extended period. Calcium carbonate-coated WFs or *in-situ* $CaCO_3$ formed WFs (CWFs) may exhibit a different aging behavior from typical WFs during the aging process. It was presumed that hydrogen peroxide bleached WFs (BWFs) may exhibit faster color changes than CWFs. An accelerated aging test was performed to reveal the effect of $CaCO_3$ coated WFs on aging. All properties of the WFs-containing sheets (WFs, CWFs, and BWFs) were also compared to those of bleached chemical pulp sheets.

2. Materials and Methods

2.1. Wood Fibers

Bleached kraft softwood pulp (mixture of Hemlock, Douglas fir, and Cedar, Canada) in 20 wt%, and hardwood pulp (mixture of Aspen and Poplar, Canada) in 80 wt% were mixed together, and refined in a valley beater until freeness (Canadian standard freeness. ISO-5267-2) reached 500 mL. We added WFs, CWFs, and BWFs to this mixed pulp in 5 and 10% to compare their properties.

2.2. Wood Flours

WFs were donated by Korea N Company that were the mixture of hardwoods imported from China. We selected this material because the biggest paper mill in Korea already used the same materials in large quantity for the manufacture of commercial paperboard. The important specification of the wood flours were the size and the composition (mainly from oak) when Korea N company supplies them to the paper mill. However, those wood flours did not make smooth surfaces for printing paper because of their large size. They were further pulverized to pass through the 200-mesh screen. To increase their brightness, hydrogen peroxide bleaching was applied to the WFs (5 wt% H_2O_2 based on dried wood flour weight, sodium silicate, pH13 initially controlled by NaOH, liquid ratio 1:10, 80 °C, 90 min. duration). These bleached WFs (called as BWFs) lost about 32.8% of their initial weight during the bleaching process [10].

To prepare the *in-situ* $CaCO_3$ formed WFs, the screened ones were first mixed with CaO by 169 and 281 wt% based on the dry weight of WFs for making 1:3 and 1:5 ratios of wood flour: $CaCO_3$, respectively, in the reaction vessel initially in 10% concentration. Second, we injected carbon dioxide to the aqueous mixture while stirring at 350 rpm until a neutral pH was reached. The pH was checked until a pH change was not detected for 3 min at neutral pH. These resultant WFs look like the coated WFs with $CaCO_3$, and we called them CWFs.

2.3. Preparation of Handsheets and Testing Methods

Handsheets of 60 g/m^2 basis weight were prepared from chemical pulp with and without WFs according to the standard method (TAPPI Test Methods T205, 1995). Composition of the handsheets are shown in Table 1. Cationic PAM (cationic polyacrylamid, +5 meq/g, CIBA Specialty Chemicals Korea, Seoul, Korea) was used for retention (0.1% based on sheet weight). The ash content (TAPPI Test Method T 413 om-93, 1993), density and bulk (TAPPI T410 om-98, T411 om-97), breaking length (ISO 1924-1, 1992), bending resistance (TAPPI T543 om-00), Bekk smoothness (TAPPI T479 cm-99), brightness (ISO 2470, 1977), and folding endurance (ISO 5626) of the handsheets were measured according to the respective standards. For tensile test, we used 20 mm/min. of loading speed and 10 specimens (15 mm width × 150 mm length) per each sample conditioned overnight at 23 °C and

50% RH before testing. The results were converted to breaking length in km. An accelerated aging treatment was performed based on the dry heat treatment method at 105 °C (ISO 5630-1:1991(e)) for 5, 10, and 20 days.

Table 1. Composition of the handsheets in percent.

Sample Name	Wood Fibers	Wood Flour 200 Mesh Passed	Wood Flour: In-situ CaCO$_3$ = 1:3	Wood Flour: In-situ CaCO$_3$ = 1:5	Bleached Wood Flour
Pulp only	100				
Wood flour, 5 (WFs)	95	5			
Wood flour, 10 (WFs)	90	10			
In-situ 1:3 WF, 5 (CWFs-3)	95		5		
In-situ 1:3 WF, 10 (CWFs-3)	90		10		
In-situ 1:5 WF, 5 (CWFs-5)	95			5	
In-situ 1:5 WF, 10 (CWFs-5)	90			10	
Bleached WF, 5 (BWFs)	95				5
Bleached WF, 10 (BWFs)	90				10

3. Results and Discussion

3.1. Morphology of the in-situ CaCO$_3$ Formed WFs

Morphologies of WFs with and without CaCO$_3$ coating are shown in Figure 1. Most of the WF surface was covered with newly formed calcium carbonate for the CWFs, and there were almost no loose CaCO$_3$ around the CWFs in Figure 1b. It seemed that most of the CaCO$_3$ formed were attached to the WFs. Using an electron microscope, it was not possible to differentiate between a 1:3 and 1:5 (WF: in-situ CaCO$_3$) ratio for the CWFs.

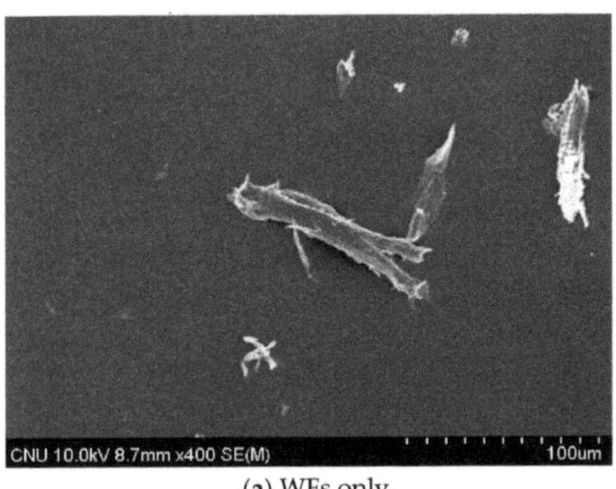

(**a**) WFs only.   (**b**) *In-situ* CaCO₃ formed WFs.

Figure 1. Morphologies of WFs with and without in-situ CaCO$_3$ coating WFs: *in-situ* CaCO$_3$ = 1:5).

3.2. Physical Properties

The brightness and smoothness of the paper containing WFs are the most critical properties in this study. Figures 2 and 3 show the changes in brightness and Bekk smoothness. Adding 5% WFs to wood pulp caused a brightness drop from 82.8% (Pulp only) to 65.1% (17.7% difference). Through peroxide bleaching of WFs, the brightness decreased by 5.4% from the 'Pulp only' sheet. The *in-situ* CaCO$_3$ formation method caused a brightness drop of 6.1% for the 1:5 case. Adding 10% CWFs caused a larger brightness drop.

The Bekk smoothness for papers containing WFs or BWFs decreased greatly. The greater the amount of WFs or BWFs, the less smoother the papers are. This indicates that WFs smaller than the ones that passed through the 200-mesh screen are required. However, the CWFs exhibited much higher smoothness than the other WF-containing sheets. We believe this is because the CWFs consist of mostly $CaCO_3$, not WFs. The CWFs consist of 75% and 83% $CaCO_3$ for the 1:3 and 1:5 ratios, respectively. WFs are bulky, rigid, and rough, but the CWFs are denser than WFs and occupy less volume inside the paper sheet at the same addition level, thereby allowing for less of a chance to lower smoothness. Furthermore, CWFs seem to be slightly deformable under the wet pressing pressure because the calcium carbonates formed on the surface of wood flours are not chemically attached, but instead adsorbed on the wood surface, resisting the turbulence of the process water. In short, CWFs can result in brightness as high as BWFs but can produce much higher smoothness than BWFs. If the improvement in smoothness and brightness of CWFs containing sheets is evident, it may be the correct approach to utilize more CWFs in printing paper than WFs or BWFs.

The bulk of the handsheets were shown in Figure 4, where we observed that WFs, whether bleached or not, resulted in very high bulk. The CWFs produced less bulk than WFs, although still much higher than the 'Pulp only' case. In summary, CWFs can result in brightness as high as BWFs but produce much higher smoothness than BWFs while still producing high bulk. The improvement in smoothness and brightness of CWFs-containing sheets is evident that we believe it is the effective approach to utilize more CWFs in printing paper.

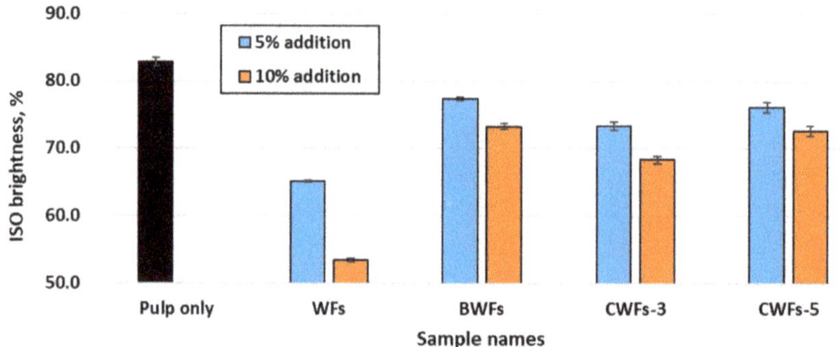

Figure 2. Brightness changes caused by the addition of pre-treated WFs (Standard deviation bars were included).

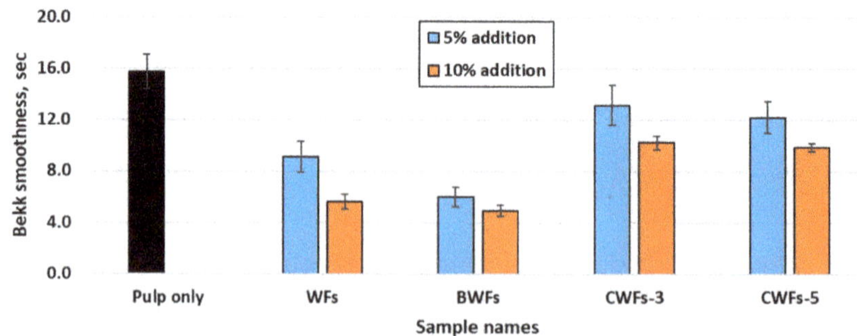

Figure 3. Bekk smoothness changes caused by the addition of pre-treated WFs (Standard deviation bars were included).

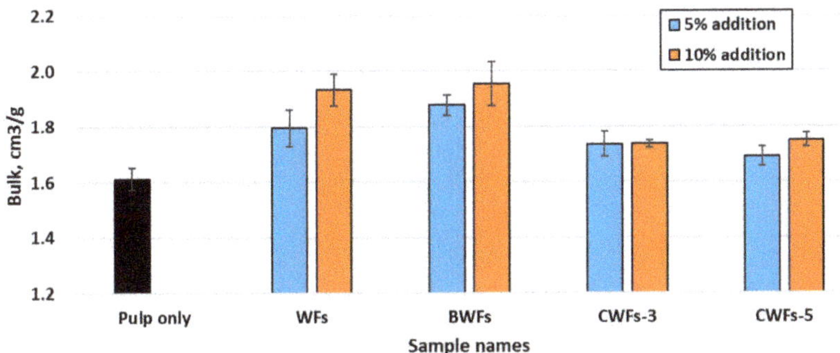

Figure 4. Bulk changes caused by the addition of pre-treated WFs (Standard deviation bars were included).

3.3. Strength Properties

The use of WFs in paper increased bulk as expected, but a decrease in strength properties is unavoidable. Breaking length, which is the measure of tensile strength after compensating for the basis weight differences of paper sheets, of the sheets were shown in Figure 5, where breaking length was higher for the CWFs than for the other WF-containing sheets. We surmise this is because CWFs have less specific surface area than the other WFs do. WFs and calcium carbonate make very little, if any, or no hydrogen bonds with wood fibers. More surface area of either WFs or CWFs per unit weight means less bonding. The WFs and BWFs have more surface area per unit weight than CWFs because the WFs and BWFs consist of wood flours only, but the CWFs both wood flours (usually wood density of 0.4–0.8 g/cm^3) and calcium carbonates (density of 1.5 g/cm^3), respectively. Figure 5 shows there were no differences in breaking length development between the WFs and their bleached ones (BWFs).

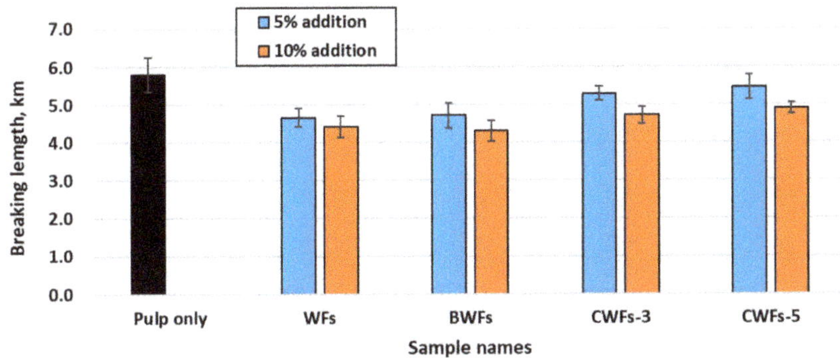

Figure 5. Breaking length changes caused by the addition of pre-treated WFs (Standard deviation bars were included).

The substitution of expensive chemical pulp with other inexpensive materials is successful only when physical properties including brightness, smoothness, and strength properties are acceptable to customers or equivalent to those of the chemical pulp sheet. Out of essential strength properties, stiffness is usually most difficult to increase, and that is why WFs are under current consideration. Figure 6 shows stiffness changes caused by the application of WFs. All WFs-containing sheets exhibited higher stiffness than 'Pulp only' sheets, but the CWFs showed lower stiffness than the other WFs-containing sheets. However, CWFs containing sheets gave 15%–20% higher stiffness than the 'Pulp only' sheet. Sheet bulk is the most important factor for developing high stiffness because stiffness is proportional to the cube of sheet thickness. The bulk is the reciprocal of the thickness.

In Figure 4, the bulk of the CWFs were lower than that of the WFs and the BWFs but higher than the 'Pulp only'.

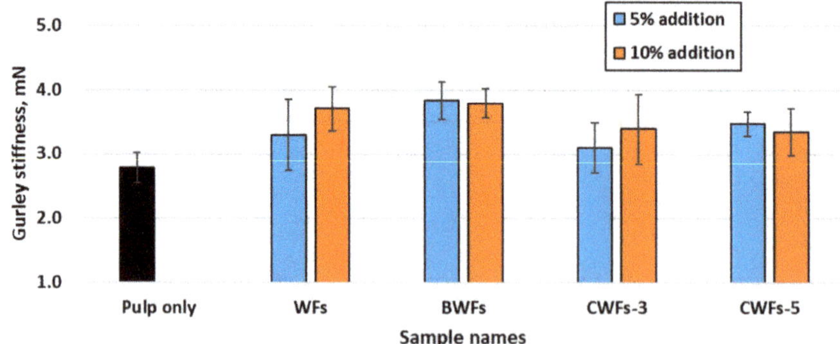

Figure 6. Stiffness changes caused by the addition of pre-treated WFs (Standard deviation bars were included).

Double folds, which is the measure of how many times a paper sheet can endure folding under a given tensile load until failure, are shown in Figure 7. Large differences between 'Pulp only' and WFs-containing sheets are evident. The application of CWFs in printing paper needs to improve folding properties and thus needs to use more long fibers in the wood pulp part.

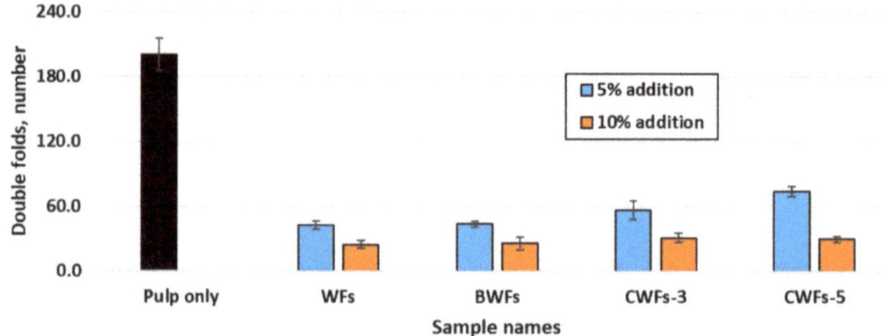

Figure 7. Double folds changes caused by the addition of pre-treated WFs (Standard deviation bars were included).

3.4. Accelerated Aging Test

The use of lignin containing materials in printing paper should cause color change or aging problems. Furthermore, strength properties of those papers deteriorate faster than wood-free paper. Accelerated aging (ISO 5630-1:1991) of paper and board by dry heat treatment at 105 °C may provide information concerning the natural changes that may occur in the material over a period of years. Figures 8–11 show the results of WFs-containing papers after accelerated aging for 20 days at 105 °C.

Brightness decreased greatly due to accelerated ageing for the WFs-containing and 'Pulp only' sheets, but decreased much more slowly for the CWF sheets (Figure 8). There might be large color changes in the WFs within the CWFs, but calcium carbonate on the surface of WFs seemed to partially conceal these color changes. After 20 days, the brightness of the 'Pulp only' sheet was close to that of the CWFs containing sheet.

The slow deterioration of breaking length (Figure 9), stiffness (Figure 10), and double folds (Figure 11) can be explained in the same way. That is, calcium carbonate has not changed its color and bonding properties by the prolonged exposure to the temperature of 105 °C, and the surface of CWFs consist of mostly calcium carbonate. However, chemical pulp and WFs lose their brightness and strength properties quickly by accelerated heat aging. Thus, the CWF sheets had slow deterioration and showed nearly identical strength properties to the 'Pulp only' sheet after 20 days of aging (Figures 9–11).

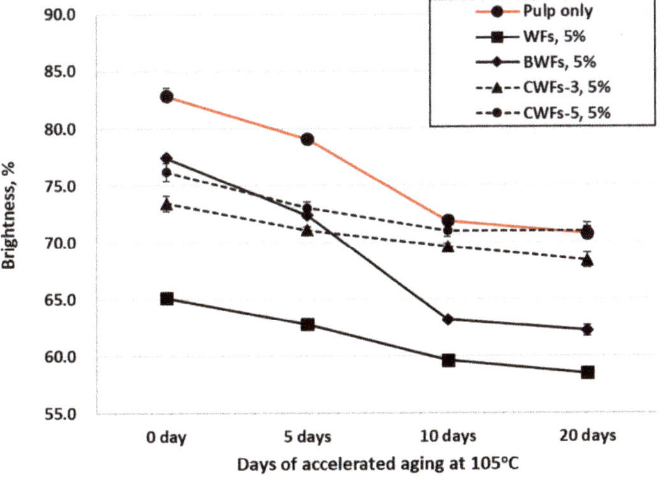

Figure 8. Brightness changes by accelerated aging for the WFs-containing sheets (Standard deviation bars were included).

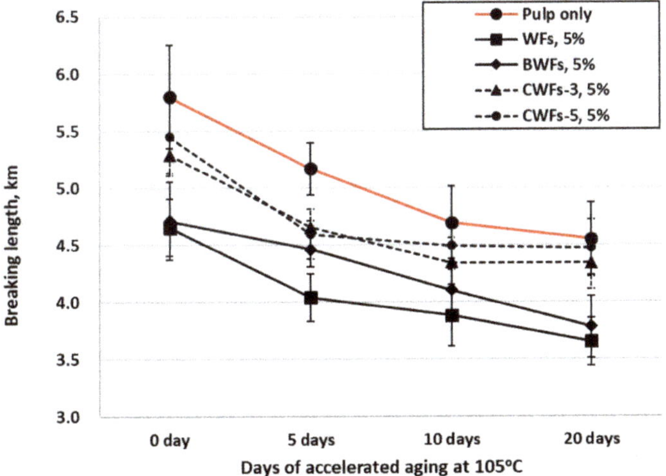

Figure 9. Breaking length changes due to accelerated aging for the WFs-containing sheets (Standard deviation bars were included).

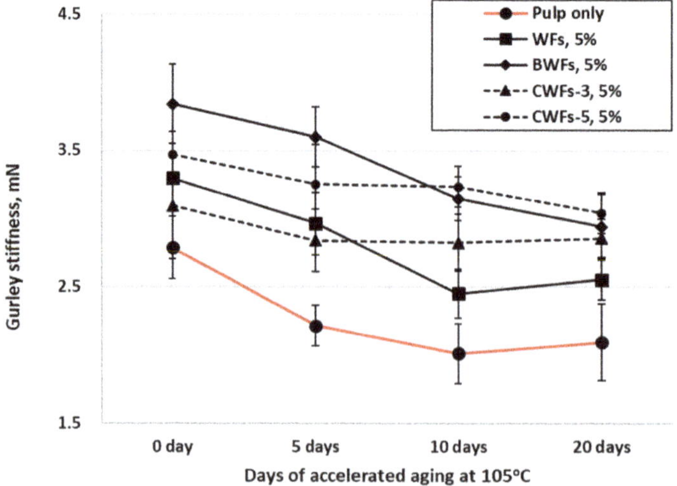

Figure 10. Stiffness changes due to accelerated aging for the WFs-containing sheets (Standard deviation bars were included).

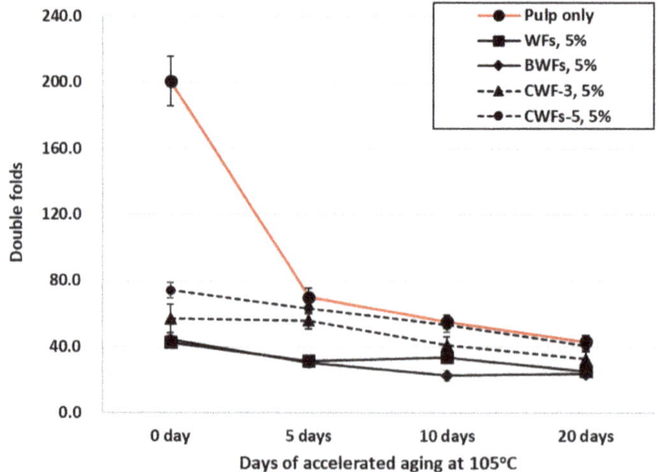

Figure 11. Double folds changes due to accelerated aging for the WFs-containing sheets (Standard deviation bars were included).

3.5. Cost Consideration of the in-situ CaCO₃ Formed WFs (CWFs)

To make bleached WFs with 5% H_2O_2 requires chemical costs, process energy, and wastewater treatment. The yield loss was about 32.8% (Seo et al., 2018). On the other hand, for the CWFs with a 1:5 ratio, yield increases 500%. The price of a 200-mesh pass WFs should be more than 2 times higher than the price of calcium carbonate. Bleached WFs should be much more expensive than unbleached ones. Therefore, the cost of preparing CWFs should be much less expensive than the WFs and BWFs used in this study. Furthermore, there is no need to treat wastewater for the CWFs.

4. Conclusions

The wood flours (WFs) that passed through a 200-mesh screen, were added to the bleached chemical pulp in 5% and 10% by weight after H_2O_2 bleaching (BWFs) or after CaCO₃ coating (CWFs) to replace the bleached chemical pulp, but turned out to be slightly inferior in brightness and strength properties other than stiffness. It is believed that the size of the 200-mesh screen-passed WFs was not small enough to develop equivalent smoothness to that of the chemical pulp. The CWFs resulted in similar brightness but much higher smoothness and could be prepared in a much less expensive way than the BWFs. Accelerated aging showed that the calcium carbonates on the surface of CWFs conceal the color change of the WFs inside, resulting in much slower deterioration of strength and optical properties in the CWF containing sheets. Compared to the BWFs, the CWFs developed better properties and rendered lower production costs. Therefore, further development of the CWFs for replacing bleached chemical pulp should be promising.

Author Contributions: D.S.K. and J.S.H. participated in the experimental design, actual experiments and data analysis. Y.B.S. conceived the idea of wood flour utilization in printing paper, experimental design, and all the writing. All authors approved read and approved the final manuscript.

Funding: This research was supported by the National Strategic Project-Carbon Upcycling of the National Research Foundation of Korea (NRF) funded by the Ministry of Science and ICT (MSIT), the Ministry of Environment (ME) and the Ministry of Trade, Industry and Energy (MOTIE). (2017M3D8A2086048).

Conflicts of Interest: The authors declare no conflict of interest

References

1. Chae, H.J.; Park, J.M. Study on Drainage and Physical Properties of KOCC Handsheet Containing Pretreated Wooden Fillers. *J. Korea Tech. Assoc. Pulp Pap. Ind.* **2011**, *43*, 21–29.
2. Lee, J.Y.; Lee, E.K.; Sung, Y.J.; Kim, C.H.; Choi, J.S.; Kim, B.H.; Lim, G.B.; Kim, J.S. Application of new powdered additives to paperboard using peanut husk and garlic stem. *J. Korea Tech. Assoc. Pulp Pap. Ind.* **2011**, *43*, 40–48.
3. Lee, J.Y.; Kim, C.H.; Choi, J.S.; Kim, B.H.; Lim, G.B.; Kim, D.M. Development of New Powdered Additive and Its Application for Improving the Paperboard Bulk and Reducing Drying Energy (I). *J. Korea Tech. Assoc. Pulp Pap. Ind.* **2012**, *44*, 58–66.
4. Karinkanta, P.; Ämmälä, A.; Illikainen, M.; Niinimäki, J. Fine grinding of wood—Overview from wood breakage to applications. *Biomass Bioenergy* **2018**, *113*, 31–44. [CrossRef]
5. Kim, S.Y.; Lee, J.Y.; Kim, C.H.; Lim, G.B.; Park, J.H.; Kim, E.H. Surface modifications of organic fillers to improve the strength of paperboard. *BioResources* **2015**, *10*, 1174–1185. [CrossRef]
6. Lee, J.Y.; Kim, C.H.; Kim, E.H.; Park, T.U.; Jo, H.M.; Sung, Y.J. Effect of rice husk organic filler and wood powder on the bulk and the drying energy of paperboard. *J. Korea Tech. Assoc. Pulp Pap. Ind.* **2017**, *49*, 69–75. [CrossRef]
7. Shin, T.G.; Kim, C.H.; Chung, H.K.; Seo, J.M.; Lee, Y.R. Fundamental study on developing lignocellulosic fillers for papermaking (I). *J. Korea Tech. Assoc. Pulp Pap. Ind.* **2008**, *40*, 21–29.
8. Kim, C.H.; Lee, J.Y.; Lee, Y.R.; Chung, H.K.; Back, K.K.; Lee, H.J.; Gwak, H.J.; Gang, H.R.; Kim, S.H. Fundamental Study on Developing Lignocellulosic Fillers for Papermaking (II). *J. Korea Tech. Assoc. Pulp Pap. Ind.* **2009**, *41*, 1–6.
9. Kim, H.H.; Kim, C.H.; Seo, J.M.; Lee, J.Y.; Kim, S.H.; Park, H.J.; Kim, G.C. Use of modified lignocellulosic fillers to improve paper properties. *Appita J. J. Tech. Assoc. Aust. N. Z. Pulp Pap. Ind.* **2011**, *64*, 338–343.
10. Seo, Y.B.; Lee, Y.H.; Jung, J.K. Feasibility study of using the in-situ $CaCO_3$ formed wood flours for printing paper. *J. Korea Tech. Assoc. Pulp Pap. Ind.* **2018**, *50*, 44–53.
11. Seo, Y.B.; Lee, Y.H.; Chung, J.K. The improvement of recycled newsprint properties by in-situ $CaCO_3$ loading. *BioResources* **2014**, *9*, 6254–6266. [CrossRef]
12. Seo, Y.B.; Ahn, J.H.; Lee, H.L. Upgrading waste paper by in-situ calcium carbonate formation. *J. Clean. Prod.* **2017**, *155*, 212–217. [CrossRef]

Article

Sulfation–Roasting–Leaching–Precipitation Processes for Selective Recovery of Erbium from Bottom Ash

Josiane Ponou [1], Marisol Garrouste [2], Gjergj Dodbiba [1], Toyohisa Fujita [3] and Ji-Whan Ahn [4,*]

[1] Graduate School of Engineering, the University of Tokyo, 7-3-1 Hongo, Bunkyo-ku, Tokyo 113-8656, Japan; ponou@sys.t.u-tokyo.ac.jp (J.P.); dodbiba@sys.t.u-tokyo.ac.jp (G.D.)

[2] MINES ParisTech—PSL Research University, Rue Claude Daunesse, F-06904 Sophia Antipolis CEDEX, B.P. 207 1 Paris, France; marisol.garrouste@mines-paristech.fr

[3] School of Resources, Environment and Materials, Guangxi University, 100 Daxue Road, Nanning, Guangxi 530004, China, fujitatoyohisa@gxu.edu.cn

[4] Korea Institute of Geoscience and Mineral Resources (KIGAM), 92 Gwahang-no, Yuseong-gu, Daejeon 305-350, Korea

* Correspondence: ahnjw@kigam.re.kr; Tel.: +82-42-868-3578

Received: 8 April 2019; Accepted: 20 June 2019; Published: 24 June 2019

Abstract: Bottom ash (BA) is mainly composed of compounds of Al, Fe, Ca, and traces of rare earth elements (REEs). In this study, the selective recovery of erbium (Er) as REEs by means of sulfation–roasting–leaching–precipitation (SRLP) using BA was investigated. A pre-treatment process of sulfation and roasting of BA was developed to selectively recover REEs using ammonium oxalate leaching (AOL) followed by precipitation. Most of the oxides were converted to their respective sulfates during sulfation. By roasting, unstable sulfates (mostly iron) decomposed into oxides, while the REE sulfates remained stable. Roasting above 600 °C induces the formation of oxy-sulfates that are almost insoluble during AOL. Dissolved REEs precipitate after 7 days at room temperature. The effects of particle size, roasting temperature, leaching time, and AOL concentration were the important parameters studied. The optimal conditions of +100–500 µm particles roasted at 500 °C were found to leach 36.15% of total REEs in 2 h 30 min and 94.24% of the leached REEs were recovered by precipitation. A total of 97.21% of Fe and 94.13% of Al could be separated from Er.

Keywords: bottom ash; sulfation; roasting; oxalic acid; rare earth; erbium; secondary raw material

1. Introduction

The demand for rare earth elements (REEs) has grown due to their wide applications in the technologies of the 21st century. In 2017, the raw material supply group of the European Commission compiled a list of 41 elements or groups of elements considered critical [1]. Among these elements, heavy and light REEs as well as scandium were highlighted because of the supply risk due to their rarity and their high impact on the global economy. The availability of technology to separate REEs from one another is also among the important factors accounting for the high cost of some REEs [2]. Therefore, new sources of REEs need to be identified to enhance the global supply. Many researchers have identified and extracted REEs from low-grade REE materials such as power plant coal, hard coal fly ash, bauxite, coal deposits, and waste fly ash or by recycling electric and electronic equipment waste using various methods [3–14]. Many processes such as subsequent solvent extraction, liquid-liquid extraction, acid leaching, etc. have been developed for rare earth extraction [10,15]. However, the presence of large amount of iron in the leached solution is still challenging for researchers [15]. Moreover, finding a relatively low-cost process, which could selectively recover REEs, is of importance in the chemical engineering field.

In this study, a combination of sulfation, roasting, and leaching processes was thoroughly examined to recover REEs from bottom ash (BA) during the final stage by precipitation in order to find out a low-cost method for recovery of REEs in low concentrated secondary material. Ammonium oxalate was used as a leaching reagent; note that simultaneously, the affinity of this acid may induce the precipitation of REEs [16,17]. Roasting temperature, leaching time, and ammonium oxalate concentration are the main parameters, which were noted to study the selectivity of the process.

2. Materials and Methods

2.1. Materials

BA was collected from a recycling company in Japan (For confidential reason, the company will be unnamed). The change in structure with increasing temperature of the sulfated BA was investigated using a thermo-gravimetric differential thermal analyzer (TG-DTA) under ambient air flow, from room temperature up to 1000 °C, with a heating rate of 10 °C/min (Thermo plus EVO TG8120/Rigaku). Sulfated BA was roasted using an electric furnace (HPM-OG/AS ONE). The concentration of REEs as well as other metals in the solution was analyzed using an inductively coupled plasma-optimal emission spectrometer (ICP-OES; Perkin Elmer Optima 5300). Characterization of the powder samples was performed using X-ray diffraction (XRD) and X-ray fluorescence (XRF).

2.2. Sulfation–Roasting–Leaching–Precipitation

Sulfation–roasting–leaching–precipitation (SRLP) is a combination of four consecutive and complementary processes. The original sample was first crushed and passed through 500 μm and 100 μm sieves to obtain two samples (+100–500 μm and −100 μm). As described by Borra et al., both dried samples were moistened (the BA-to-deionized water mass ratio was 1:0.4) to ensure homogeneous mixing with acid, and thus aid in the conversion of metal oxides to sulfates [5,18]. After moistening, the samples were mixed with 64% sulfuric acid (sample-to-acid mass ratio was 1:1), as this ratio was found to ensure good dissolution of REEs while not excessively increasing the dissolution of iron [5]. The mixtures were dried at 70 °C for 24 h to ensure complete sulfation. During this step, metals are converted into their respective sulfates. After drying, the samples were roasted at selected temperatures in an electric furnace. Some of these sulfates, the iron sulfates in particular [19], are unstable at elevated temperatures, and consequently decompose during the roasting step into water-insoluble oxides [5]. Rare earth sulfates, conversely, remain stable at high temperature and are water-soluble [20]. After roasting, the samples were ground to obtain homogeneous powders. Ammonium oxalate leaching experiments were conducted at 25 °C and 300 rpm and the sample dosage was maintained at 67 g/L. Unless specified, the concentration of the ammonium oxalate was 50 g/L. After leaching, the solutions were filtered with 0.20 μm filter paper. The obtained filtrate was analyzed using an ICP-OES.

During the subsequent leaching step, the metal oxides remain in the solid residue, whereas the rare earth sulfates dissolve in the solution. Typically used to precipitate REEs, ammonium oxalate is, in this study, used for both leaching and selective precipitation of REEs [16,17] (its affinity to rare earth elements may confer a property of good leaching reagent for REEs). Therefore, after reaching the optimal conditions to extract as many REEs as possible, the leached solution was kept for seven days until precipitation occurred, and the remaining solution was filtered and analyzed. In addition, to identify the optimal temperature at which the iron sulfate decomposition may be completed, sulfated samples were subjected to thermal decomposition from room temperature to 1000 °C using TG-DTA. All the experiments were repeated several times to avoid discrepancies in the results. The flowsheet of the experimental procedure is shown in Figure 1.

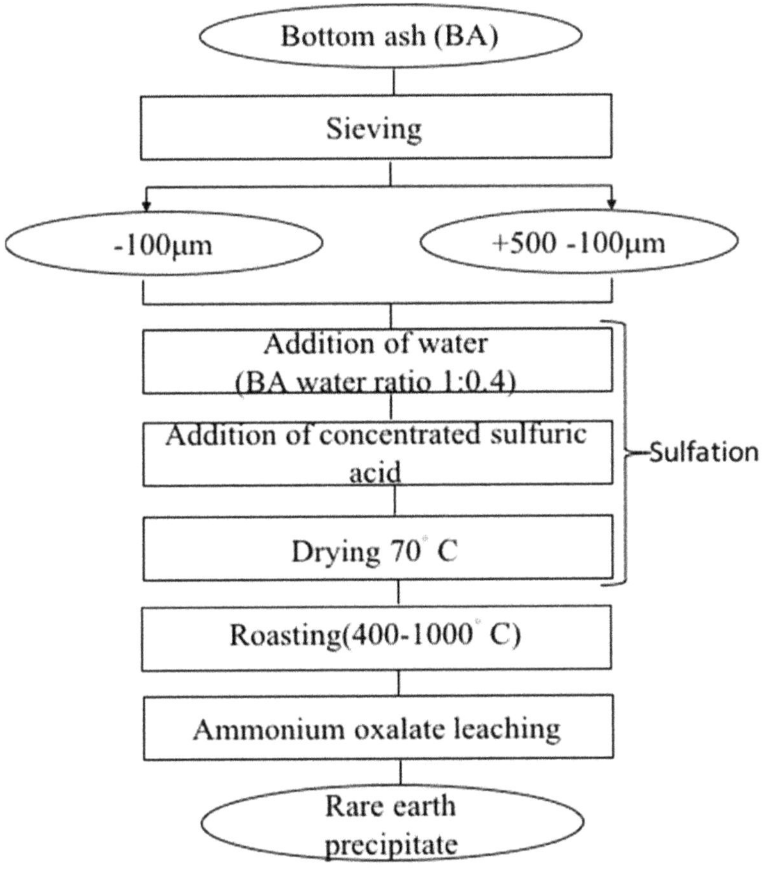

Figure 1. Experimental procedure flowsheet.

3. Results and Discussion

3.1. Sample Characterization

As-received BA was ground using a planetary mill and sieved to produce two different size fractions, i.e., −100 µm and +100–500 µm. To compare the elemental composition of the samples, each size fraction was analyzed using XRF, and the REE composition was detected using an ICP-OES. The results of the present study were compared to previous studies as shown in Table 1. Compared to previous low-grade materials, the BA used in the present work is more concentrated in REEs, particularly erbium (Er). Aluminum oxide, calcium oxide, iron oxide, and silica account for nearly 83% of the total mass of the +100–500 µm BA, and are therefore the dominant metals. Copper, chrome, and zinc oxide are less dominant. The REEs are more concentrated in the size fraction +100–500 µm with erbium as the dominant element. Therefore, analysis in the present work will consider aluminum, calcium, and iron oxides as the dominant metals, chromium and copper oxides as less dominant metals, and erbium (Er) as REEs. Furthermore, sample with the size fraction of +100–500 µm was used.

Table 1. Quantitative analysis of the as-received sample.

	Metals	BA-100 μm	BA +100–500 μm	Bauxite Residue	Lignite Ash	Refuse Ash	Biomass Ash
Metal concentration, (%)	MgO	0.53	0.63	NA	2.60	1.90	5.20
	Al_2O_3	8.18	7.87	23.60	29.80	20.30	12.40
	SiO_2	14.21	18.20	10.20	48.40	50.40	54.40
	P_2O_5	1.84	2.12	NA	0.40	1.00	2.20
	SO_3	1.60	1.26	NA	2.80	0.90	4.00
	Cl	1.19	0.84	NA	NA	NA	NA
	K_2O	1.86	2.18	NA	0.20	0.90	8.30
	CaO	49.02	38.78	11.20	7.90	3.30	9.20
	TiO_2	3.76	3.57	5.70	1.40	1.90	0.10
	Cr_2O_3	0.26	0.33	NA	NA	NA	NA
	MnO	0.46	0.56	NA	NA	NA	NA
	Fe_2O_3	12.58	19.25	44.60	5.40	19.30	2.20
	CuO	1.14	1.09	NA	NA	NA	NA
	ZnO	2.30	2.04	NA	NA	NA	NA
Rare earth concentration, (ppm)	Sc	5.00	4.00	121.00	28.00	297.00	10.00
	Y	5.00	6.00	76.00	1.20	0.30	0.20
	La	14.00	13.00	114.00	134.00	58.00	32.00
	Ce	32.00	30.00	368.00	295.00	114.00	61.00
	Pr	-	-	28.00	2.00	0.54	0.30
	Nd	-	-	99.00	152.00	43.00	25.00
	Sm	-	-	21.00	33.00	9.30	5.20
	Eu	-	-	5.00	8.10	1.80	1.00
	Gd	-	-	22.00	20.20	5.80	3.20
	Tb	-	-	3.00	3.70	1.00	0.60
	Dy	-	-	17.00	22.80	6.20	3.30
	Ho	-	-	4.00	4.60	1.20	0.60
	Er	1467.00	1607.00	13.00	1.30	0.30	0.20
	Tm	-	-	2.00	2.60	0.70	0.40
	Yb	4.00	4.00	14.00	9.40	2.60	1.20
	Lu	4.00	4.00	2.00	1.50	0.40	0.20
Sources		Present work		Borra et al., 2016		Singh et al., 2011	

NA = Not available

3.2. Thermo-Gravimetric Differential Analysis of the Sulfurized Sample

As suggested by Borra et al. (2016) [5], +100–500 μm sieved BA was sulfurized with 64% sulfuric acid at a mass ratio of 1:1 and dried for one day. The obtained sample was termed "sulfurized sample" or "70 °C". Prior to roasting, the sulfurized sample was subjected to thermal decomposition at a rate of 10 °C/ min using TG-DTA. The results are shown in Figure 2. The TGA thermogram indicates the weight loss in the sample and that of the DTA shows the phase changes in the sample with increasing temperature. Until 300 °C, the sample loses approximately 25% of its original weight and some peaks appears in the DTA thermogram. Meaning that most sulfates lose their physical and chemical bound water molecules until 300 °C [5]. From 300–650 °C, no significant changes were observed. From 650–750 °C, the sample once again lost 10% of its weight followed by the appearance of a significant peak at 750 °C. Borra et al. (2016) attributed these changes to the decomposition of iron sulfate into iron(III) oxide [5].

3.3. Roasting Process

To understand the progressive changes occurring with increasing roasting temperature and their effect on the leaching process, sulfurized bottom ash was roasted at 400 °C, 500 °C, 600 °C, 700 °C, 800 °C, 900 °C, and 1000 °C, respectively, for an hour in the electric furnace. With increasing temperature, a progressive change in color is observed, which may indicate a phase changes in the material. The sample color shifts from white to brown from 600 °C to 700 °C, indicating the partial decomposition of iron sulfate into iron(III) oxide. As shown by the XRD patterns (Figure 3), iron sulfate is totally decomposed into iron(III) oxide from 800 °C. These results are in accordance with those observed using TG-DTA. Calcium sulfate and aluminum sulfate were formed during the early stage and remained almost stable. REEs were not observed because their concentration was too low to be detected by the instrument.

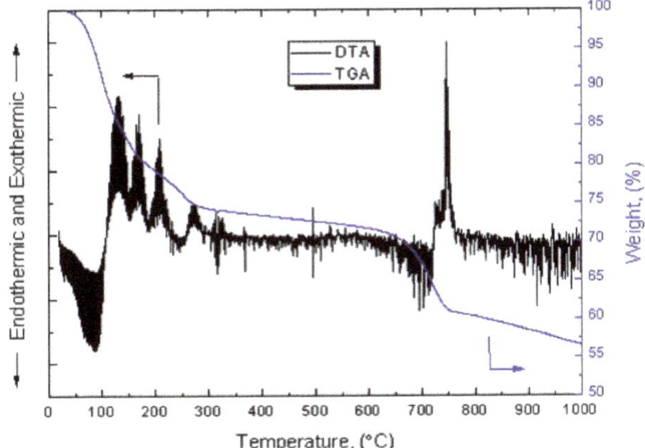

Figure 2. TG-DTA analysis of the sulfated and dried +100–500 μm bottom ash. Experimental conditions: size fraction: +100–500 μm; deionized-water-to-sample mass ratio: 1:0.4; 64% sulfuric-acid-to-sample mass ratio 1:1; drying: 70 °C; drying time 24 h.

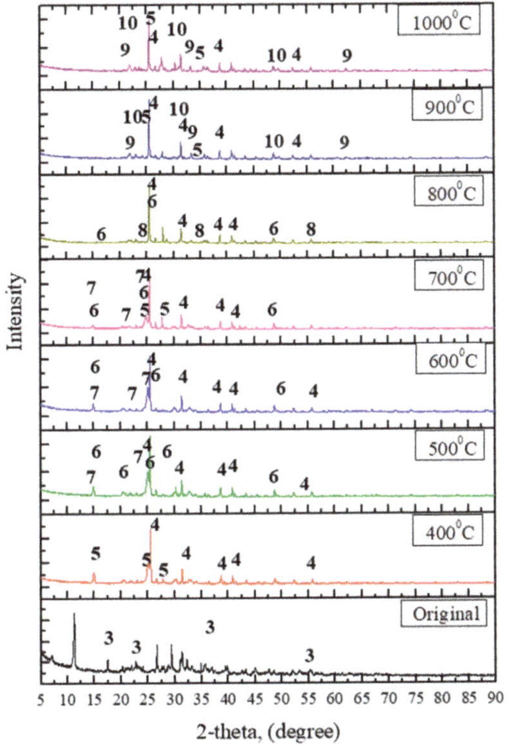

Figure 3. XRD patterns of roasted sulfurized samples. $3 \rightarrow$ Katoite $Ca_{2.93}\ Al_{1.97}\ (Si.64O_{2.56})\ (OH)_{9.44}$; $4 \rightarrow$ Calcium Sulfate $CaSO_4$; $5 \rightarrow$ Chromium Sulfate $Cr_2(SO_4)_3$; $6 \rightarrow$ Millosevichite, syn $Al_2(SO_4)_3$; $7 \rightarrow$ Mikasaite $Fe_2(SO_4)_3$; $8 \rightarrow$ Iron(III) oxide, α-Fe_2O_3; $9 \rightarrow$ Hematite, syn Fe_2O_3; $10 \rightarrow$ Albite, syn Na $(AlSi_3O_8)$.

3.4. Leaching Process

Ammonium oxalate is among the reagents that shows a high affinity to precipitate REEs. In this study, ammonium oxalate was used for the lixiviation of REEs during the first step. Leaching of sulfurized samples using deionized water was also tested to compare both results. During the second step, the leached solution was kept at room temperature for a period of time to ensure the precipitation of REEs, particularly erbium. Sulfurized samples are indicated in the graphs as 70 °C. The concentration of ammonium oxalate as well as the leaching time as a function of roasting temperature were the main parameters investigated. Since ammonium oxalate has particular affinity to REEs, REEs are selectively leached and precipitate as solid oxalate salts as shown in equation (1).

$$2REEs^{3+} + 3C_2O_4^{2-} \rightarrow (REEs)_2(C_2O_4)_3 \tag{1}$$

3.4.1. Effect of Roasting Temperature on the Leaching Process

To note the optimal roasting temperature at which Er (the most abundant rare earth element in the material) and other metals will reach their optimal leaching conditions, samples were roasted at 400 °C, 500 °C, 600 °C, 700 °C, 800 °C, 900 °C, and 1000 °C, respectively, and the sulfurized sample (70 °C) and the original were used. The ideal conditions for selectivity were to obtain a high concentration of Er and a low concentration of metals in the solution. Experiments were conducted for 2 h. Ammonium oxalate leaching as well as water leaching were both done. Er and metal concentrations in one gram of the sample were reported in Figure 4. Figure 4a shows the leaching results using ammonium oxalate, and Figure 4b indicates the leaching results using the deionized water function of roasting temperatures compared to the original sample.

Figure 4a displays the results suggesting that without sulfuric acid pre-treatment, the leaching ratio is very low, which highlights the importance of the pre-treatment process. The concentration of Er as well as other metals considerably increased during the sulfation process. Sulfation itself increases the dissolution rate of the chemical elements without selectivity. Roasting at 400 °C or 500 °C slightly decreases the metal rate in solution and increases the Er concentration with temperature, showing the affinity of ammonium oxalate with REEs. The dissolution of Er decreases from 600 °C to 700 °C, temperatures at which the decomposition of iron sulfate into iron oxide is ongoing. The dissolution of both REEs and metals is basically non-existent above 700 °C. Below 700 °C, water-insoluble forms of Fe, Al, and Cu are formed, explaining the observed low yield. It was concluded that the entire stability of the metals by roasting, which addresses the water-insoluble forms, is not favorable for the leaching process. These results are in accordance with those reported by Borra et al., (2016) which showed that a higher roasting temperature over a longer duration has a negative effect on the dissolution of REEs because of the increasing number of sulfates that are decomposed to water-insoluble oxy-sulfates and oxides [5,18]. Hence, the decrease in the Er yield from 600 °C may be related to the oxidization of rare earth oxides. In conclusion, 500 °C may be the optimal temperature for REE leaching.

Furthermore, the water dissolution results of Er and metals are shown in Figure 4b. Er leaching decreased with increasing roasting temperature. The sulfation process increased the yield of Er in water, but the roasting temperature had a negative effect on water leaching compared to ammonium oxalate leaching, even when the roasting temperature was less than 700 °C. This result may be due to the sulfates losing their chemical water and becoming increasingly insoluble in water. Therefore, an unroasted sulfurized sample (70 °C) shows the optimal dissolution yield in water.

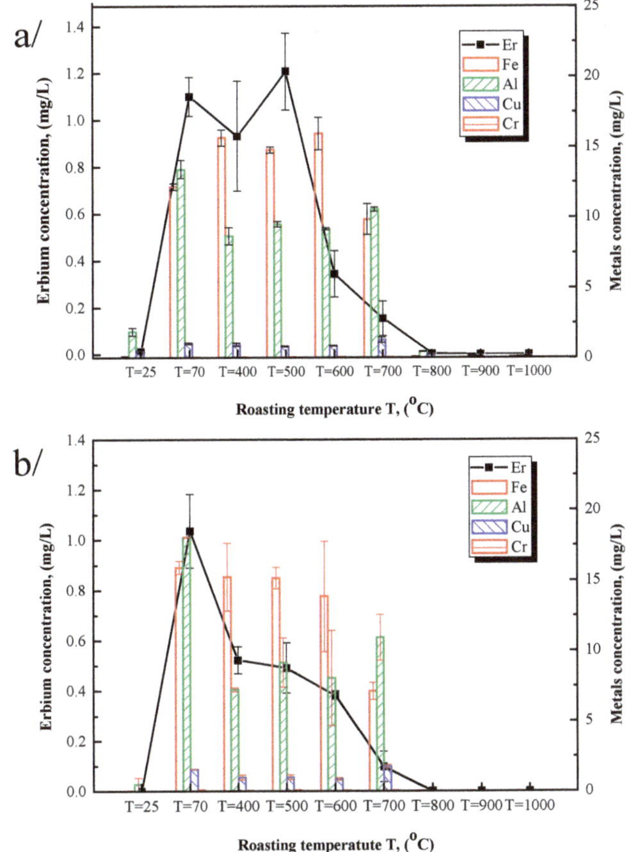

Figure 4. Effect of roasting temperature on the leaching process. (**a**) Acetic acid leaching; (**b**) Water leaching. Experimental conditions: size fraction: +100–500 μm; ammonium oxalate concentration 50 g/L; leaching time 2 h; sample dosage 67 g/L.

3.4.2. Effect of Leaching Time on REE Recovery

The leaching time for a 500 °C roasted sample was evaluated every 30 min. The results are shown in Figure 5a. Er concentration started decreasing after 2 h 30 min of leaching while the metal concentration was still increasing. This phenomenon may be explained by lixiviation followed by precipitation of erbium after 2 h 30 min. Ammonium oxalate is known to show a strong affinity to REEs and is used for REE recovery via precipitation [16]. Therefore, 2 h 30 min may be the optimal time to leach a 500 °C roasted sample to recover as much as possible REEs by precipitation.

Comparatively, the water dissolution results (Figure 5b) indicated that the concentration of Er in solution does not vary with time. However, the lowest metal yield was observed after 2 h of contact time. Ammonium oxalate has strong lixiviation and precipitation forces vs. REEs.

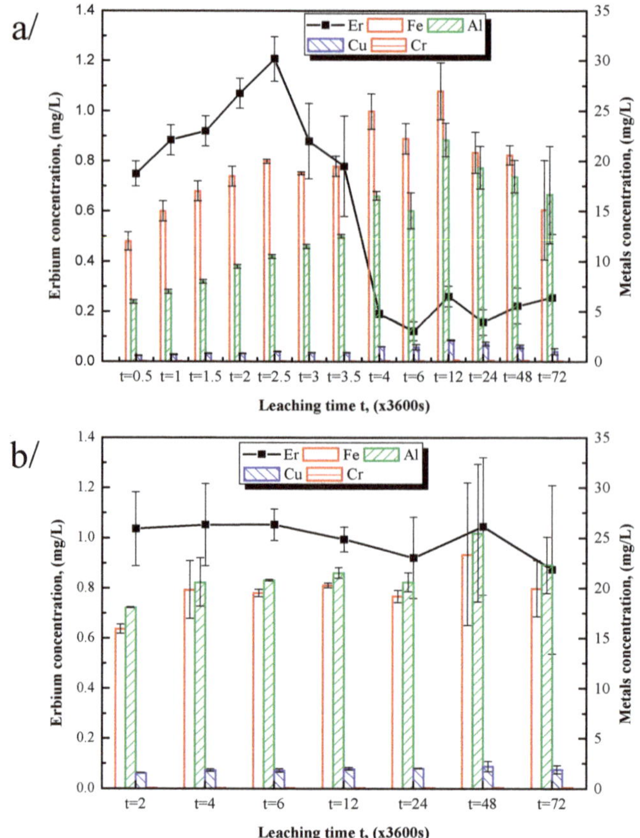

Figure 5. Leaching time effect: (**a**) Ammonium oxalate leaching: Sample 500 °C roasted sample; (**b**) Water leaching: Sample 70 °C dried sample. Experimental conditions: size fraction: +100–500 μm; ammonium oxalate concentration 50 g/L; sample dosage 67 g/L.

3.4.3. Effect of Leaching Acid Concentration

Leaching experiments were conducted at various concentrations of ammonium oxalate under the optimal conditions previously found: a +100–500 μm particle size fraction, a roasting temperature of 500 °C, and 2.5 h of leaching time. Figure 6 shows the effect of the ammonium oxalate concentration on Er or other metal recoveries. The ammonium oxalate concentration shows no significant effect on the metal concentration in the solution, but highly affects Er leaching. Er concentrations in solution using 50 g/L or 75 g/L of acid are quite similar, indicating an equilibrium at approximately 50 g/L.

3.5. Precipitation and Recovery of REEs

The prepared sample, under the obtained optimal conditions for ammonium oxalate was maintained, at room temperature, to investigate the precipitation design of erbium as well as other metals in the solution. Figure 7 shows the concentration of erbium and metals in solution on the day of the experiment and after 7 days. As shown previously in Figure 5a, Er remained in the solution after 3 days of leaching. Therefore, a longer time is required to completely precipitate Er. According to Figure 7, the original concentration of Er in solution is approximately 1198 ppm, which represents 36.15% of the total erbium present in the 500 °C roasted sulfurized BA (referring to aqua regia leaching).

When 94.24% of the Er in the solution precipitated, 97.21% of Fe, 94.13% of Al, and 70.74% of Cu remained in the solution. Therefore, a high purity of Er sulfate precipitate could be separated from Fe, Al, and Cu. A longer maintenance time may increase the concentration of Fe in the precipitate. The selective recovery of Er was effective by sulfation–roasting–leaching–precipitation process.

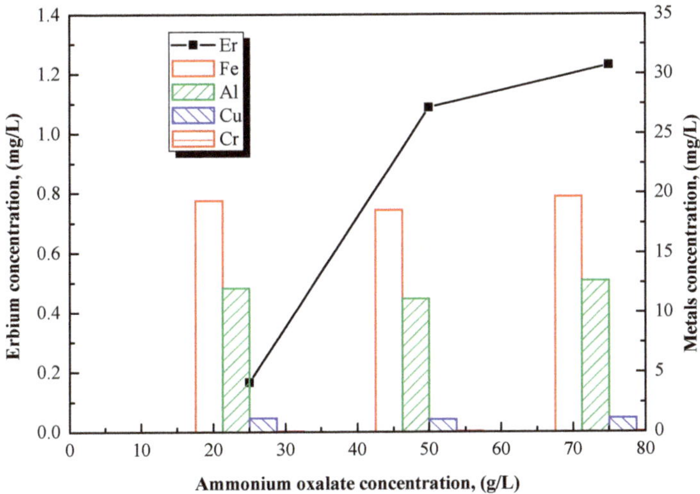

Figure 6. Ammonium oxalate concentration effect. Experimental conditions: Size fraction: +100–500 μm; sulfurized sample roasted at 500 °C; leaching time 2 h 30 min; sample dosage 67 g/L.

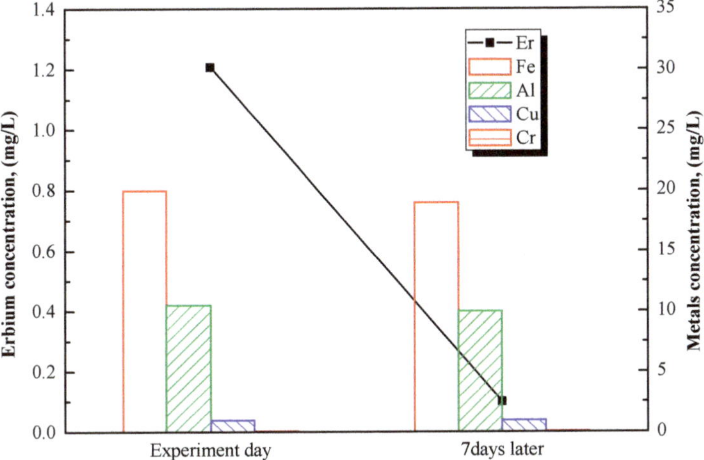

Figure 7. Concentration of erbium and other metals in the leached solution as a function of keeping time using an ammonium oxalate leaching solution. Experimental conditions: Size fraction: +100–500 μm; sulfurized sample roasted at 500 °C; ammonium oxalate concentration 50 g/L; leaching time 2 h 30 min; sample dosage 67 g/L.

4. Application, Limitation, and Perspectives

Bottom ash is an important secondary source of metals. In Japan, 80% of bottom ash is still landfilled without any recycling process [21] due to the fact that the ash contains a high concentration of heavy and other metals, which do not fit the Japanese industrial standards to be reuse as cement kiln, etc. Therefore, in addition to previous authors' works [22,23] concerning the removal of heavy metals from ashes, erbium recovery from bottom ash gives hope to the REE industry as well as the waste management field. Erbium which finds its main application in optical fiber fabrication is really on high demand in REE market. The present work findings could be an alternative to pre-concentrate erbium from bottom ash. However, only 36.15% of the erbium containing, in the bottom ash could be leached. Repeating experiments could enhance the leaching rate. 7 days are enough to recover REE precipitates with fewer metals content. Beyond 7 days, metals especially iron in solution precipitates progressively and a complex of rare earth iron may be formed. To avoid this phenomenon, the precipitation duration should be strictly respected. Furthermore, more purification processes of the rare earth precipitate need to be undergone to obtain a high purity erbium powder which could be reused in high technologies.

5. Conclusions

Sulfation, roasting leaching, and precipitation were applied to BA to selectively recover rare REEs, particularly Er. XRF analysis of the BA showed a high concentration of aluminum oxide, iron oxide, and calcium oxide. Further analysis indicated some trace of REEs, of which Er was abundant. The sulfation process induced the formation of sulfate molecules of Al, Fe, and Cu, which were transformed into their most stable forms with an increasing roasting temperature up to 600 °C. Iron(III) oxide in 800 °C roasted sample was observed in the XRD patterns.

Analysis of the effect of Er leaching using ammonium oxalate on the roasting temperature pointed out 500 °C as the optimal temperature to selectively recover Er. Roasting above 600 °C considerably decreased the leaching rate due to the formation of oxy-sulfates or oxide forms of the metals, which are insoluble in water and almost insoluble in ammonium oxalate.

Moreover, the leaching of the 500 °C roasted sample showed a 36.15% Er yield compared to aqua regia leaching of the same sample. The leaching time of 2 h 30 min with ammonium oxalate showed a progressive precipitation of Er without a pronounced effect on other metals in the solution. When 94.24% of Er in the solution precipitated, 97.21% of Fe, 94.13% of Al, and 70.74% of Cu remained in the solution. Therefore, a high purity of concentrated Er sulfate could be obtained by precipitation. The present study could be an alternative for a secondary source of erbium.

Author Contributions: Conceptualization, J.P.; investigation, M.G., J.P.; writing-original draft preparation, J.P.; writing-review and editing, G.D.; supervision, T.F., J.-W.A.

Funding: This research was funded by the Carbon Mineralization Flagship Center, the Ministry of Science and ICT, the Ministry of Trade, Industry and Energy, and the Ministry of Environment of the Korea.

Acknowledgments: This work was joint research between the University of Tokyo and the Korean Institute of Geoscience and Mineral Resources (KIGAM). The financial support and assistance of the Sponsor, the Carbon Mineralization Flagship Center, the Ministry of Science and ICT, the Ministry of Trade, Industry and Energy, and the Ministry of Environment of the Korea are gratefully acknowledged. Special thanks to the smart lab of Tokyo University for conducted the XRD analysis.

Conflicts of Interest: The authors declare no conflict of interest.

References

1. European commission. Communication from the commission to the European parliament. In Proceedings of the Council, the European economic and Social Committee and the Committee of the Regions, Brussels, Belgium, 13 September 2017; pp. 1–8. Available online: https://eur-lex.europa.eu/legal-content/EN/TXT/?uri=CELEX:52017DC0490 (accessed on 21 May 2018).
2. Mineral Prices. Available online: http://mineralprices.com (accessed on 21 May 2018).
3. Singh, S.; Ram, L.C.; Mastro, R.E.; Verma, S.K. A comparative evaluation of minerals and trace elements in the ashes from lignite, coal refuse and biomass fired power plants. *Int. J. Coal Geol.* **2011**, *87*, 112–120. [CrossRef]

4. Ponou, J.; Wang, L.P.; Dodbiba, G.; Okaya, K.; Noda, M.; Fujita, T. Recovery of rare earth elements from aqueous solution obtained from Vietnamese clay minerals using dried and carbonized parachlorella. *J. Environ. Chem. Eng.* **2014**, *2*, 1070–1081. [CrossRef]
5. Borra, C.R.; Mermans, J.; Pontikes, Y.; Binnemans, K.; Gerven, T.V. Selective recovery of rare earths from bauxite residue by combination of sulfation, roasting and leaching. *Miner. Eng.* **2016**, *92*, 151–159. [CrossRef]
6. Ponou, J.; Dodbiba, G.; Anh, J.W.; Fujita, T. Selective recovery of rare earth elements from aqueous solution obtained from coal power plant ash. *J. Environ. Chem. Eng.* **2016**, *4*, 3761–3766. [CrossRef]
7. Rivera, R.M.; Ulenaers, B.; Ounoughene, G.; Binnemans, K.; Gerven, T.V. Extraction of rare earths from bauxite residue (red mud) by dry digestion followed by water leaching. *Miner. Eng.* **2018**, *119*, 82–92. [CrossRef]
8. Koller, A.; Scott, C.; Hower, J.C.; Vazquez, J.A.; Lopano, C.L.; Dai, S. Distribution of rare earth elements in coal combustion fly ash, determined by SHRIMP-RG ion microprobe. *Int. J. Coal Geol.* **2017**, *184*, 1–10.
9. Huang, Z.; Fan, M.; Tiand, H. Coal and coal byproducts: A large and developable unconventional resource for critical materials-Rare earth elements. *J. Rare Earth* **2018**, *36*, 337–338. [CrossRef]
10. Fulford, G.D.; Lever, G.; Sato, T. Recovery of Rare Earth Elements from Bayer Process Red Mud. U.S. Patent 5030424, 9 July 1991.
11. Ochsenkühn-Petropulu, M.; Lyberopulu, T.; Ochsenkühn, K.M.; Parissakis, G. Recovery of lanthanides and yttrium from red mud by selective leaching. *Anal. Chim. Acta* **1996**, *319*, 249–254. [CrossRef]
12. Qu, Y.; Lian, B. Bioleaching of rare earth and radioactive elements from red mud using Penicillium tricolor RM-10. *Bioresour. Technol.* **2013**, *136*, 16–23. [CrossRef] [PubMed]
13. Smirnov, D.I.; Molchanova, T.V. The investigation of sulphuric acid sorption recovery of scandium and uranium from the red mud of alumina production. *Hydrometallurgy* **1997**, *45*, 249–259. [CrossRef]
14. Yatsenko, S.P.; Pyagai, I.N. Red mud pulp carbonization with scandium extraction during alumina production. *Theor. Found. Chem. Eng.* **2010**, *44*, 563–568. [CrossRef]
15. Logomerac, V.G. Complex utilisation of red mud by smelting and solvent extraction. *Trav. Com. Int. Etude Bauxites Alum. Alum.* **1979**, *15*, 279–285.
16. Taran, M.; Aghaie, E. Designing and optimization of separation process of iron impurities from kaolin by oxalic acid in bench-scale stirred-tank reactor. *Appl. Clay Sci.* **2015**, *107*, 109–116. [CrossRef]
17. Jorjani, E.; Shahbazi, M. The production of rare earth elements group via tributyl phosphate extraction and precipitation stipping using oxalic acid. *Arab. J. Chem.* **2012**. [CrossRef]
18. Guo, X.; Li, D.; Park, K.H.; Tian, Q.; Wu, Z. Leaching behavior of metals from a limonitic nickel laterite using a sulfation-roasting-leaching process. *Hydrometallurgy* **2009**, *99*, 144–150. [CrossRef]
19. Onal, M.A.R.; Borra, C.R.; Guo, M.; Blanpain, B.; Gerven, T.V. Recycling of NdFeB magnets using sulfation, selective roasting and water leaching. *J. Sustain. Metall.* **2015**, *1*, 199–215. [CrossRef]
20. Bergman, A.; Gardestrom, P.; Ericson, I. Release and refixation of ammonium during photorespiration. *Int. J. Plant Biol.* **1981**, *53*, 528–532.
21. Tanigaki, N. *Latest Results of Bottom Ash Handling in Japan, Japan Environmental Facilities Manufacturers Association (JEFMA)*; Nippon Steel & Sumikin Engineering Co., Ltd.: Chiyoda-ku, Japan, 2015.
22. Pöykiö, R.; Mäkelä, M.; Watkins, G.; Nurmesniemi, H.; Dahl, O. Heavy metals leaching in bottom ash and fly ash fractions from industrial-scale BFB-boiler for environmental risks assessment. *Trans. Nonferrous Met. Soc. China* **2016**, *26*, 256–264. [CrossRef]
23. Patra, S.; Whaung, S.T.; Kwan, W.L. Analysis of heavy metals in incineration bottom ash in Singapore and potential impact of pre-sorting on ash quality. *Energy Procedia* **2017**, *143*, 454–459. [CrossRef]

 *sustainability*

Article

Development of a New Clean Development Mechanism Methodology for the Quantification of Greenhouse Gas in Calcium Sulfoaluminate Cement

Eun-don Jeon, Kyu-uk Lee and Chung-kook Lee *

Korea Research Institute on Climate Change, Chuncheon-si, Gangwon-do 24239, Korea;
Jeoned@Kric.re.kr (E.-d.J.); Kyuuklee@Kric.re.kr (K.-u.L.)

Received: 31 January 2019; Accepted: 5 March 2019; Published: 11 March 2019

Abstract: The purpose of this research was to probe beyond the scope of the "National Strategy Project on Carbon Mineralization" to develop a "United Nations Framework Convention on Climate Change, Clean Development Mechanism" (UNFCCC CDM) methodology that enables reduction of greenhouse gas (GHG) by "green cement" under the calcium sulfoaluminate (CSA) cement technologies. The findings will be utilized as the foundations and developed into the UNFCCC CDM project. There were two existing methodologies related to cement, but neither was applicable for CSA cement. The existing methodologies are applicable only when there is one clinker, but CSA cement utilizes more than one clinker. Through this research, we developed methodologies to use waste-based material for avoiding emission leakage and utilized more than one clinker to calculate GHG emissions and reduction. For this purpose, we utilized the CSA cement methodology for calculating GHG reduction compared to Portland cement and found that CSA cement allowed for a reduction of 0.281 tCO_2-eq/ton above the reduction enabled by Portland cement. We are presently preparing to register the CSA cement methodology for UNFCCC CDM methodology approval. With the technology transfer and support for this CSA cement technology and methodology, developing countries will be able to achieve their national GHG reduction targets and gain carbon credits. Thus, CSA cement technology could serve as an important tool to deal with GHG emissions and climate change.

Keywords: green cement; climate change; CO_2; clean development mechanism; CDM methodology; carbon credit; emission trading scheme

---

1. Introduction

Under the Paris Agreement, the international community has set national greenhouse gas reduction targets that are to be met by 2030. Korea has announced that it will set its national greenhouse gas reduction target at 37% against the 2030 emission estimate (Business as Usual) through the INDC (intended nationally determined contribution). Through the 2030 National Greenhouse Gas Reduction Roadmap, 37% is divided into 32.5% domestic reductions and 4.5% overseas (international) reductions. Within the INDC, only the total reduction targets are specified and the targets for overseas reductions are not specified; however, additional preparations are necessary to achieve the reduction targets. As of 2016, Korea's total industrial greenhouse gas emissions amounted to 51,456 million tCO_2-eq and the cement sector accounted for 49.6%, at 25,543 million tCO_2-eq [1]. Thus, in order to reduce greenhouse gases in the industrial sector, a lot of efforts are needed in the cement sector.

Korean cement companies are designed and managed by the ETS (Emission Trading Scheme) and TMS (Target Management System) considering the large amount of greenhouse gas emissions. Companies have studied a variety of ways to reduce greenhouse gases. Among them, some cement companies have developed CSA (calcium sulfoaluminate) cement with a lower rate of greenhouse gas emissions compared to conventional OPC (ordinary Portland cement) as a means to reduce greenhouse gas emissions. It is now necessary to develop a new methodology for the United Nations Framework Convention on Climate Change, Clean Development Mechanism (UNFCCC CDM) that meets CSA cement technology at an international level. The methodology will serve as a means for CSA cement developers to quantify greenhouse gas reductions. As of now, there is a "cement production methodology," which is a registered methodology in the CDM, but this cannot be applied to CSA cement technology. Therefore, this study aimed to develop a new methodology for CDM that can quantify greenhouse gas reductions of mixed cement products using two or more clinkers with CSA cement technology.

2. Theoretical Background

The Clean Development Mechanism (CDM) is a greenhouse gas reduction program under the Tokyo Protocol. The reduced amount of greenhouse gas in non-annex countries (countries that have ratified or accepted UNFCCC but are not obligated to reduce greenhouse gas emissions) can be verified through the third-party verification agency and utilized as an accomplished goal or a trading type of carbon credit. In order to receive the verification of carbon credit, a project should proceed with the CDM methodology certified by UNFCCC. The CDM methodology means an applicable standards of additionality, baseline, and monitoring methods for the registration of the CDM project and the verification of a greenhouse reduction amount regarding the greenhouse reduction project.

A quantitative method for the emission amount of greenhouse gas by cement has been internationally presented in "CO_2 Accounting and Reporting Standard for the Cement Industry" [2] issued by The World Business Council for Sustainable Development (WBCSD) and the "2006 IPCC Guidelines for National Greenhouse Gas Inventories" [3]. In Korea, its method has been utilized in TMS and ETS with an emission amount of greenhouse gas computed using "instruction for the goal management and operation of greenhouse gas energy" [4]. Previous studies have the same theoretical computational equation for the emission amount of greenhouse gas and the differences in the boundary of operation, the kinds of greenhouse gas to report, and the consideration of indirect emission by the purpose (e.g., a report for the emission amount of greenhouse gas by a company, an establishment of a national inventory for greenhouse gas, etc.).

In the case of the CDM methodology, for the purpose of a computation of the reduced amount of greenhouse gas, only the emission amount of greenhouse gas by the facility or process shall be computed with the application of the reduction technology since the greenhouse gas differs from the computation method of previous studies. The rest shall be excluded for simplification. Therefore, life cycle assessment (LCA) is a powerful tool for computing the amount of greenhouse gas emissions, which has not been considered as it is different from the concept of CDM methodology in the present study.

The CDM project should be promoted with the CDM methodology certified by UNFCCC. There are two cases of methodologies for the cement production registered in the UNFCCC, but to apply them to the manufacturing process of CSA cement, a new methodology should be developed. The contents of the new methodology must include an equation to compute the reduction amount of greenhouse gas in CSA cement accurately. An applicability must be set for completed and consistent computation.

3. Methods

In this research, UNFCCC CDM methodologies related to cement were analyzed, and the applicable conditions for CSA cement production were determined. However, there were no methodologies suitable for CSA cement, and hence we developed new methodologies for quantification of GHG reduction by means of using CSA cement. For developing methodologies, we analyzed standard process for two methodologies, their applicability, and method of GHG emissions calculation for the validation of the new methodology. We developed new CSA cement methodologies and are presently preparing to register at UNFCCC CDM for the approval process. Also, we have provided limited information on this research for privacy reasons. In summary, we provide the estimated GHG reduction achieved by utilizing the new CSA cement methodology.

4. Results

4.1. Demand for the New CDM Methodology

To be issued carbon credits from the UNFCCC CDM, it is mandatory to register the CDM project via the approved methodology from the UNFCCC. In UNFCCC CDM, there are 216 registered methodologies and 2 related cement methodologies (Table 1).

Table 1. Methodologies on UNFCC CDM (Cement Production).

Methodology	Title
ACM0005	Increasing the blend in cement production
ACM0015	Emission reductions from raw material switch in clinker production

In the ACM (approved consolidated methodology) 0005 [5] methodology, GHG reduction is mainly achieved through increasing the "ratio of mixed material" in the process of cement mixing/grinding to reduce the clinker production. This enables: (1) reduction of CO_2 in the process emission in the kiln process (decarboxylation of limestone), and (2) reduction of fossil fuel usage during the kiln process.

The ACM0015 [6] methodology's main principle is to replace limestone with non-carbonate material so as to reduce the usage of limestone and fossil fuel in the kiln process and CO_2 emissions in the process.

In CDM, 17 projects applying these two methodologies were approved by the UNFCCC. The total GHG reduction per year for these projects was approximately 2,559,000 tCO$_2$-eq.

In this research, CSA cement was used for: (1) replacing the limestone in the "raw material input" for reduction of CO_2 emission and fossil fuel usage in the kiln process, and (2) adding mixed material for reducing usage of a clinker that in turn brought down the CO_2 emission in the cement mixing/grinding process (Figure 1). Therefore, for achieving those conditions, we needed to compound those two methodologies, viz. ACM0005 and ACM0015.

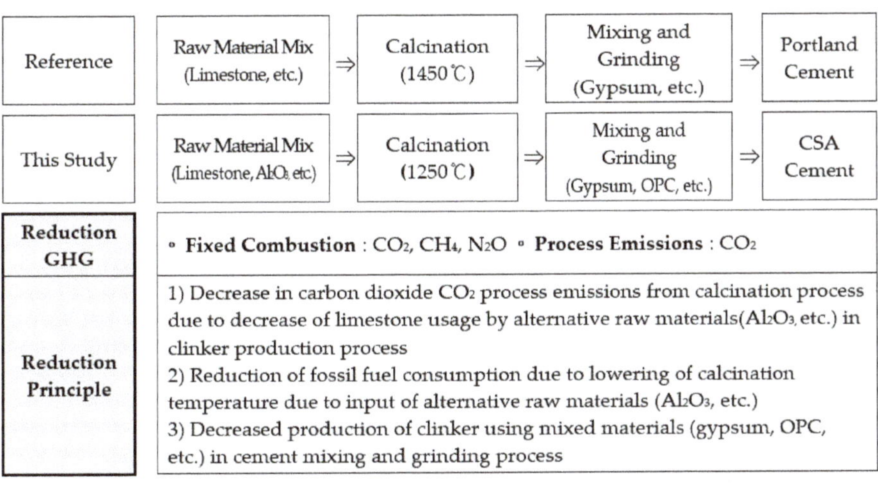

Figure 1. GHG reduction principle in CSA cement.

Also, ACM0005 and ACM0015 only allow utilization of one clinker for calculations of GHG reduction. For this reason, these two methodologies cannot be applied to CSA cement that has to mix OPC clinkers (more than one) for maintaining the quality.

4.2. Development of New CDM Methodology

As previously mentioned, the registered CDM methodologies were not applicable for utilization in the CSA cement process. For obtaining approval for the GHG reduction effect of CSA cement, it is necessary to apply the registered CDM methodologies to the CSA cement methodology. Within the two methodologies, GHG reduction happens during the "mixing and grinding process" of ACM0005, whereas it is the "raw material mix process" that enables GHG reduction under ACM0015. However, for CSA cements, GHG reduction happens during both of the processes. These were the reasons why it was necessary to develop a new methodology for the CSA cement process (Figure 2).

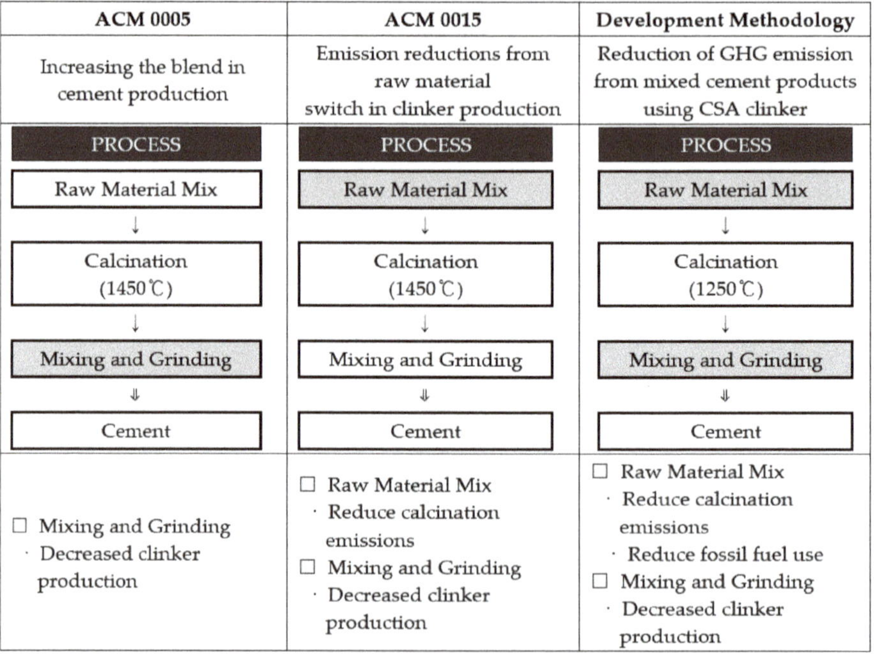

ACM 0005	ACM 0015	Development Methodology
Increasing the blend in cement production	Emission reductions from raw material switch in clinker production	Reduction of GHG emission from mixed cement products using CSA clinker
PROCESS	PROCESS	PROCESS
Raw Material Mix	Raw Material Mix	Raw Material Mix
↓	↓	↓
Calcination (1450 ℃)	Calcination (1450 ℃)	Calcination (1250 ℃)
↓	↓	↓
Mixing and Grinding	Mixing and Grinding	Mixing and Grinding
⇓	⇓	⇓
Cement	Cement	Cement
☐ Mixing and Grinding · Decreased clinker production	☐ Raw Material Mix · Reduce calcination emissions ☐ Mixing and Grinding · Decreased clinker production	☐ Raw Material Mix · Reduce calcination emissions · Reduce fossil fuel use ☐ Mixing and Grinding · Decreased clinker production

Figure 2. GHG reduction principle in CSA cement.

4.2.1. Applicability

Applicability of the new CDM methodology is presented in Table 2.

Related to the applicability of the CSA cement export parts, GHG reduction here happens during the production phase of the technology, and therefore the export part is excluded from the applicability conditions. In addition, GHG reduction for CSA cement tech mainly happens during the "clinker process" and thus can be categorized as a reduction in "process emissions," since we created the block condition that does not fit in the new methodologies. Also, we presented methodologies to ensure quality and compare quality of the produced cement during the baseline and project scenario with support from the applicable third-party certification institutions.

Especially, we enabled utilization of only waste materials, excluding mineral materials, as alternative material for the CSA cement production process. The reason for excluding mineral materials was to prevent emission leakage during the mining and transport process and thereby curb the total emission from the CSA cement production process.

Table 2. Applicability of the development methodology.

Applicability
1. This methodology can be applied to cement mixing processes sold in the target country. However, the emission reductions achieved during the process exclude reductions in greenhouse gas emissions after cement shipment.
2. All clinkers used in the project should be produced at the cement plant located within the business boundary, and the whole cement production process must be performed at the cement plant located within the project boundary.
3. There must be a component and quality standard for each type of cement certified by a country or a third-party certification body. It should be possible to compare the quality of cement produced in this project with the quality of cement produced in the baseline scenario.
4. Baseline clinker production raw materials should include $CaCO_3$ and/or $MgCO_3$, and the project should use raw materials (such as fly ash) that can be partially or completely replaced.
5. The additional effect of clinker production using alternative raw materials may extend over the lifetime, but the objective of the project should not be to increase the capacity and extend the life of the equipment.
6. The quality of the produced cement must meet the cement quality standards of the country concerned.
7. It should be demonstrated that alternative raw materials and additive materials did not result in increased emissions in other areas.
8. Alternative raw materials related to the production of CSA clinkers should be waste-based and not mineral-based.

4.2.2. GHG Emission Calculation Method

In the methodology, the GHG emission takes place during the clinker unit's and cement unit's development process. The project boundary is a cement production factory, facility factory (if there is one), or electricity grid. Also, indirect emission is equivalent to the loss of a power plant or electricity grid when purchasing the power. The alternative material and additives' transport emission is included apart from these, but the transport emission for the raw material for producing the clinker is excluded as a conservative simplification. The GHG emissions for the CSA cement process could be divided into: (1) kiln emissions, (2) fossil fuel and power use for clinker production, and (3) power used for mixed grinding to prepare alternative raw materials and additives.

During the production of CSA cement clinker, OPC is added for equalizing the quality, but while utilizing the registered methodologies there are no ways to apply the CSA cement clinker (Equation (1)). Given this constraint, we developed a method for utilizing more than one clinker for emission calculations (Equation (2)):

$$PE_{clinker,\ y} = PE_{calcin,\ y} + PE_{fossil\ fuel,\ y} + PE_{ele,\ grid,\ CLNK,\ y} + PE_{ele,\ sg,\ CLNK,\ y} \tag{1}$$

where:

$PE_{clinker,\ y}$ = CO_2 emissions per ton of clinker in the project activity plant in year "y" (t CO_2/t clinker)

$PE_{calcin,\ y}$ = Emissions per ton of clinker due to calcination of calcium carbonate and magnesium carbonate in year "y" (t CO_2/t clinker)

$PE_{fossil\ fuel,\ y}$ = Emissions per ton of clinker due to combustion of fossil fuels for clinker production in year "y" (t CO_2/t clinker)

$PE_{ele,\ grid,\ CLNK,\ y}$ = Grid electricity emissions for clinker production per ton of clinker in year "y" (t CO_2/ t clinker)

$PE_{ele,\ sg,\ CLNK,\ y}$ = Emissions from self-generated electricity per ton of clinker production in year "y" (t CO_2/t clinker)

$$PE_{clinker, y} = \sum \{(PE_{calcin, i, y} + PE_{fossil\ fuel, i, y} + PE_{ele, grid, CLNK, i, y} + PE_{ele, sg, CLNK, i, y}) \times P_{CLNK, i, y}\} \quad (2)$$

where:

$PE_{clinker, y}$ = CO_2 emissions per ton of clinker in the project activity plant in year "y" (t CO_2/t clinker)

$PE_{calcin, i, y}$ = Emissions per ton of clinker i due to calcinations of calcium carbonate and magnesium carbonate in year "y" (t CO_2/t clinker)

$PE_{fossil\ fuel, i, y}$ = Emissions per ton of clinker i due to combustion of fossil fuels for clinker production in year "y" (t CO_2/t clinker)

$PE_{ele, grid, CLNK, i, y}$ = Grid electricity emissions for clinker i production per ton of clinker in year "y" (t CO_2/t clinker)

$PE_{ele, sg, CLNK, i, y}$ = Emissions from self-generated electricity per ton of clinker i production in year "y" (t CO_2/t clinker)

$P_{CLNK, i, y}$ = Percentage of clinker i in year "y" (t clinker/t clinker)

4.3. Calculation of GHG Reduction

We utilized the newly developed methodologies for calculating reduction in GHG for CSA cement. The calculated data was from Korea H Cement Company's "Production Basic Unit of Portland Cement" and "Projection Basic Unit of CSA Cement (Experimental Value)." The applied data is on Table 3. The amount of electricity used for clinker production, cement grinding, raw material, and additive preparation was assumed to be the same.

Table 3. Data for calculating GHG reductions.

List	Baseline (OPC)	Project (CSA)	Unit
Amount of fossil fuel used for clinker production	900	641	Kcal/kg-clinker
Electricity used in clinker production	28.11	28.11	KWh
Free CaO Content	0.004485	0.004848	%
CaO content of clinker produced	0.623	0.44	%
Free MgO Content	0.000142	0.000251	%
MgO content of clinker produced	0.0308	0.0182	%
Amount of electricity used for cement grinding	43.41	43.41	KWh
Amount of electricity used to prepare raw materials and additives	25.43	25.43	KWh
Clinker mixing ratio of cement	0.889	0.8	%
Additive mixing ratio of cement	0.111	0.2	%

Given these assumptions, the GHG reductions were calculated as shown in Table 4. The CSA cement production process achieved a reduction of 0.281 tCO_2-eq/ton above the reduction realized for Portland cement, which was the baseline. As determined during the research, the GHG reduction point was the "kiln process," where it was possible to curb the process emissions by reducing the fossil fuel usage and also by changing the ratio of the clinker content. In the latter, GHG reductions were calculated per ton of cement.

As for the "National Strategy Project of Carbon Mineralization," which is the fundamental R&D(research and development) for this research, the third phase plan is to produce 30,000 ton/year of CSA cement and thereby realize 8430 tCO_2-eq/year of GHG reduction. The resulting data of GHG reduction is based on the experimental values. If the CSA cement is produced by the generally considered method, the GHG reduction could be less than the values presented here.

Table 4. GHG reduction calculation results.

List	Baseline (OPC)	Project (CSA)	Unit
GHG Emissions in the Clinker	0.876	0.621	$tCO_2/tClinker$
GHG Emissions from Calcination	0.519	0.361	$tCO_2/tClinker$
GHG Emissions from Fossil Fuel Use	0.342	0.243	$tCO_2/tClinker$
GHG Emissions from Electricity Use	0.015	0.016	$tCO_2/tClinker$
GHG Emissions from Cement Grinding and Mixture Preparation	0.032	0.032	$tCO_2/tCement$
Total GHG Emissions Unit	0.810	0.529	$tCO_2/tCement$
GHG Reduction	**0.281**		$tCO_2/tCement$

5. Conclusions

In this research, we attempted to develop a method for quantification of the reduction in GHG using a CSA cement process and developed CDM methodologies to register the CSA cement process as a CDM project in the UNFCCC. There are some registered methodologies related to cements, but those methodologies do not satisfy the conditions of CSA cement. Hence, we developed a new methodology for CSA cement. The new methodology describes how to calculate GHG reduction during the CSA cement process when more than one type of clinker is utilized. Using these calculations, it was found that the GHG reduction was 0.2851 tCO_2-eq/ton while producing 1 ton of CSA cement.

We believe that these new methodologies, once registered, will potentially have an impact on the following situations.

- Impact on developing countries cement technologies transfer and support for industrial parts.
- GHG reduction when utilizing and producing cement and thereby the chance for creating carbon credits (emissions).
- Will allow countries to choose material and technologies that enable reduction of GHG in their industrial sectors for their development.

This CSA cement technology could be utilized in developing countries' infrastructure construction sectors and will also realize a greater reduction in GHG compared to general cement. In addition, development of this new CDM methodology contains the possibility of allowing CDM principles after POST2020 under the Paris Agreement. For this, there has been opportunity for developing countries to consider the technologies under CDM. We also developed a type of Greenfield project for developing countries to develop their own CSA cement and achieve their GHG reduction targets.

Author Contributions: Conceptualization, E.-d.J.; Investigation, E.-d.J.; Supervision, C.-k.L.; Writing—Original draft, E.-d.J.; Writing—Review and editing, E.-d.J. and K.-u.L.

Funding: This research was funded by [Ministry of Science and ICT (MSIT); Ministry of Environment (ME); and Ministry of Trade, Industry and Energy (MOTIE)] grant number [NRF-2017M3D8A2085273] And The APC was funded by [NRF-2017M3D8A2085273].

Acknowledgments: This research was supported by the National Strategic Project-Carbon Upcycling of the National Research Foundation of Korea (NRF) funded by the Ministry of Science and ICT (MSIT); the Ministry of Environment (ME); and the Ministry of Trade, Industry and Energy (MOTIE), (NRF-2017M3D8A2085273).

Conflicts of Interest: The authors declare no conflict of interest.

References

1. Greenhouse Gas Inventory Research Center. *National Greenhouse Gas Inventory Report of Korea*; Greenhouse Gas Inventory Research Center: Seoul, Korea, 2018.
2. World Business Council for Sustainable Development. *CO2 Accounting and Reporting Standard for the Cement Industry*; WBCSD: Seoul, Korea, 2011.
3. IPCC. *2006 IPCC Guidelines for National Greenhouse Gas Inventories*; IPCC: Geneva, Switzerland, 2006.

4. Ministry of Environment. *Instruction for the Goal Management and Operation of Greenhouse Gas Energy*; Ministry of Environment: Seoul, Korea, 2016.
5. UNFCCC CDM Executive Board. *ACM 0005: Increasing the Blend in Cement Production*; Version 07.1.0; UNFCCC: Bonn, Germany, 2012.
6. UNFCCC CDM Executive Board. *ACM 0015: Emission Reductions from Raw Material Switch in Clinker Production*; Version 04.0.; UNFCCC: Bonn, Germany, 2014.

 *sustainability*

Article

Utilization of Lime Mud Waste from Paper Mills for Efficient Phosphorus Removal

Hong Ha Thi Vu [1], Mohd Danish Khan [2], Ramakrishna Chilakala [3], Tuan Quang Lai [2], Thriveni Thenepalli [1], Ji Whan Ahn [1], Dong Un Park [4] and Jeongyun Kim [1,*]

[1]   Center for Carbon Mineralization, Mineral Resources Division, Korea Institute of Geoscience and Mineral Resources, 124 Gwahak-ro, Gajeong-dong, Yuseong-gu, Daejeon 34132, Korea; hongha@kigam.re.kr (H.H.T.V.); thenepallit@rediffmail.com (T.T.); ahnjw@kigam.re.kr (J.W.A.)

[2]   Resources Recycling Department, University of Science and Technology, 217, Gajeong-ro, Yuseong-gu, Daejeon 34113, Korea; danish0417@ust.ac.kr (M.D.K.); tuanlai@ust.ac.kr (T.Q.L.)

[3]   Department of Bio-Based Materials, School of Agriculture and Life Science, Chungnam National University Daejeon-34132, Korea; chilakala_ramakrishna@rediffmail.com

[4]   Center for Climate Technology Cooperation, 17th Floor, Namsan Square Bldg., 173, Toegye-ro, Jung-gu, Seoul 04554, Korea; dongun.park@gtck.re.kr

*   Correspondence: kooltz77@kigam.re.kr

Received: 8 February 2019; Accepted: 9 March 2019; Published: 13 March 2019

Abstract: In this study, we utilized lime mud waste from paper mills to synthesize calcium hydroxide $(Ca(OH)_2)$ nanoparticles (NPs) and investigate their application for the removal of phosphorus from aqueous solution. The NPs, composed of green portlandite with hexagonal shape, were successfully produced using a precipitation method at moderately high temperature. The crystal structure and characterization of the prepared $Ca(OH)_2$ nanoparticles were analyzed by field emission scanning electron microscopy, Fourier transform infrared spectroscopy, and X-ray diffraction. The effects of $Ca(OH)_2$ NP dosage and contact time on removal of phosphorus were also investigated. The results show that the green portlandite NPs can effectively remove phosphorus from aqueous solution. The phosphorus removal efficiencies within 10 min are 53%, 72%, 78%, 98%, and 100% with the different mass ratios of $Ca(OH)_2$ NPs/phosphorus (CNPs/P) of 2.2, 3.5, 4.4, 5.3, and 6.2, respectively. Due to the efficient phosphorus removal, the calcium hydroxide nanoparticles (CNPs) could be a potential candidate for this application in domestic or industrial wastewater treatment.

Keywords: calcium hydroxide; lime mud; nanoparticles; phosphorus removal

---

1. Introduction

Pulp and paper industries are producing an enormous amount of pulp annually to meet the ever-growing demand for papers and packaging materials [1–3]. Statistics also revealed that the global paper and cardboard production rose from 391.2 to 410.9 million metric tons from the year 2008 to 2016 [4]. Countries such as China, the United States, and Japan account for more than 50% of total paper and cardboard production. China alone produces more than 111.3 million metric tons (as of 2016) and currently is the largest producer of pulp and cardboard, followed by the United States, Japan, and South Korea [5]. However, organic and inorganic wastes (dregs and ash) are produced in huge quantities as byproducts, causing severe ecological and environmental issues. Lime mud is one such byproduct and its estimated outcome accounts for approximately 0.47 m^3 tons^{-1} of pulp produced [6]. Lime mud is produced during the wood chips-to-pulp conversion process for paper production. The pulp is extracted from those wood chips through sodium hydroxide treatment and sodium carbonate is formed as a byproduct. For the recovery of sodium hydroxide, calcium oxide (quicklime) is then added to sodium carbonate slurry and calcium carbonate is formed, which is termed as 'lime mud'. Trace amounts of other elements such as magnesium, potassium, sulfur, and boron can also be found in lime mud.

Lime mud is classified as a toxic industrial waste mainly due to its high alkalinity and typical mineralogical characteristics and therefore requires proper treatment before discharge [7]. Reutilization of lime mud is very limited in industries and accounts for only 30% of lime mud produced, while the remainder is disposed of in landfills. However, many serious environmental issues are associated with disposal in landfills: (a) large landfill area occupation with high disposal cost; (b) landfill leachates moving into groundwater and rivers; (c) adverse impacts on nearby plants and soils; (d) toxic fine dust generation in dry and windy conditions [8]. Furthermore, landfills are also suggested to be an unsuitable option for lime mud disposal [9]. Therefore, there is an utmost requirement for cost-effective reutilization and valorization of lime mud.

Lime mud is used in many applications, such as building materials (bricks and cements) [10], wastewater treatment [11], and agricultural soils [12]. However, high costs involved in pretreatment processes such as drying, desalting, or dewatering makes lime mud a secondary option. Therefore, lime mud valorization can be a possible option to yield a more value-added product. Lime mud is mostly composed of calcium carbonate (~95–96%), and with a few steps of pretreatment, it can be converted into calcium carbonate ($CaCO_3$), calcium oxide (CaO), calcium hydroxide ($Ca(OH)_2$), ceramsite, and bioceramics [13–15]. This valorization can produce CaO and $Ca(OH)_2$, which has the capacity to effectively remove phosphate. The nanoscale size of produced product provides enormous surface area and activated surface for effective adsorption of phosphorus, fluoride, and many other heavy metals such as Pb, Cu, As, and Cd [16–18]. Numerous studies have been performed on bioceramics, lime sludge, and red mud (~46 wt. % of CaO), which have similar compositions to lime mud, for the removal of phosphorus through surface adsorption and wet chemical precipitation processes [13,19–21], whereas utilization of lime mud for phosphorus removal has not yet been documented, to the best of our knowledge.

Phosphorus is a very important element for the industries of fertilizers [22], pharmaceuticals [23], detergents [24], and batteries [25]. These industries inevitably discharge a significant amount of phosphorus-bearing wastes (mostly in the form of phosphates) in effluent streams, which ultimately causes serious environmental issues. Phosphorus is the key element responsible for the occurrence of eutrophication as it encourages the growth of algae [26,27], which then adversely affects the overall aquatic life. Severe oxygen depletion due to higher eutrophication and biological oxygen demands are some of the major consequences [28]. This directly affects the water quality and hampers aquatic life. Therefore, removal and recovery of this useful element is highly desirable to minimize any environmental impact, particularly in nearby regions of urban areas.

Conventional phosphorus treatment methods include adsorption, biological treatments, precipitation, floatation, and crystallization [29]. The reduced removal efficiencies and complex operations make most of the mentioned methods unsuitable for commercial purposes. Chemical precipitation technique is widely accepted and is also capable of removing more than 99% of phosphorus from wastewater. Moreover, precipitation of phosphorus in the form of calcium phosphate is an important physicochemical process. This further promotes the efficient and economical route for the recovery of phosphorus [28].

Numerous studies have already been conducted on the treatment of phosphorus from various chemicals in wastewater through the precipitation method [28–33]. Among them, calcium-based compounds were very effective and could even reach up to 99% phosphorus removal efficiency [32]. In particular, calcium hydroxide has some merits over other metal salts, including that it does not induce metal ions such as Al^{3+} or Fe^{3+} and anions such as SO_4^{2-} or Cl^- in the treated water. In the present work, an attempt has been made to develop green nano-calcium hydroxide from lime mud without using any toxic chemicals or energy-intensive processes.

The objectives of the present work are to valorize lime mud into a value-added product, i.e., green nano-calcium hydroxide, and to determine the optimum dosage of green nano-calcium hydroxide and residence time required for the maximum removal of phosphorus from wastewater.

2. Materials and Methods

2.1. Green Ca(OH)$_2$ Preparation

Lime mud was collected from the Moorim paper mill in Ulsan, Republic of Korea. The composition of raw lime mud was predominantly calcium carbonate with small amounts of other oxides containing Al_2O_3, SiO_2, SO_3, P_2O_5, Na_2O, Fe_2O_3, and MgO. Hydrochloric acid (HCl, 35–37% concentration) and sodium hydroxide (NaOH, 97% purity) were purchased from Junsei Chemicals, Republic of Korea. All chemicals were used as received.

The green Ca(OH)$_2$ nanoparticles were prepared by a precipitation method involving the following chemical reaction:

$$CaCO_{3\,(solid)}\text{ (lime mud)} + 2HCl_{(aqueous)} \rightarrow CaCl_{2\,(aqueous)} + CO_{2\,(gas)} + H_2O_{(aqueous)}$$

$$CaCl_{2\,(aqueous)} + 2NaOH_{(aqueous)} \rightarrow Ca(OH)_{2\,(solid)} + 2NaCl_{(aqueous)}$$

Firstly, the lime mud was ground and screened through a sieve of 100 μm. Then, 12.5 g of fine powder lime mud was dissolved in 250 mL of 1 M HCl under vigorous stirring. In order to remove residues, the solution was filtered to 0.45 μm particle size by paper filter and syringe filter. After filtering, a transparent solution was obtained. Then, the transparent solution was heated at 90 °C due to the minimum solubility of CO_2 in water. Moreover, the calcium hydroxide nanoparticles (CNPs) should have a perfect shape if the temperature is around 90 °C [34,35]. When the temperature of the transparent solution reached 90 °C, 200 mL of 1 M NaOH solution was added dropwise into the resultant solution under vigorous magnetic stirring while keeping the temperature of the mixed solution around 90 °C. Subsequently, the transparent mixture became white in color within 5 min. Finally, the white solution was filtered, and then the residue was washed by deionized water several times to remove remaining impurities and dried in an oven at 100 °C for one day.

2.2. Phosphorus Removal Study

The potassium dihydrogen phosphate (KH_2PO_4) (extra pure reagent, Daejung Chemicals & Metal Co., Ltd, Korea) was dissolved in DI water to prepare 15 mg L^{-1} phosphorus solution. Different mass ratios of Ca(OH)$_2$ NPs/phosphorus (CNPs/P) (2.2, 3.5, 4.4, 5.3, 6.2) were added into different beakers containing 150 mL of phosphorus solution with fixed concentration of 15 mg L^{-1} (pH 7) under 300 rpm magnetic stirring at room temperature and with difference contact times of 1, 5, 10, 20, and 60 min. A slight change in pH (~0.5) was observed upon addition of CNPs. The solution in the treatment beaker was directly filtered through a 0.45 μm syringe filter to separate the nanoparticles from the phosphorus solution. The filtrate was then analyzed for remaining concentration of phosphorus in the solution by a spectrophotometer (HS 3300, HUMAS) using the ascorbic acid method (3000 TP-L program). The phosphorus removal efficiency E% was determined using the following equation:

$$E\% = \frac{C_0 - C}{C_0} \times 100 \tag{1}$$

where C_0 is the primary concentration of total phosphorus (mg L^{-1}) and C is the concentration of total phosphorus after treatment (mg L^{-1}).

2.3. Physical Characterization

The crystal structure study and the identification for mineral phases of lime mud waste and the prepared Ca(OH)$_2$ sample were examined by X-ray diffraction (XRD; BD2745N) with an X-ray source of Cu Kα (λ = 0.15406 nm) in the scan range of diffraction angle 2θ from 20° to 80°. In order to identify characteristic functional groups and structure properties, the samples were measured by Fourier transform infrared spectroscopy (FT-IR; 6700 FTIR) with the scan range from 400 to 4000 cm^{-1}. The samples were directly analyzed by FT-IR without any further sample preparation. The morphological features of lime mud sample were investigated by scanning electron microscopy (SEM; JSM-6380). The Ca(OH)$_2$ NPs micro images were recorded by field emission scanning electron microscopy (FE-SEM; Hitachi-S-4800) to find out the morphologies and size of CNPs.

For analysis concentration of phosphorus before and after treatment, the solutions were measured by the water analyzer and spectrophotometer (HUMAS, HS-3300) at 880 nm using the ascorbic acid colorimetric method.

3. Results

3.1. Characteristics of Green Ca(OH)$_2$

Regarding the characterization of crystal structure of the raw lime mud and CNPs, the samples were examined by XRD, as shown in Figure 1. The raw lime mud (black line) is highly crystalline. The resultant peaks are in good agreement with the mineralogical phase of rhombohedral calcite (CaCO$_3$), having space group R-3c (space group No. 167. PDF card No. 00_081_2027). The XRD peaks at 2θ = 23.1°, 29.3°, 35.9°, 39.4°, 43.1°, 47.1°, 47.5°, 48.4°, 56.5°, 57.3°, 60.6°, and 64.6° were assigned to the (012), (104), (110), (113), (202), (024), (018), (116), (211), (122), (214), and (300) planes of the calcite phase, respectively. For Ca(OH)$_2$ nanoparticles (red line), the major diffraction peaks matched very well with the characteristic peaks of the hexagonal portlandite phase of space group P-3m1 (Space Group No. 164, PDF Card No. 00-087-0673) [34]. The peaks at 2θ = 28.6°, 34.0°, 47.0°, 50.6°, 54.2°, 62.5°, 64.1°, and 71.7° corresponded to the (100), (101), (102), (110), (111), (021), (013), and (002) planes of the portlandite phase, respectively. Besides, a minor calcite peak was found at 2θ = 29.3° due to reaction of portlandite and carbon dioxide from air. Moreover, the crystallite size (d) of produced Ca(OH)$_2$ NP powder was calculated based on the Debye–Scherrer's equation:

$$d = \frac{K\lambda}{\beta \, \cos\theta} \tag{2}$$

where K is Debye–Scherrer's constant, equal to 0.090, λ is the wavelength of X-ray radiation used (λ = 0.15406 nm), θ is the Bragg diffraction angle, and β is the half-width diffraction peak. The estimated mean crystal size of portlandite NPs for the plane (101) (with the highest diffraction peak) was approximately 24.8 nm. The lattice strain ε of crystal at the plane was calculated with the following equation:

$$\varepsilon = \frac{\beta}{4tan\theta} \tag{3}$$

The lattice strain was found to be 4.775 × 10^{-3} for the plane (101) [36–38].

To observe the structural transformation from lime mud to portlandite nanoparticles, SEM and Field Emission Scanning Electron Microscopy (FESEM) analyses were conducted. Figure 2a,b shows the scanning electron microscope images of a lime mud sample. The micrograph shows that the lime mud has irregular shape and agglomerate units. The size range of lime mud is from 300 nm to 10 μm. After preparation, the Ca(OH)$_2$ NP structure revealed irregular shape to hexagonal shape features with varying sizes, as shown in Figure 2c. Figure 2d clearly shows the hexagonal nanoplate of Ca(OH)$_2$ nanoparticles with size of approximately 550–700 nm. Furthermore, the impurity in the form of calcite in the prepared Ca(OH)$_2$ nanoparticles sample was also observed through the FESEM image. The calcite CaCO$_3$ revealed a nano-fibrous shape and nano-needle shape (Figure 2c). Other studies also verified the formation of nano-portlandite with similar morphologies and size. Darroudi et al. reported the preparation of calcium hydroxide nanoparticles with hexagonal structure and average particle size range of 600–650 nm by a facile sol–gel method in aqueous gelatin media [39]. Pereira et al.

introduced Ca(OH)$_2$ nanoplates with average dimensions of 100–300 nm and hexagonal shape [40]. Madrid et al. produced high-purity hexagonal nano-calcium hydroxide with 200–600 nm size by a precipitation method in N$_2$ atmosphere and at different temperatures [35]. Ca(OH)$_2$ nanoplates of size ranging from 350 nm to 450 nm were also developed from waste oyster shells by Khan et al. [34].

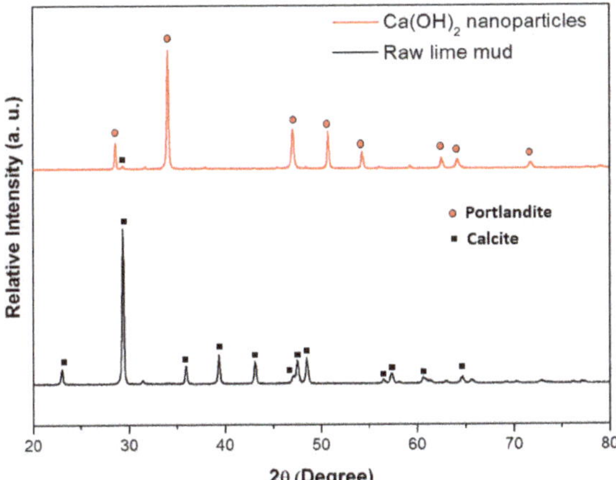

Figure 1. The XRD patterns of raw lime mud (black line) and Ca(OH)$_2$ nanoparticles (red line).

(a)

(b)

(c)

(d)

Figure 2. SEM analysis of lime mud samples and FESEM analysis of calcium hydroxide nano particles. (**a**) Low- and (**b**) high-magnification SEM images of raw lime mud; (**c**) low- and (**d**) high- magnification Field Emission Scanning Electron Microscopy (FESEM) of Ca(OH)$_2$ nanoparticles.

To identify the characteristic functional groups, the FTIR spectra of samples were recorded. Figure 3 presents the FTIR patterns of (a) raw lime mud and (b) $Ca(OH)_2$ nanoparticles. The FTIR spectra of raw lime mud indicated that the major composition of lime mud is calcite, as shown in Figure 3a. The broad stretching absorption and sharp peaks at 712, 871, and 1426 cm^{-1} denote v_4 (in-plane bending mode), v_2 (out-of-plane bending mode), and v_3 (antisymmetric stretching mode) of the CO_3^{2-} group of the calcite [41]. For green portlandite, the absorption peak at 3641.7 cm^{-1} was assigned to the characteristic hydroxyl group (OH^-) stretching vibration in the portlandite phase, as shown in Figure 3b [42]. Furthermore, the weak absorption bands at 871 cm^{-1} (v_2) and 1426 cm^{-1} (v_3) correspond to the stretching of the carbonate (CO_3^{2-}) group of the calcite phase, which was also confirmed in XRD result.

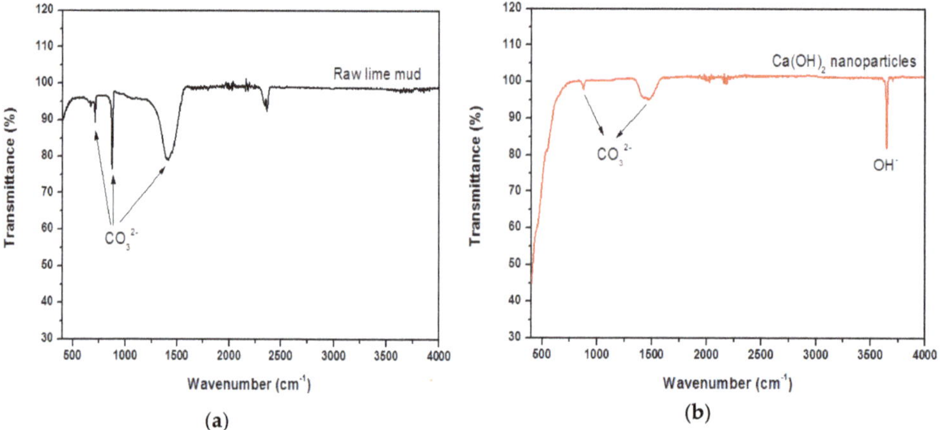

Figure 3. The FTIR patterns of (a) raw lime mud and (b) $Ca(OH)_2$ nanoparticles.

3.2. Phosphorus Treatment Application

The effect of contact time on removal of phosphorus was investigated. The experimental data were collected within 60 min to reach chemical equilibrium in the solution. The initial phosphorus concentration was fixed at 15 mg L^{-1}. The phosphorus removal efficiency of green calcium hydroxide versus contact time is presented in Figure 4. Initially, within 10 min, the phosphorus removal rate increased rapidly and then reduced gradually until equilibrium was attained. The highest removal efficiency of phosphorus was obtained within 20 min, as the system approaches equilibrium. The adsorption rate was rapid within the initial 10 min due to the presence of active sites on the surface of CNPs. However, due to continuous adsorption, those vacant sites started obtaining a saturation state, leading to limited contact between the phosphorus ions and surface area of the absorbent. Therefore, the phosphorus removal rate was lowered. The reaction reached chemical equilibrium when the surface of the CNPs was fully filled out by the phosphorus ions [43].

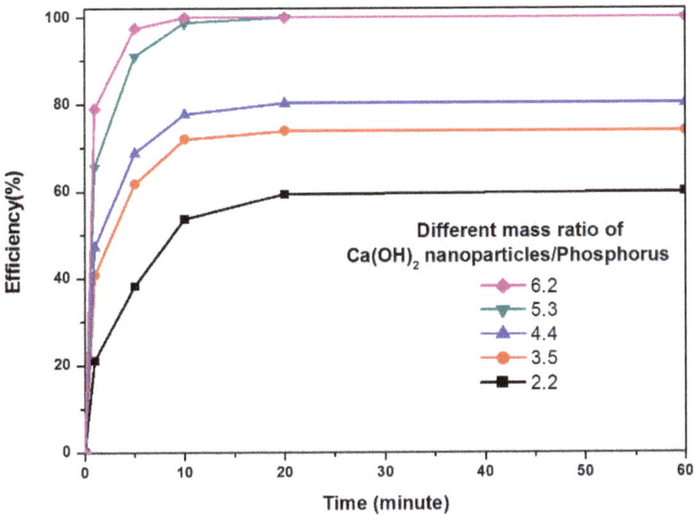

Figure 4. Effect of reaction time on phosphorus removal efficiency.

Figure 5 shows the effect of Ca(OH)$_2$ NP dosage on phosphorus removal within 10 min. The results show that the efficiency of phosphorus removal increased with the increased mass ratio of adsorbent green calcium hydroxide nanoparticles/phosphorus due to more vacant sites and surface area of the adsorbent particles for adsorbing phosphorus ions. When the mass ratio of CNPs/P was increased from 2.2 to 6.2, the phosphorus removal efficiency also increased from 53.5% to 100% within 10 min.

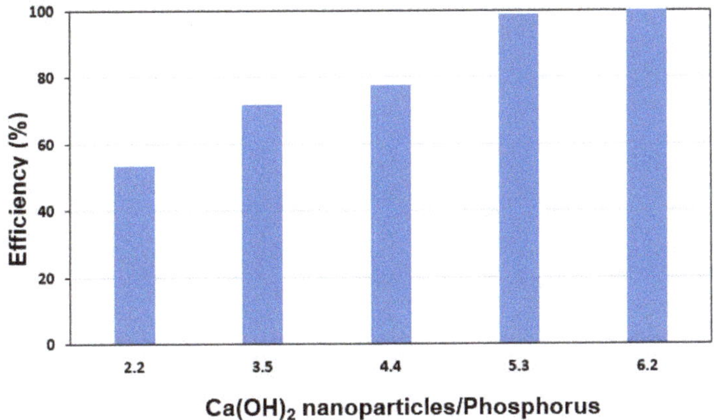

Figure 5. Effect of Ca(OH)$_2$ dosage (mass ratio) on phosphorus removal within 10 min.

Figure 6 illustrates the effect of absorbent dosage on removal of phosphorus with different contact time. The results indicated that the percent of phosphate adsorption increased when the mass ratio of CNPs/P and contact time were increased. When the mass ratio of CNPs/P were 2.2, 3.5, 4.4, 5.3, and 6.2, the phosphorus removal efficiencies were 59.2%, 73.8%, 80.2%, 100%, and 100%, respectively (t = 20 min). Other previous studies have also confirmed the enhancement of phosphorus removal efficiency with increasing adsorbent dosage and contact time. Torit et al. reported 80% phosphorus removal from domestic wastewater within 2 h with 5 g of calcinated eggshell in 1.7 mg L^{-1} of primary phosphorus concentration, which means the mass ratio of eggshell ash/phosphorus was about 2941 [43]. Deng et al. introduced a mass ratio of recycled concrete aggregate/phosphorus of 100, and about 95% phosphorus removal efficiency with 24 h contact time was achieved [44]. Nawar et al. determined that the removal efficiency of phosphorus ions from drinking water was 97% with a mass ratio of absorbent doses, alum sludge/phosphorus, of 50 within 30 min [45]. In this study, we reported 100% phosphorus removal efficiency with a mass ratio of CNPs/P of only 6.2 within 10 min.

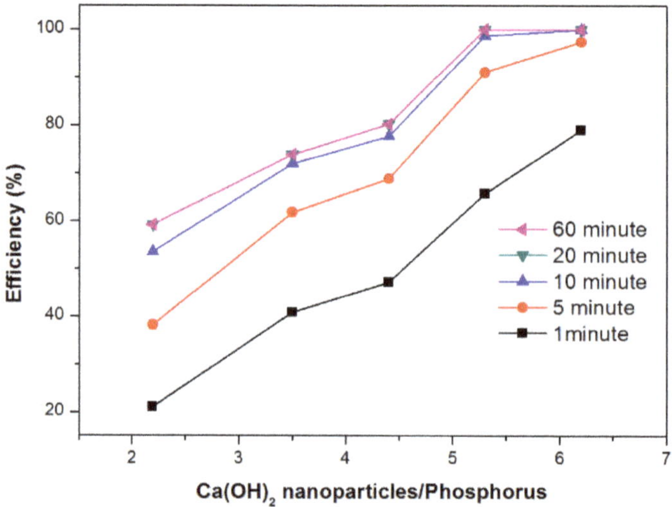

Figure 6. Effect of Ca(OH)$_2$ dosage on removal of phosphorus with different contact time.

Figure 7 presents the phosphorus adsorption rate and shows that phosphorus removal capacity increased until chemical equilibrium. The graph shows that the best effective adsorption capacity was achieved at a mass ratio of 6.2 of Ca(OH)$_2$ nanoparticles/phosphorus. The adsorption rate reached the maximum (adsorption capacity at equilibrium = 160.7 mg/g, efficiency = 100%) within 10 min.

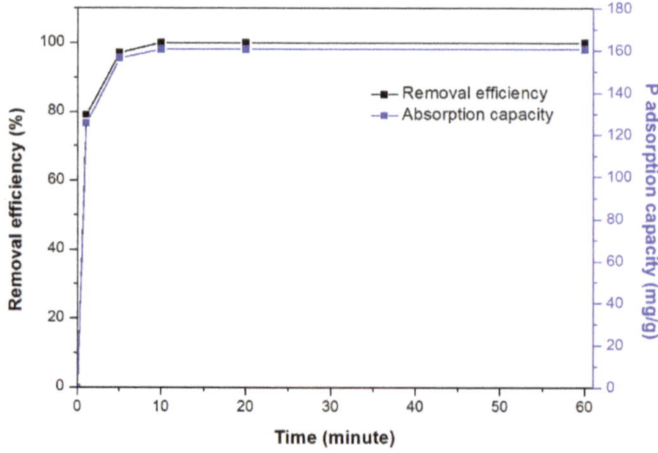

Figure 7. Kinetic studies of phosphorus adsorption.

4. Conclusions

In this work, green hexagonal portlandite NPs were successfully synthesized from waste lime mud by a precipitation method. The crystal structure and hexagonal shape of calcium hydroxide nanoparticles were determined through XRD and FESEM. The average particle size was found to be about 550–700 nm. In addition, the prepared $Ca(OH)_2$ nanoparticles were introduced as an absorbent able to remove phosphorus. A perfect phosphorus removal efficiency of 100% was achieved within 10 min when a CNPs/P mass ratio of 6.2 was employed to remove phosphorus in 150 mL of 15 mg L^{-1} phosphorus solution. Due to the efficient phosphorus removal results, the nano-portlandite (CNPs) is strongly suggested as an appropriate absorbent in phosphorus treatment from domestic and industrial wastewater. Moreover, green portlandite production from lime mud has tremendous environmental and economic benefits with reduced residue storage.

Author Contributions: H.H.T.V., M.D.K., and R.C. planned and designed the experiment; H.H.T.V., T.Q.L., and C.R. carried out the experiments; H.H.T.V. and M.D.K. analyzed the data and wrote the paper; T.T., D.U.P., J.W.A., and J.K. reviewed and revised the manuscript.

Funding: This research was supported by the National Strategic Project—Carbon Mineralization Flagship Center of the National Research Foundation of Korea (NRF) funded by the Ministry of Science and ICT (MSIT), the Ministry of Environment (ME), and the Ministry of Trade, Industry, and Energy (MOTIE) (2017M3D8A2084752).

Conflicts of Interest: The authors declare no conflict of interest.

References

1. He, J.J.; Lange, C.R.; Dougherty, M. Laboratory study using paper mill lime mud for agronomic benefit. *Process Saf. Environ.* **2009**, *87*, 401–405. [CrossRef]
2. Nurmesniemi, H.; Poykio, R.; Keiski, R.L. A case study of waste management at the Northern Finnish pulp and paper mill complex of Stora Enso Veitsiluoto Mills. *Waste Manag.* **2007**, *27*, 1939–1948. [CrossRef] [PubMed]
3. Zhou, Y.T.; Zhao, H.; Bai, H.L.; Zhang, L.P.; Tang, H.W. Papermaking effluent treatment: A new cellulose nanocrystalline/polysulfone composite membrane. *Proc. Environ. Sci.* **2012**, *16*, 145–151. [CrossRef]
4. Paper Industry—Statistics & Facts. Available online: https://www.statista.com/topics/1701/paper-industry/ (accessed on 30 January 2019).
5. Production Volume of Paper and Cardboard in Major Countries from 2009 to 2016 (in 1000 Metric Tons). Available online: https://www.statista.com/statistics/240598/production-of-paper-and-cardboard-in-selected-countries/ (accessed on 30 January 2019).
6. Wirojanagud, W.; Tantemsapya, N.; Tantriratna, P. Precipitation of heavy metals by lime mud waste of pulp and paper mill. *Songklanakarin J. Sci. Technol.* **2004**, *26*, 45–53.
7. Ren, X. Cleaner production in China's pulp and paper industry. *J. Clean. Prod.* **1998**, *6*, 349–355. [CrossRef]
8. Liu, Y.J.; Naidu, R.; Ming, H. Red mud as an amendment for pollutants in solid and liquid phases. *Geoderma* **2011**, *163*, 1–12. [CrossRef]
9. Monte, M.C.; Fuente, E.; Blanco, A.; Negro, C. Waste management from pulp and paper production in the European Union. *Waste Manag.* **2009**, *29*, 293–308. [CrossRef]
10. Madrid, M.; Orbe, A.; Carre, H.; Garcia, Y. Thermal performance of sawdust and lime-mud concrete masonry units. *Constr. Build. Mater.* **2018**, *169*, 113–123. [CrossRef]
11. Hartley, W.; Edwards, R.; Lepp, N.W. Arsenic and heavy metal mobility in iron oxide-amended contaminated soils as evaluated by short- and long-term leaching tests. *Environ. Pollut.* **2004**, *131*, 495–504. [CrossRef]
12. Bellaloui, A.; Chtaini, A.; Ballivy, G.; Narasiah, S. Laboratory investigation of the control of acid mine drainage using alkaline paper mill waste. *Water Air Soil Pollut.* **1999**, *11*, 57–73. [CrossRef]
13. Correa, T.H.A.; Toledo, R.; Silva, N.S.; Holanda, J.N.F. Novel nano-sized biphasic calcium phosphate bioceramics (β-CPP/β-TCP) derived of lime mud waste. *Mater. Lett.* **2019**, *243*, 17–20. [CrossRef]
14. Zhang, J.; Zheng, P.; Wang, Q. Lime mud from papermaking process as a potential ameliorant for pollutants at ambient conditions: A review. *J. Clean. Prod.* **2015**, *103*, 828–836. [CrossRef]
15. Qin, J.; Cui, C.; Cui, X.; Hussain, A.; Yang, C. Preparation and characterization of ceramite from lime mud and coal fly ash. *Constr. Build. Mater.* **2015**, *95*, 10–17. [CrossRef]

16. Yang, X.; Wan, Y.; Zheng, Y.; He, F.; Yu, Z.; Huang, J.; Wang, H.; Ok, Y.S.; Jiang, Y.; Gao, B. Surface functional groups of carbon-based adsorbents and their roles in the removal of heavy metals from aqueous solutions: A critical review. *Chem. Eng. J.* **2019**, *366*, 608–621. [CrossRef]

17. Chaudhary, M.; Maiti, A. Defluoridation by highly efficient calcium hydroxide nanorods from synthetic and industrial wastewater. *Colloids Surf. A. Physicochem. Eng. Asp.* **2019**, *561*, 79–88. [CrossRef]

18. Hu, H.; Li, X.; Huang, P.; Zhang, Q.; Yuan, W. Efficient removal of copper from wastewater by using mechanically activated calcium carbonate. *J. Envron. Manag.* **2017**, *203*, 1–7. [CrossRef]

19. Krishna, K.C.B.; Niaz, M.R.; Sarkar, D.C.; Jansen, T. Phosphorus removal from aqueous solution can be enhanced through the calcination of lime sludge. *J. Envron. Manag.* **2017**, *200*, 359–365. [CrossRef]

20. Baraka, A.M.; El-Tayieb, M.M.; Shafai, M.E.; Mahamed, N.Y. Sorptive removal of phosphate from wastewater using activated red mud. *Aust. J. Basic Appl. Sci.* **2012**, *6*, 500–510.

21. Ni, F.; He, J.; Wang, Y.; Luan, Z. Preparation and characterization of a cost-effective red mud/polyaluminum chloride composite coagulant for enhanced phosphate removal from aqueous solutions. *J. Water Process Eng.* **2015**, *6*, 158–165. [CrossRef]

22. Kruger, O.; Adam, C. Phosphorus in recycling fertilizers—Analytical challenges. *Environ. Res.* **2017**, *155*, 353–358. [CrossRef]

23. Eussen, S.R.; Verhagen, H.; Klungel, O.H.; Garssen, J.; van Loveren, H.; van Kranen, H.J.; Rompelberg, C.J. Functional foods and dietary supplements: Products at the interface between pharma and nutrition. *Eur. J. Pharmacol.* **2011**, *668*, S2–S9. [CrossRef] [PubMed]

24. Puijenbroek, P.J.T.M.V.; Beusen, A.H.W.; Bouwman, A.F. Datasets of the phosphorus content in laundry and dishwasher detergents. *Data Brief* **2018**, *21*, 2284–2289. [CrossRef] [PubMed]

25. Ott, C.; Degg, A.; Walke, P.; Reiter, F.; Nilges, T. Influence of copper on the capacity of phosphorus-anode in sodium-ion-batteries. *J. Solid State Chem.* **2019**, *270*, 636–641. [CrossRef]

26. Harrison, R.M. *Pollution: Causes, Effects, and Control*, 3rd ed.; The Royal Society of Chemistry: London, UK, 1996.

27. Kiely, G. *Environmental Engineering*; McGrawHill: New York, NY, USA, 1997.

28. Hosni, K.; Ben, M.S.; Chachi, A.; Ben Amor, M. The removal of PO_4^{3-} by calcium hydroxide from synthetic wastewater: Optimisation of the operating conditions. *Desalination* **2008**, *223*, 337–343. [CrossRef]

29. Yang, M.; Shi, J.; Xu, Z.; Zhu, S.; Cui, Y. Phosphorus removal and recovery from fosfomycin pharmaceutical wastewater by the induced crystallization process. *J. Environ. Manag.* **2019**, *231*, 207–212. [CrossRef]

30. Zhang, J.; Bligh, M.W.; Liang, P.; Waite, T.D.; Huang, X. Phosphorus removal by in situ generated Fe(II): Efficacy, kinetics and mechanism. *Water Res.* **2018**, *136*, 120–130. [CrossRef]

31. Namasivayam, C.; Sakoda, A.; Suzuki, M. Technical Note: Removal of phosphate by adsorption onto oyster shell powder-kinetic studies. *J. Chem. Technol. Biotechnol.* **2011**, *80*, 356–358. [CrossRef]

32. Xiong, J.; Quin, Y.; Islam, E.; Yue, M.; Wang, W. Phosphate removal from solution using powdered freshwater mussel shells. *Desalination* **2011**, *276*, 317–321. [CrossRef]

33. Penn, C.J.; Bryant, R.B.; Callahan, M.P.; McGrath, J.M. Use of Industrial By-products to Sorb and Retain Phosphorus. *Commun. Soil Sci. Plant Anal.* **2011**, *42*, 633–644. [CrossRef]

34. Khan, M.D.; Ahn, J.W.; Nam, G. Environmental benign synthesis, characterization and mechanism studies of green calcium hydroxide nano-plates derived from waste oyster shells. *J Environ. Manag.* **2018**, *223*, 947–951. [CrossRef]

35. Madrid, J.A.; Lanzon, M. Synthesis and morphological examination of high-purity Ca(OH)$_2$ nanoparticles suitable to consolidate porous surfaces. *Appl. Surf. Sci.* **2017**, *424*, 2–8. [CrossRef]

36. Bindu, P.; Thomas, S. Estimation of lattice strain in ZnO nanoparticles: X-ray peak profile analysis. *J. Theor. Appl. Phys.* **2014**, *8*, 123–134. [CrossRef]

37. Venkateswarlu, K.; Sandhyarani, M.; Nellaippan, T.A.; Rameshbabu, N. Estimation of crystallite size, lattice strain and dislocation density of nanocrystalline carbonate substituted hydroxyapatite by V-ray peak variance analysis. *Proc. Mater. Sci.* **2014**, *5*, 212–221. [CrossRef]

38. Shafi, P.M.; Bose, A.C. Impact of crystalline defects and size on X-ray line broadening: A phenomenological approach for tetragonal SnO$_2$ nanocrystals. *AIP Adv.* **2015**, *5*, 057137. [CrossRef]

39. Darroudi, M.; Bagherpour, M.; Hosseini, H.A.; Ebrahimi, M. Biopolymer-assisted green synthesis and characterization of calcium hydroxide nanoparticles. *Ceram. Int.* **2016**, *42*, 3816–3819. [CrossRef]

40. Pereira, M.R.N.; Salviano, A.B.; Medeiros, T.P.; Santos, M.R.D.; Cibaka, T.E.; Andrade, M.H.C.; Porto, A.O.; Lago, R.M. Ca(OH)$_2$ nanoplates support on activated carbon for the neutralization/removal of free fatty acids during biodiesel production. *Fuel* **2018**, *221*, 469–475. [CrossRef]
41. Makreski, P.; Jovanovski, G. Minerals from Macedonia IX. Distinction between some rhombohedral carbonates by FT IR spectroscopy. *Bull. Chem. Technol. Macedonia* **2003**, *22*, 25–32.
42. Asikin-Mijan, N.; Taufiq-Yap, Y.H.; Lee, H.V. Synthesis of clamshell derived Ca(OH)$_2$ nano-particles via simple surfactant-hydration treatment. *Chem. Eng. J.* **2015**, *262*, 1043–1051. [CrossRef]
43. Torit, J.; Phihusut, D. Phosphorus removal from waste water using eggshell ash. *Environ. Sci. Pollut. Res. Int.* **2018**, *25*, 1–9.
44. Deng, Y.; Wheatley, A. Mechanisms of phosphorus removal by recycled crushed concrete. *Int. J. Environ. Res. Public Health* **2018**, *15*, 357. [CrossRef]
45. Nawar, N.; Ahmad, M.E.; El Said, W.M.; Moalla, S.M.N. Adsorptive removal of phosphorus from wastewater using drinking water treatment-alum sludge (DWT-AS) as low cost adsorbent. *Am. J. Chem. Appl.* **2015**, *2*, 79–85.

Case Report

A Case Study of Environmental Policies and Guidelines for the Use of Coal Ash as Mine Reclamation Filler: Relevance for Needed South Korean Policy Updates

Hanna Cho [1], Sang-woo Ji [2], Hee-young Shin [2] and Hwanju Jo [2,*]

[1]   Environmental Assessment Group, Korea Environmental Institute, Bldg. B, 370, Sicheong-daero, Sejong 30147, Korea
[2]   Mineral Resources Division, Korea Institute of Geoscience and Mineral Resources, 124, Gwahak-ro, Yuseong-gu, Daejeon 34132, Korea
*   Correspondence: chohwanju@kigam.re.kr; Tel.: +82-42-868-3473

Received: 30 April 2019; Accepted: 30 June 2019; Published: 2 July 2019

Abstract: The South Korean government is pursuing a national project to use the complex carbonates found in coal ash to capture CO_2 and promote coal ash recycling. One possible approach is the use of coal ash as fill material in mine reclamation, but environmental concerns have so far blocked the implementation of this procedure, and no relevant regulations or guidelines exist. In this study, we review international approaches to the environmental management of coal ash recycling and consider how the lessons learned can be applied to South Korea. Each studied country was proactively using coal ash for beneficial uses under locally suitable conditions. The United States, European Union, United Kingdom, Australia, and Japan are all putting coal ash to beneficial use following thorough analyses of the environmental impact based on several considerations, including bulk concentration, coal ash leachate concentration, field inspections, and water quality monitoring. Our findings can contribute to the development of proper regulations and policies to encourage the use of recycled coal ash in South Korea as an approach to managing carbon emissions and climate changes. There are currently no relevant regulations in South Korea, so we consider the adoption of the strictest standards at each stage of the other cases at the time of introduction. Based on our findings, detailed and appropriate management guidelines can be developed in the future. Establishing management plans for complex carbonates, verifying their environmental stability, and using them as fill material will provide clear benefits for South Korea in the future.

Keywords: South Korea; mine reclamation; coal ash; complex carbonates; environmental policy

1. Introduction

Despite commitments and efforts to increase the share of renewable energy in the global energy supply, coal remains one of the world's biggest energy sources, generating considerable amounts of coal fly ash [1]. Globally, many efforts are being made to recycle the 7.8 billion tons of coal ash currently being generated annually, only 53.3% of which is currently recycled. For example, in 2016, the United States reused 60.2 million tons of coal combustion products (CCPs) out of 107.4 million tons produced. Although the rate of ash utilization thus increased from 52% to 56%, the total volume of material utilized stayed about the same as production declined. Even though the coal ash production volume has declined 7% since 2015 as coal's share of the total energy generation shrank in response to environmental regulations and competition from other energy sources, the coal ash utilization volume remained approximately level with that of the prior year [2]. Coal ash has been successfully used for many years in a wide range of applications, including in building material, asphalt, concrete pavement, soil stabilization, road base, structural fill, embankments, mine reclamation, mineral filler, and fertilizer, as well as in small-scale applications, such as the production of zeolites and geopolymers [3]. The United States and Australia, which have large land areas and many developed mines, also recycle coal ash as landfill material and filling.

South Korea generated about 9 million tons of coal ash in 2016; this figure is expected to reach 14 million tons in 2021 and increase about 1.6 times over the next 10 years. The construction of circulating fluidized bed combustion (CFBC) power plants generates coal ash containing 40% CaO or more. The capacity of current South Korean disposal facilities is nearing saturation, and some recently built power plants are targeting 100% recycling of generated coal ash [4]. Recently, South Korea has been developing complex carbonate minerals that use coal ash to capture carbon dioxide, an important approach to carbon dioxide reduction and coal ash recycling. Although these materials are intended for use as a high-function green cement and mine fill, they cannot be legally used for the latter purpose at present. In South Korea, coal ash (the main component of complex carbonates) can only be legally used for 15 applications, not including fill material in mine reclamation sites. There is a movement to legalize the use of coal ash at mine reclamation sites, but this effort is on hiatus due to concerns regarding the environmental impact, including the potential contamination of soil, ground surfaces, and groundwater resulting from the presence of soluble metal species in ash leachate [5]. Therefore, in order to safely use coal ash at mine reclamation sites in South Korea, its environmental impacts must first be assessed and predicted.

Currently, the United Kingdom, Japan, the European Union, the United States, Australia, and Canada all make beneficial use of coal ash as a mine fill material. Therefore, in this study, we analyzed these entities' relevant regulations and environmental management guidelines to assess the best practices for verifying and managing environmental stability when using coal ash for this purpose. We also considered the relevant regulations and guidelines for domestic waste recycling and mine fill in different countries. Based on these case studies, we then developed the necessary environmental management guidelines for the use of complex carbonates as fill material for mine reclamation sites in South Korea.

2. Case Studies

2.1. United States (US)

Code of Federal Regulations (CFR) Title 40 contains regulations related to environmental protection; those dealing with waste are 40 CFR 257 (Criteria for Classification of Waste Disposal Facilities and Practices) [6] and 40 CFR 261 (Identification and Listing of Hazardous Waste) [7]. These rules do not apply when coal combustion residuals (CCRs) are used in construction material or fill; instead, the Surface Control and Reclamation Act (SMCRA) [8], state regulations, and water quality standards must be satisfied.

In the United States, laws and regulations on coal ash are applied differently from state to state. We focused on such management in Pennsylvania, where coal ash is actively used for mine reclamation projects. Here, the use of coal ash was legalized following the revision of the 1986 Solid Waste Management Act in July 1992, after which the "Guidelines for Beneficial Use of Coal Ash at Coal Mines" [9] were developed in accordance with the regulations. Further revisions were made in 1997 and 1998 to provide guidance on certification standards and procedures for the use of coal ash in mining sites. In December 2010, the addition of Title 25, Chapter 290 of the state code, "Beneficial Use of Coal Ash" [10], codified all the requirements for the use, certification, and management of coal ash.

Certification follows two steps: certification of the ash itself and certification at the mine site. The first ensures that coal ash does not cause pollution and meets the minimum physical and chemical standards for further use. This requires defining the coal of origin, the combustion method, and the collection and storage method. For example, the concentration criteria for leaching and bulk conditions of coal materials are given in Tables 1 and 2, respectively [11]. The former criteria have not been finalized in the regulations, because the potential for contamination and the impact on health can vary by circumstances. The parameters presented here are based on the state's Department of Environmental Protection (DEP)'s experience and the results of studies by the National Research Council. The hydraulic conductivity, density, and acid neutralization capacity should also be measured. Even if the criteria are met, the generator should continue to prove that the coal ash meets the certification criteria through regular monitoring reports.

Table 1. Parameters required for leaching tests and their maximum acceptable leaching limits in the state of Pennsylvania in the United States (US) [11].

Parameter	Leaching Limit (mg/L)	Parameter	Leaching Limit (mg/L)
Ag	2.5	Mg	*
Al	5.0	Mn	2.5
As	0.25	Mo	4.375
B	15	Na	*
Ba	50	NH$_3$	30
Be	0.1	Ni	2.5
Ca	*	NO$_2$	1.0
Cd	0.125	NO$_3$	10
Cl	250	Pb	0.375
Co	17.5	SO$_4$	2500
Cr	2.5	Sb	0.15
Cu	25	Se	0.5
F	*	Tl	0.05
Fe	7.5	V	6.125
Hg	0.05	Zn	50
K	*	pH	7 or above

* Limit not established.

Table 2. Parameters required for bulk chemical analysis and their loading limits in the US state of Pennsylvania [11].

Parameter	Loading Rate (lbs/acre)	Parameter	Loading Rate (lbs/acre)
Al	*	Mg	*
Sb	*	Mn	*
As	36	Hg	15
Ba	*	Mo	16
Be	*	Ni	370
B	60	K	*
Cd	34	Se	88
Ca	*	Ag	*
Cr	2672	Na	*
Co	*	S	*
Cu	1320	Ti	*

Table 2. *Cont.*

Fe	*	V	*
Pb	264	Zn	2464

* Limit not established.

Certified coal ash is not necessarily suitable at all mine sites, requiring second site-based certification based on local characteristics, such as groundwater hydrology, mine waste chemistry, and local rivers. Groundwater assessments generally include conditions in wells, ponds, mine waste leachates, and abandoned mine shafts. Upgradient groundwater should also be studied, especially if pollution is already present. In addition, a minimum of three downgradient groundwater sites should be assessed.

Coal ash use can be implemented once it has passed these certification steps. The use of coal ash at mine sites also requires water quality monitoring plans to be submitted to the state's DEP. This should be planned so that the impact of coal ash use can be clearly determined based on an in-depth analysis of groundwater and surface water flows within mine site and the surrounding area. The water quality monitoring points should be representative of the overall water quality based on adequate samples and include both upgradient and downgradient locations. The reported values should meet all standards for drinking water in the state (Table 3). If measured values exceed these standards, then the DEP should be notified, and the water should be purified until stabilization is achieved.

Table 3. Water quality standards for monitoring coal ash use at mine reclamation sites in the US state of Pennsylvania [11].

Parameter	Limit (mcl)	Parameter	Limit (mcl)
Ag	*	Mg	*
Al	*	Mn	*
As	0.01	Mo	*
B	*	Na	*
Ba	2	NH_3	*
Be	0.004	Ni	*
Ca	*	NO_2	*
Cd	0.005	NO_3	*
Cl	*	Pb	0.015
Co	*	SO_4	*
Cr	0.1	Sb	*
Cu	1.3	Se	0.005
F	4	Tl	0.002
Fe	*	V	*
Hg	0.002	Zn	*
K	*		

* Limit not established.

2.2. European Union (EU)

The use of CCPs in Europe is being influenced by standardization and environmental legislation. In December 2010, a new Waste Framework Directive took effect in the EU, by which each member state has to define CCPs as byproducts and determine when a material is removed from waste status.

The producers of CCPs are required to register their products according to the Registration, Evaluation, Authorisation and Restriction of Chemicals (REACH) regulation as a precondition for placing a product on the market [12,13]. The new law's legal framework was established to improve waste management, promoting the prevention, reuse, and recycling of waste while exempting certain wastes from waste laws. It redefines byproducts as not being waste if they can be appropriately handled and put to beneficial use without potential negative influence on the environment and human health while conforming to general environmental regulations and certifications.

The law also sets criteria for end of waste (EoW) to prevent the discarding of certain materials having other applications that can be achieved in environmentally safe and economically beneficial ways. Examples include construction waste, coal ash, slag, scrap metal, glass, wastepaper, and tires. To be considered as EoW, the material should be clearly discernable during the recovery process and should avoid producing negative environmental impacts. Criteria were set for each potential material by the Institute for Prospective Technological Studies (IPTS) and the Joint Research Centre (JRC) based on a study performed at the request of the European Council [14]. In summary, the EU has been actively pursuing the beneficial use of CCPs under the condition that their environmental impact is thoroughly taken into consideration. In line with this, every member country has been independently devising regulations on the use of industrial byproducts.

United Kingdom (UK)

The UK sets standards for the production and use of byproducts from specific wastes through quality protocols (QPs) that define how to recover waste and turn it into a usable product [15]. These are a key mechanism in the national guidelines for waste recycling, which define the processes for using mass-produced materials, the product quality to be maintained, the point at which waste ceases to be waste, and when it can be used as a regular product (without the need for waste management controls). The UK's Environment Agency's Waste Protocols Project has established nine EoW QPs with another nine under development [16]. Waste types already covered include biodegradable waste compost, blast furnace slag, wood waste, and non-packaging plastics, which are recycled to produce products for use in agriculture, construction, energy, and manufacturing. In addition, the EQual project encouraged businesses to recycle waste into new products in ways that protected human health and the environment [17]. EQual helped businesses decide whether their recycled waste met regulatory standards when used in new products and provided technical guidance for regulating waste sites.

For example, the UK's Building Research Establishment has suggested guidelines for the environmental monitoring of fly ash grouting using two methods for groundwater monitoring: large-scale lysimeter experiments and collection from wells and piezometers in nearby areas. These are recommended prior to, during, and after construction. Groundwater quality should be determined using the stricter threshold value of either the Drinking Water Standard (DWS) or the Environmental Quality Standard (EQS), because the effects of some pollutants on humans and organisms are variable [18].

2.3. Australia

Similar to the US, Australia's Environmental Protection Act and management regulations apply different standards to the use of coal ash in each state: material with identical characteristics may be classified as inert waste in one state but as specified waste in another state, leading to different handling requirements. Water quality monitoring generally follows the Australian and New Zealand Guidelines for Fresh and Marine Water Quality produced by the Australian and New Zealand Environment Conservation Council (ANZECC), with each state further devising its own regulations [19].

2.3.1. Queensland

In Queensland, coal ash use is regulated by the 2011 Waste Reduction and Recycling Act [20] and is divided into three categories (bound, unbound, and soil ameliorant) with varying general conditions for the use of each that limit the harmful environmental effects of coal ash use. In addition, certifications for management and use of coal ash are required for the generator, transporter, and user. These regulations were revised in 2019 to state that coal combustion products are not waste and should be used as a resource based on Article 159 of the original law, following the relevant requirements and conditions. This reflects Queensland's vision of leading improved waste management in Australia by preventing unnecessary consumption or waste by adopting innovative resource restoration methods and managing all products and materials as valuable and finite resources.

In order for coal ash to be considered as EoW, the producer should monitor and analyze the resources, submitting annual reports using the criteria shown in Table 4 [21]. The concentration of pollutants should fall within the difference between the mean and standard deviation (at least 95% confidence interval), and statistical analysis should be performed to prove that no anomalies are present. Otherwise, analyses should be performed every six months; if actual or potential changes in resource composition are detected, further monitoring should be performed and appropriate measures should be taken [22]. Such monitoring and analysis should be performed by individuals with appropriate qualifications. Resource producers should document and store records of the source, date of dispatch, contact details of the recipient, and quantity of supply of the resource.

Table 4. Resource quality criteria in Queensland, Australia (total maximum concentration) [22].

Parameter	Bound Applications (mg/kg)	Unbound Applications (mg/kg)	Soil Ameliorant and Land Applications (mg/kg)
As (total)	*	20	20
Be	*	*	60
B	*	100	10 [1]
Cd	*	1	1
Cr (total)	*	100	100
Cr (III)	*	*	100
Cr (VI)	*	1.5	1
Co	*	100	100
Cu	*	100	100
Pb	*	50	50
Hg	*	1	1
Mo	*	*	10
Ni	*	60	60
Se	*	10	5
Zn	*	200	200
Electrical conductivity (us/cm)	*	*	10,000
pH	*	5–12.5	5–12.5

* No specified level (monitoring is still required), [1] Measured using the hot $CaCl_2$ method.

Resource users applying for unbound status must not use the resource below the groundwater table, within 50 m of a water supply bore, or in areas with a pH greater than 8 [22].

Water quality management in Queensland should primarily satisfy the Queensland Water Quality Guidelines (QWQ 2009), based on national standards [23,24], which aim to protect environmental values (EVs) through management approaches that vary by EV. The selection of parameters for evaluating EV protection levels depends on the evaluation purpose. The selected parameters should be highly specific to the given situation, and the most appropriate and cost-effective parameters should be selected, rather than simply monitoring a series of conventional parameters. The development of corresponding guideline values aids in this process [23].

2.3.2. New South Wales

In this state, the 2014 Coal Ash Exemption was applied as a Resource Recovery Exemption under Part 9, Clauses 91 and 92 of the Environment Operations (Waste) Regulations. Coal ash use should primarily satisfy the conditions given in Table 5 [25,26]. This exemption states that the use of coal ash or coal ash mixture is permitted if a consumer meets certain exemption requirements according to the 1997 Environment Operations Act. Here, coal ash refers to fly ash and floor ash obtained by burning coal in Australia and not to brine-conditioned or treated ash.

Table 5. Assessment procedures and criteria for coal ash use in New South Wales, Australia [25].

Parameter	Procedure	Criteria (mg/kg)	Frequency
Total metals (Cd, Pb, Hg)	US EPA 200.2	Cd (10) Pb (100) Hg (5)	Annual
Leachable TCLP metals	US EPA 1311	various	Annual

The exemption requires that coal ash satisfies all the requirements for chemicals and other substances and that it is mixed with a substance satisfying the requirements of the Resource Recovery Order before being used as a soil ameliorant for plant cultivation, as a cement admixture such as concrete, or in underground or road construction contexts as pipe bedding material; selected backfill adjacent to structures; road pavement, base, and sub-base structures; composite filler in asphalt pavements; rigid and composite pavement structures; working platforms atop earthworks; or fill for reinforced soil structures (including geo-grid applications). Prior to use, producers should prepare a plan that clearly describes the procedures for coal ash sample preparation and storage and then perform sampling and testing as required by the relevant regulations. The specified conditions are designed to minimize the potential hazards to the environment, human health, or agriculture. The coal ash supplier should evaluate whether the material is appropriate for its intended use and whether such use would produce any harmful effect. Coal ash must not be supplied if the average or maximum value of any sampled attribute exceeds the relevant standard (Table 6).

Table 6. Compositional requirements for coal ash use in New South Wales, Australia [26].

Parameters	Maximum Average Concentration for Characterization	Maximum Average Concentration for Routine Testing	Absolute Maximum Concentration
Hg	0.5	*	1
Cd	0.5	0.5	1
Pb	25	25	50
As	10	*	20
B	75	*	150 for engineering uses; 60 for soil amendments
Cr (total)	25	25	50
Cu	20	*	40
Mo	10	*	20
Ni	25	25	50
Se	10	10	20
Zn	35	35	70
Electrical conductivity [1]	*	*	None for engineering uses; 4 for soil amendments
pH in non-cementitious mixes [2]	7–12.5	7–12.5	6–13
pH in cementitious mixes	*	*	*

[1] Although thresholds are not provided for electrical conductivity, this must be tested, and records must be kept.
[2] Ranges given for pH are minimum to maximum acceptable. * Not required.

New South Wales requires follow-up monitoring based on ANZECC guidelines and the state's own WQO (water quality objectives) that consider EVs and long-term goals. The criteria, restrictions, and conditions in the regulations are not intended to be directly applied, but rather considered by the industry, community, or planning/regulating authority when a situation arises that exerts an impact on the present and future environment [27].

2.4. Japan

In Japan, technologies are being developed for the practical application of coal ash recycling in diverse areas, including evaluating the conversion of coal ash into useful material and developing construction methods using such material. Japan is also focused on ensuring the quality and environmental stability of a reliable coal ash supply to improve recycling efficiency.

A variety of quality evaluation and management methods have been established. For example, for construction applications, quality responsibilities are clearly delineated into three areas: the original coal ash, the coal ash product, and the construction usage. Each source (i.e., power plants) independently carries out a quality assurance process for ash provided to construction projects, consultations are carried out prior to the start of construction, and technical guidance is provided during construction.

Coal ash environmental safety criteria are independently managed based on the Environmental Standards of Soil Pollution, the strictest standards in Japan. Coal ash is applied to public construction projects only when the leachate from hardeners such as concrete and processed products has been confirmed to satisfy the standard values. For low-strength ameliorated soil, where coal ash is mixed with soil, only coal ash without any leachate concerns is selected for use to ensure safety. For use in ameliorated soil in public construction projects, a public institute carries out leaching tests to verify safety. For example, Shikoku Electric Power carries out a two-step leaching test to ensure coal ash safety by conducting leaching tests on (1) the original coal ash and (2) the product incorporating the coal ash. The environmental safety responsibilities are distributed among the producer, the intermediate contractors, and the final user [28].

A substantial proportion of Japanese coal ash recycling occurs through coastal development projects including embankments, beachfront parks, river banks, beach mounding, breakwater foundations, revetments and quays, ground reinforcements, and land development. As such applications imply direct contact with sea water, the concerns regarding environmental stability are inevitably greater.

When coal ash mixtures are used in civil works and road construction, where they are differentiated from the surrounding soil, they are excluded from soil pollution laws, and soil environmental standards do not apply. As a result, to ensure a separate and appropriate environmental safety standard, regulations were established based on basic ideas on environmental safety quality and test methods for circulated materials. Tests are performed in two steps: mixture test and environmental safety model inspection and construction and environmental safety level test [29]. The criteria for environmental safety stipulate that the environmental catalyst surrounding the area where coal ash mixture is used must satisfy environmental standards. Whether a leaching test should be carried out and the method used depend on the micropollutant leachate path. Test selection is based on the potential application of the Soil Pollution Plans Act and the probability of direct ingestion.

Japanese environmental safety standards differ between general applications (identical to Soil Pollution Plans Act regulations) and port applications (which have been altered). As port facilities are semi-permanent, their monitoring period is different, and the environmental standard reference values for coal ash mixtures used in this context are less stringent than for general applications (Table 7). As the contribution of coal ash to elevated levels of various substances leaching into sea water in port facilities has been shown to be negligible, reference values in this setting have been set as three or more times the typical soil pollution standards. Standards for fluorine and boron, which have higher background concentrations in sea water, are 20 times the typical value.

Sampling involves collecting the coal ash mixture used in the investigated area in a sufficient amount to allow for retesting if necessary. Environmental safety inspections are carried out once prior to construction, but additional investigations may occur within the same site if (1) there is a possibility of increasing micropollutants due to significant changes in the quality of coal ash being supplied as the source for the coal ash mixture, (2) there has been a change in the facilities or processes producing the coal ash mixture, or (3) the mixing conditions are new. However, if such changes involve only the mix ratio between coal ash and other materials, additional investigations may be omitted.

Table 7. Coal ash leachate standards for general and port applications in Japan [29].

Parameter	General Applications (mg/L)	Port Applications (mg/L)	Contents (mg/kg)
Cd	0.003	0.009	150
Pb	0.01	0.03	150
Cr^{6+}	0.05	0.15	250
As	0.01	0.03	150
Ag	0.0005	0.0015	15
Se	0.01	0.03	150
F	0.8	15	4000
B	≤1	20	4000

2.5. Republic of Korea

In South Korea, coal ash is defined as a specified byproduct based on the Act on the Promotion of Saving and Recycling of Resources. This refers to any byproduct that specially requires the recycling of the whole or a part of the substance for efficient use. Coal ash recycling methods follow the Waste Control Act and the Act on the Promotion of Saving and Recycling of Resources, while specific recycling and construction methods are defined by the Ministry of Environment's Guidelines on the Recycling of Steel Slag and Coal Ash by the Discharging Entity [30].

The latter rule stipulates that coal ash recycling may only be applied in the following 15 uses: in concrete admixture such as ready-mixed concrete; as a source of cement; as a lightweight aggregate; as a secondary product source for cement; as an aggregate for mounding or molding or roads; as a source of wood adhesive; as an alternative source for manufacturing cement clinker; as an aggregate for a drainage layer; as a source for ceramics production; as a fill material for rubber and plastic; as a source for paint, abrasives, or insulators; as a source for iron and steel manufacturing (metal recovery); and as a source of upper soil fertilizer (only as a floor material). In order for coal ash to be used as a fill material for mine reclamation, this regulation must be expanded.

Recycling beyond the defined uses is provisionally allowed by the Recycling Environmental Assessment so as to promote waste recycling [31]. This may be useful in the environmental assessment of general waste recycling, but in the case of applications to specific fields, such as mine reclamation and large waste quantities, tests for total pollutant amounts or long-term environmental impacts are limited. Moreover, this procedure includes many general environmental assessments that consider neither the purpose of recycling nor the characteristics of the relevant area while testing for unnecessary aspects. The criteria for environmental assessment are also ambiguous, as the properties of media and their purpose for use are not clearly stated, although the criteria are based on the process test standards for soil pollutants, the leachate test standards for upflow permeability, the process test standards for waste, and common global test standards. In order to properly assess the actual use of complex carbonates sourced from coal ash as fill material for mine reclamation in South Korea, far more specific methods of environmental management should be developed.

3. Discussion

The environmental management of coal ash recycling varies by country. In the US, where regulations and use vary by state, Pennsylvania actively pursues coal ash recycling and uses it as mine fill with regulations based on coal ash bulk concentration, coal ash leachate concentration, field inspection, and water quality monitoring. In Europe, there is a general effort to use waste proactively as long as it does not have any negative influence on the environment and human health while conforming to environmental certifications and general environmental regulations. In the UK, efforts are taken to use waste proactively through various projects based on the substances produced, relying on QP and water quality monitoring. In Australia, similar to the US, each state has its own management regulations. Queensland defines coal ash bulk concentration as the management focus, whereas New South Wales focuses on both this and leachate concentration. In terms of water quality monitoring, both states primarily consider the Australian and New Zealand Guidelines for Fresh and Marine Water Quality but each state has individual EV standards and follow them. Japan promotes coal ash recycling in diverse areas, clearly distinguishing between coal ash itself, coal ash products, and construction applications. The coal ash environmental safety criteria in Japan are independently managed and based on the Environmental Standards of Soil Pollution—the strictest standards in Japan.

In South Korea, current regulations regarding the purpose of coal ash recycling do not specify its use as fill material in mine reclamation, preventing its use for that purpose. As part of efforts to expand the scope of waste recycling, the Recycling Environmental Assessment provisionally allows coal ash use but allows only two categories—media-contact type and non-media-contact type—making it difficult to carry out purpose-driven environmental assessments and thus placing a limitation on the use of coal ash. To overcome this limitation and to simultaneously assess and verify the environmental stability of complex carbonates, new environmental management methods should be developed for this application.

We compared the coal ash leachate standards that permit its use in mine reclamation in each country (Table 8) and the related water quality monitoring standards (Table 9). Although environmental assessment standards cannot be standardized, as each country uses distinct approaches reflecting their regional characteristics and environments, there are currently no relevant regulations in South Korea. Therefore, we propose the adoption of the strictest standards. Moreover, each step of the environmental management strategy, from the generation of coal ash in the future and the production of complex carbonates to the construction and use of coal ash as a filler material in abandoned mines and the respective monitoring, must be discussed.

Table 8. The coal ash leachate standards of the countries.

Parameter (mg/kg)	US (Pennsylvania)	Australia (New South Wales)	Japan	South Korea	EPA-TCLP Hazardous Waste Limit	Environmental Management Proposal (SPLP)
Al	5	*	*	*	*	5
As	0.25	0.5	0.01	1.5	5	0.25
B	50	*	*	*	*	50
Ba	15	10	≤1	*	100	10
Be	0.1	*	*	*	*	0.1
Cd	0.125	0.05	0.003	0.3	1	0.05
CN	*	*	*	1	*	1
Cr	2.5	*	*	*	5	2.5
Cr+6	*	*	0.05	1.5	*	*
Cu	25	10	*	3	*	25

* Not defined.

Table 8. *Cont.*

Fe	7.5	*	*	*	*	7.5
Hg	0.05	0.01	0.0005	0.005	0.2	0.01
Mn	2.5	*	*	*	*	2.5
Pb	0.375	0.5	0.01	3	5	0.375
Se	*	0.1	*	*	1	0.1
Tl	0.05	*	*	*	*	0.05
Zn	50	50	*	*	*	50

* Not defined.

Table 9. Water quality monitoring criteria of the countries.

Parameter (mg/L)	US Pennsylvania (Drinking Water)	UK (Drinking Water)	South Korea (Drinking Water)	South Korea (Groundwater)	Environmental Management Proposal
Al	*	*	0.2	*	0.2
As	0.01	0.01	0.01	0.05	0.01
B	*	*	1	*	1
Ba	2	*	*	*	2
Be	0.004	*	*	*	0.004
Cd	0.005	0.005	0.005	0.01	0.005
CN	*	0.05	0.01	0.01	0.05
Cr	0.1	0.05	0.05	*	0.05
Cr+6	*	*	*	0.05	
Cu	1.3	*	*	*	1.3
F	4	1.5	1.5	*	1.5
Fe	*	*	0.3	*	0.3
Hg	0.002	0.001	0.001	0.001	0.001
Mn	*	*	0.3	*	0.3
Pb	0.015	0.01	0.01	0.1	0.01
Se	0.05	0.01	0.01	*	0.01
Tl	0.002	*	*	*	0.002
Zn	*	*	3	*	3

* Not defined.

4. Conclusions

A national project in South Korea is being pursued to develop the use of complex carbonates to capture CO_2 in coal ash, thus capturing carbon emissions and allowing for the large-scale recycling of coal ash. One intended approach is to use complex carbonates as fill material in mine reclamation, but environmental concerns, such as the environmental safety of coal ash, have thus far prevented their use for this purpose in South Korea.

In several other countries, coal ash—the main raw material for compound carbonate—is used as a mine filling material, and suitable environmental management has been implemented. We examined environmental regulations and guidelines on the use of recycled coal ash in other countries to assess its potential application as a fill material in mine reclamation in South Korea. Each studied country was proactively using coal ash for beneficial uses under locally suitable conditions. The environmental assessment standards cannot be standardized, as each country uses distinct approaches reflecting their regional characteristics and environments. Thus, we propose managing each step under the strictest conditions.

This study analyzed the regulations and policies needed to utilize complex carbonate as mine filler. In the future, it will be studied as well as environmental assessment verification based on experimental data and economic assessments. Our findings, which are based on case studies of other countries, can guide the future of coal ash use in South Korea in a safe and wise manner that will overcome opposition to its use based on incorrect presumptions of negative impacts and unnecessary restrictions. Hopefully, the safe use of complex carbonates will contribute to climate change mitigation and help create a sustainable society through recycling resources.

Author Contributions: conceptualization, H.-y.S., S.-w.J., and H.J.; investigation, H.C. and H.J.; writing—original draft preparation, H.C.; writing—review and editing, H.J.; funding acquisition, S.-w.J.

Funding: This study was conducted by Korea Environment Institute on "A study on the assessment and verification of environmental safety about abandoned mine filler (2018)". This study was funded by the Ministry of Science and ICT, Ministry of Environment, Ministry of Trade, Industry and Energy, and the National Strategy Project for the National Research Foundation of Korea—Carbon Upcycling (NRF-2017M3D8A2085336).

Conflicts of Interest: The authors declare no conflict of interest.

References

1. Izquierdo, M.; Querol, X. Leaching behaviour of elements from coal combustion fly ash: An overview. *Int. J. Coal Geol.* **2012**, *94*, 54–66. [CrossRef]
2. Coal Ash Recycling Reached Record 56 Percent Amid Shifting Production and Use Patterns. American Coal Ash Association (ACAA), 2017. Available online: https://www.acaa-usa.org/Portals/9/Files/PDFs/News-Release-Coal-Ash-Production-and-Use-2016.pdf (accessed on 22 April 2019).
3. Yao, Z.T.; Ji, X.S.; Sarker, P.K.; Tang, J.H.; Ge, L.Q.; Xia, M.S.; Xi, Y.Q. A comprehensive review on the applications of coal fly ash. *Earth Sci. Rev.* **2015**, *141*, 105–121. [CrossRef]
4. *Study on the CO₂/Environmental Monitoring Standardization*; Korea Institute of Geoscience and Mineral Resources (KIGAM): Daejeon, Korea, 2018; p. 104.
5. Praharaj, T.; Powell, M.A.; Hart, B.R.; Tripathy, S. Leachability of elements from sub-bituminous coal fly ash from India. *Environ. Int.* **2002**, *27*, 609–615. [CrossRef]
6. 40 CFR 257, U.S. Government Publishing Office. Available online: https://www.govinfo.gov/app/details/CFR-2011-title40-vol25/CFR-2011-title40-vol25-part257 (accessed on 22 April 2019).
7. 40 CFR 261, U.S. Government Publishing Office. Available online: https://www.govinfo.gov/app/details/CFR-2012-title40-vol27/CFR-2012-title40-vol27-part261 (accessed on 22 April 2019).
8. Surface Control and Reclamation Act (SMCRA). U.S. Department of the Interior. Available online: https://www.osmre.gov/lrg.shtm (accessed on 22 April 2019).
9. Guidelines for Beneficial Use of Coal Ash at Coal Mines. Pennsylvania Department of Environmental Protection, 2016. Available online: http://www.depgreenport.state.pa.us/elibrary/GetDocument?docId=7891&DocName=GUIDELINES%20FOR%20BENEFICIAL%20USE%20OF%20COAL%20ASH%20AT%20COAL%20MINES.PDF%20 (accessed on 22 April 2019).
10. Beneficial Use of Coal Ash. Pennsylvania Code, Chapter 290. Available online: https://www.pacode.com/secure/data/025/chapter290/chap290toc.html (accessed on 22 April 2019).
11. *Coal Ash Monitoring Parameters and Certification Standards*; Pennsylvania Department of Environmental Protection: Harrisburg, PA, USA, 2017.
12. Feuerborn, H.J. Coal combustion products in Europe—An update on production and utilisation, standardisation and regulation. In Proceedings of the World of Coal Ash Conference, Denver, CO, USA, 9–12 May 2011.
13. Waste Framework Directive (2008/98/EC). European Commission. Available online: http://ec.europa.eu/environment/waste/framework/ (accessed on 22 April 2019).
14. End of Waste Criteria. Joint Research Centre of the European Commission. Available online: http://susproc.jrc.ec.europa.eu/activities/waste/index.html (accessed on 22 April 2019).
15. Quality Protocols: Converting Waste into Non-Waste Products. U.K. Government. Available online: https://www.gov.uk/government/collections/quality-protocols-end-of-waste-frameworks-for-waste-derived-products (accessed on 22 April 2019).

16. Waste Protocols Project. U.K. Environment Agency. Available online: https://uk.practicallaw.thomsonreuters.com/3-503-1368?transitionType=Default&contextData=(sc.Default)&firstPage=true&comp=pluk&bhcp=1 (accessed on 22 April 2019).

17. EQual: Ensuring Quality of Waste-Derived Products to Achieve Resource Efficiency. U.K. Government. Available online: https://www.gov.uk/government/groups/equal-ensuring-quality-of-waste-derived-products-to-achieve-resource-efficiency (accessed on 22 April 2019).

18. *Stabilising Mine Workings with Pfa Grouts: Environmental Code of Practice*; BRE Construction Division: Newcastle, UK, 2006; ISBN 1-86081-909-5.

19. Australian and New Zealand Guidelines for Fresh and Marine Water Quality. Australian Government. Available online: http://www.waterquality.gov.au/guidelines/anz-fresh-marine (accessed on 22 April 2019).

20. Waste Legislation. Queensland, Australia Government. Available online: https://www.qld.gov.au/environment/pollution/management/waste/recovery/legislation (accessed on 22 April 2019).

21. End of Waste (EOW) Framework. Queensland, Australia Government. Available online: https://environment.des.qld.gov.au/waste/end-of-waste-framework.html (accessed on 22 April 2019).

22. *End of Waste Code, Coal Combustion Products (ENEW07359717)*; Department of Environment and Science: Queensland, Australia, 2019. Available online: https://environment.des.qld.gov.au/assets/documents/regulation/wr-eowc-approved-coal-combustion-products.pdf (accessed on 22 April, 2019).

23. Water Quality Guidelines. Queensland Government. Available online: https://environment.des.qld.gov.au/water/guidelines/#document_availability (accessed on 22 April 2019).

24. Shen, B.; Poulsen, B.; Luo, X.; Qin, J.; Thiruvenkatachari, R.; Duan, Y. Remediation and monitoring of abandoned mines. *Int. J. Min. Sci. Technol.* **2017**, *27*, 803–811. [CrossRef]

25. *CCP Environment Monitoring Program 2007/8*; Ash Development Association of Australia (ADAA): Port Kembla NSW, Australia, 2007; Available online: http://www.adaa.asn.au/uploads/default/files/adaa_emp_2007.pdf (accessed on 22 April 2019).

26. *Resource Recovery Exemption Under Part 9, Clauses 91 and 92 of the Protection of the Environment Operations (Waste) Regulation 2014*; New South Wales EPA: Sydney NSW, Australia, 2014. Available online: https://www.epa.nsw.gov.au/-/media/3F3155BCFE174DBB9201025C7CE4964A.ashx?la=en (accessed on 22 April 2019).

27. Water Quality. New South Wales Office of Environment & Heritage. Available online: https://www.environment.nsw.gov.au/water/waterqual.htm (accessed on 22 April 2019).

28. *Effective Utilization of Coal Ash*; Japan Society of Civil Engineers: Tokyo, Japan, 2003. (In Japanese)

29. Kojima. Challenges and regulation for international trade in recyclable waste. *Econ. Stud.* **2014**, *63*, 71–84. (In Japanese)

30. *Guidelines of Application of Steel Slag and Coal Combustion Products (CCPs)*; Korea Ministry of Environment: Sejong, Korea, 2016. (In Korean)

31. *Regulations on the Procedures and Methods for the Evaluation of Recycling Environment*; Korea Ministry of Environment: Sejong, Korea, 2016. (In Korean)

 *sustainability*

Project Report

Strategy of Developing Innovative Technology for Sustainable Cities: The Case of the National Strategic Project on Carbon Mineralization in the Republic of Korea

Jongyeol Lee [1], Changsun Jang [1], Kyung Nam Shin [1] and Ji Whan Ahn [2,*]

[1]   Division of Climate Technology Cooperation, Green Technology Center (GTC), Seoul 04554, Korea
[2]   Center for Carbon Mineralization, Mineral Resources Division, Korea Institute of Geoscience and Mineral Resource (KIGAM), Daejeon 34132, Korea
*   Correspondence: ahnjw@kigam.re.kr

Received: 30 April 2019; Accepted: 26 June 2019; Published: 1 July 2019

Abstract: Technology cooperation, including technology transfer, development of projects, and establishment of international networks, is an important instrument for attaining greenhouse gas mitigation and the sustainable development of a global society. In this context, carbon mineralization technology has received attention because of its high potential for carbon sequestration, environmental conservation, and economic market value. This project report introduces a national top-down approach for developing and implementing international technology cooperation in the Republic of Korea, focusing on carbon mineralization. The Ministry of Science and Information and Communication Technology (MSIT) leads international technology cooperation, identifies prominent climate technologies, and addresses scientific agendas to presidential meetings. The inter-ministerial bodies established the climate technology roadmap and masterplan for a climate change response. With the support of these inter-ministerial efforts, a National Strategic Project on carbon mineralization was developed by a presidential-level decision as a top-down approach. Furthermore, the demonstration of this technology was emphasized to enhance the possibility of success in commercialization. This project also includes demonstration of a pilot, sequestering 6000 tons of CO_2 and manufacturing 30,000 tons of carbonate. This successive and holistic approach, comprising of a range of hierarchical levels of government, is recommended for deriving a high impact on global society of prominent climate technology.

Keywords: National Strategic Project; carbon mineralization; carbon utilization; National Designated Entity (NDE); climate cooperation

---

1. Introduction

As the greenhouse gas (GHG) mitigation targets of each country were set after the Paris Agreement, technology transfer was considered a crucial measure for the implementation of international GHG mitigation in the context of a New Climate Regime [1]. The technology mechanism established by the United Nations Framework Convention on Climate Change (UNFCCC) in 2010 became a fundamental instrument of international climate technology cooperation. In addition to this systemic basis of technology cooperation, the identification and development of prominent climate technologies are crucial tasks for the mitigation of global climate change and environmental problems.

The solution of environmental problems is also a timely issue in the context of Sustainable Development Goals (SDGs). Particularly, waste management and recycling are common concerns of all countries because the generation of urban waste has been rapidly increasing [2]. Furthermore, an increasing demand for electricity is pushing developing countries to utilize more coal-fired power plants, generating a substantial amount of coal ash waste and related environmental problems [3–5]. Recently, the implementation of circular economy—the concept of facilitating the circulation of material and energy by improving the level of recycling—has been attempted ambitiously to establish this conceptual model in the real world [6].

The Republic of Korea has also devoted itself to sustainability and climate change mitigation. The roadmap of GHG mitigation was developed in 2016 to attain a 37% reduction in GHG emissions compared to business-as-usual (BAU) levels in the context of the Paris Agreement. In this roadmap, GHG mitigation was imposed in each sector, including electricity, industry, waste, and transport. However, the energy efficiency of the industry sector in the Republic of Korea is already at the highest level. Carbon capture and storage (CCS) technology, which captures the outlet of CO_2 and stores it in stable geological and marine structures, has the potential for GHG mitigation. However, the applicability of this technology is not sufficient in the Republic of Korea because of the geological limitations and high cost [7,8].

Meanwhile, several environmental problems threatening sustainability are still serious despite strict environmental regulations. Abandoned mines, which were operating during the Republic of Korea's period of rapid economic growth, cause several problems in local communities, such as the heavy metal contamination of soil and water ecosystems, and the risk of geological subsidence [9–11]. Accordingly, innovative technology that can shift the process and structure of society is essential to the Republic of Korea.

Carbon mineralization has received attention in the backdrop of climate change mitigation and sustainability. Carbon mineralization technology converts CO_2 gas into valuable inorganic compounds [12–14]. Carbon utilization, including carbon conversion and carbon mineralization, is expected to generate a market size of 800 billion dollars yr^{-1}, and sequester 7 billion metric tons of CO_2 yr^{-1}, at maximum, by 2030, in the case of strategic actions at the global scale [15]. In particular, utilizing CO_2 for manufacturing building materials (aggregates of carbonate and concrete) showed the largest potentials of market size and CO_2 sequestration: 550 billion dollar yr^{-1} and 5 billion tons of CO_2 yr^{-1}, respectively [15]. In addition to these direct benefits, the externality of environmental problems from coal ash and industrial waste can be solved by carbon mineralization technology. Therefore, carbon mineralization might be a realistic solution for CO_2 sequestration and environmental problems, while also encouraging the participation of the private sector.

Recently, the government of the Republic of Korea has devoted itself to the development of climate technology and international cooperation, including carbon mineralization technology. Developing countries have encountered complexities in economic growth and environmental regulation. Sharing the similar experience of rapid economic growth and environmental conservation in the Republic of Korea—which were possible on the basis of science and technology with education—could contribute to solutions for developing countries. Thus, establishing an international network, implementing systems of technology cooperation, and determining a strategy for technology development were led by the government of the Republic of Korea with contributions from ministerial to presidential levels. Finally, nine National Strategic Projects, organizing a consortium of the national research institutes and private corporations, were developed to enhance the already high potential of prominent technologies. Carbon mineralization was also selected as prominent technology, and the project of carbon mineralization includes technology development, progressive demonstration, environmental monitoring and standardization, and methodology development of the Clean Development Mechanism (CDM).

A detailed analysis of carbon mineralization in the Republic of Korea can provide a national roadmap of climate and environmental technology policy for other countries. Accordingly, this article describes the strategy and process of climate technology development and international cooperation, led at the national level. The component and consortium of the National Strategic Project on carbon mineralization of the Republic of Korea is also introduced. Then, a new model of climate technology development and cooperation is suggested to contribute to global GHG mitigation and environmental solutions.

2. The National Strategic Project on Carbon Mineralization

The National Strategic Project on carbon mineralization was developed to obtain a packaged technology including 1) the recycling of industrial wastes and CO_2 gas from factories and power plants; 2) the manufacturing of green cement and carbonate for abandoned mine reclamation; and 3) the development of a CO_2 sequestration project methodology. In particular, carbon mineralization technology was applied to the production of carbonate. The duration of the project (2017–2022) consists of a 1st phase (demonstration of mini-pilot), 2nd phase (demonstration of pilot), and 3rd phase (operation of demonstration plant and achievement of outcome), with a total governmental grant of 13 million dollars. This research project group consists of six teams: demonstrating low-concentration CO_2 treatment; demonstrating carbonate production; demonstrating production of green cement and construction of waterproof layer; demonstrating aggregate production and construction for abandoned mine reclamation; standardizing CO_2 and environmental monitoring system; and researching the methodology of the carbon mineralization CO_2 sequestration project (e.g., CDM) (Table 1).

Table 1. The consortium of the National Strategic Project on carbon mineralization.

Research Team	Task
Pre-treatments of low-concentration CO_2 gas and raw material	- Establishment of a database for utilizing low-concentration CO_2 from the exhaust gas of power plants and factories - Developing a packaged technology of collecting resources
Production of carbonate	- Enhancement of CO_2 fixation to inorganic waste (e.g., coal and industrial residue) - Development of continuous and rapid process technology
Demonstration of green cement production and waterproof layer construction	- Enhancement of waterproof cement - Development of optimal technology of mixing carbonate and cement for waterproof layer
Production and demonstration on abandoned mine reclamation	- Development of scale-up package technology with considerations of mine tunnels and geological stability - Design of optimal material for mine reclamation
CO_2/environmental monitoring and standardization	- Development of integrative CO_2/environmental monitoring technology on materials of mine reclamation - Development of a standardization technology for the domestic and international standardization of products for mine reclamation
Development of Clean Development Mechanism (CDM) methodology for the carbon mineralization project	- Development of new CDM methodology in accordance with mechanisms under the United Nations Framework Convention on Climate Change (UNFCCC)

Low-concentration CO_2 is captured from outlet of power plants and factories. After the pre-treatment of filtering impurities, this captured CO_2 is utilized in the manufacturing process of carbonate by a chemical reaction between CO_2 and metal oxide (e.g., calcium oxide and magnesium oxide). This carbon capture and utilization (CCU) fixes CO_2 emission to the atmosphere. Then, aggregate for abandoned mine reclamation is produced with the carbonate. Accompanying this is construction of a waterproof layer that prevents the outlet of leachate from the surface of abandoned mines after reclamation. From this, a construction standard for abandoned mine reclamation with these products and environmental monitoring systems would be established. Finally, the effect of CO_2 sequestration by the successive packaged technologies is quantified by the CDM project.

The demonstration of the carbon mineralization technology in the reclamation of abandoned mines is planned in Gangwon and Chungcheong provincial regions, where most power plants and cement and mining corporations exist (Figure 1). There are more than 2500 abandoned mines in the Republic of Korea, and approximately 480 of these abandoned mines are located in Gangwon Province. In the process of the demonstration, power plants and cement factories are major components, providing raw materials and manufacturing building materials for mine reclamation. Furthermore, the support of public and private sectors will be provided in the demonstration process. The central government supports the organization and operation of the project. The municipal authorities support the administrative processes by permitting the demonstration. The participating companies cooperate by providing low-concentration CO_2 and the testbed of abandoned mines for the demonstration. This project is expected to be a pilot demonstration, annually manufacturing 30,000 tons of carbonate with sequestering 6000 tons of CO_2 (fixation rate of 20%) by the demonstration of abandoned mine reclamation.

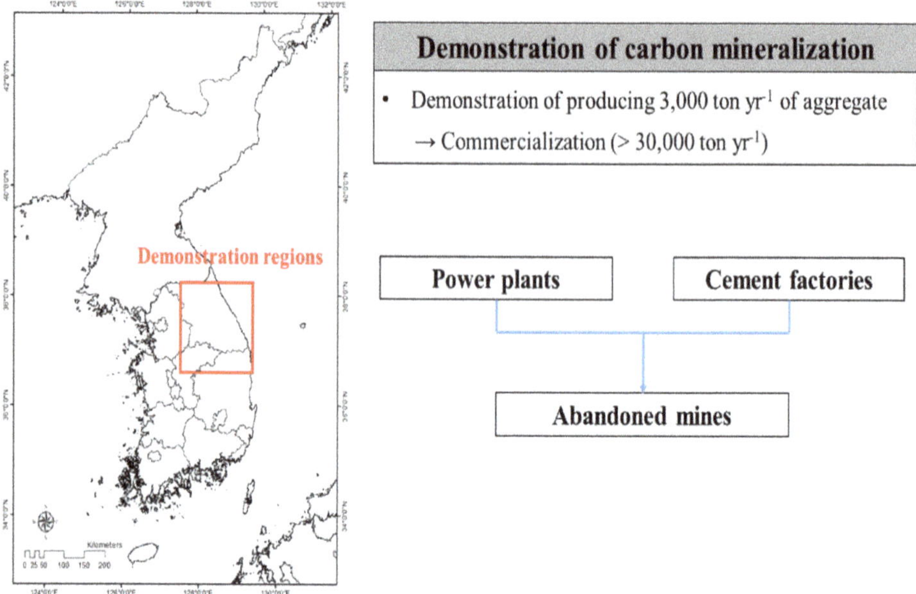

Figure 1. Demonstration plan of the National Strategic Project on carbon mineralization in Gangwon and Chungcheong provincial regions in the Republic of Korea.

3. Process of Developing the National Strategic Project on Carbon Mineralization

The National Strategic Project on carbon mineralization technology was developed by successive processes at the national scale, comprising of a range of hierarchical levels (Figure 2). The inter-ministerial approach, led by the Ministry of Science and Information and Communication Technology (MSIT), established a framework for developing climate technology and international technology cooperation. The national top-down approach was led by the presidential councils with inter-ministerial support. Each process, identified by three phases (initial, intermediate, and final) is described in detail in the following subsections.

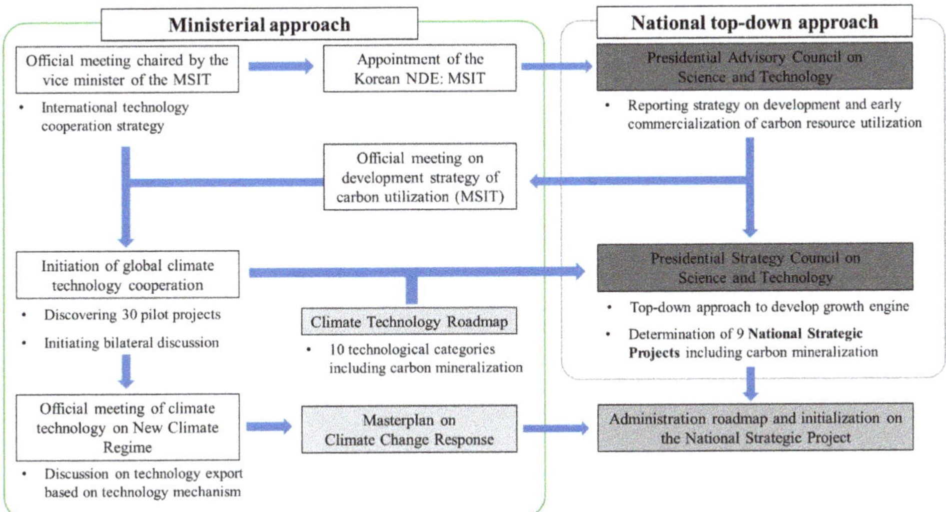

Figure 2. Schematic diagram of the governmental processes to develop the National Strategic Projects. The white, light-gray, and dark gray boxes represent the activities of the Ministry of Science and Information and Communication Technology (MSIT), inter-ministerial bodies, and presidential bodies, respectively.

3.1. The Initial Phase

The official meeting on the mitigation of national GHG emission and developing new industries, held by the vice minister of the MSIT in September 2015, initiated the international technology cooperation strategy of the Republic of Korea. This meeting included discussions about attaining the national GHG mitigation goal and developing new environmental industries with the consideration of three strategies: 1) acting as a main player in the international technology cooperation mechanism; 2) expanding international technology cooperation projects; and 3) establishing inter-ministerial cooperation to promote the technology cooperation. Particularly, measures for facilitating the export of Korean climate technologies to developing countries by global technology cooperation were also conferred.

The technology and financial mechanisms were established under the UNFCCC in order to mitigate GHG emission and enhance the capacity of adaptation to climate change (Figure 3). The technology mechanism is responsible for primary cooperation in the transfer of technology among countries. The financial mechanism amplifies the effect of the transferred technology through financing. The MSIT was appointed as the National Designated Entity (NDE) of the Republic of Korea in December 2015, which is responsible for a national channel for the technology cooperation. The Ministry of Economy and Finance was appointed as the National Designated Authority (NDA) as a part of the global financial mechanism.

Unlike most other countries, the Republic of Korea appointed the MSIT as NDE, rather than the Ministry of Environment. This implies a strong will towards leading international cooperation by facilitating innovative technologies under the technology mechanism. This Korean NDE contributes to the export of Korean technologies and creating new industries in the context of the New Climate Regime. Consequently, a Climate Technology Cooperation Team was established under the MSIT in March 2016 in response to the appointment of the NDE. This team arranges domestic, UNFCCC, and global cooperation (Figure 4).

Figure 3. The technology and financial mechanisms under the UNFCCC.

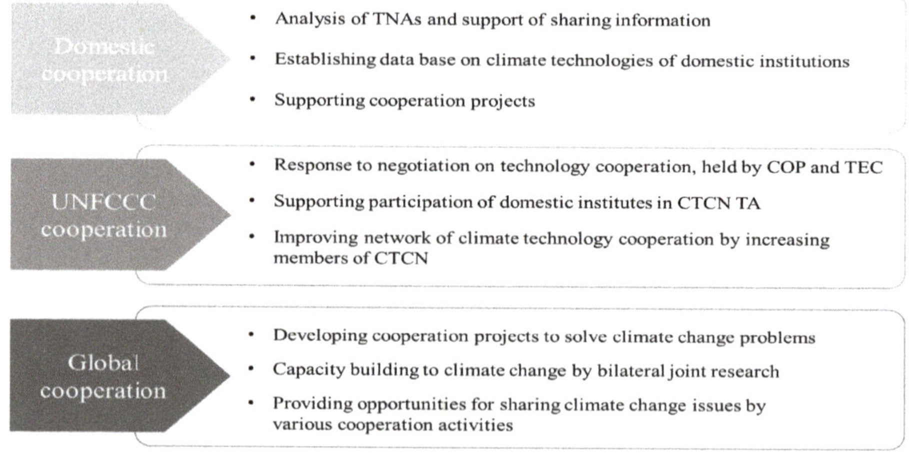

Figure 4. Functions of the cooperation team of the National Designated Entity (NDE). Abbreviations: Technology Needs Assessment (TNA), Conference of Parties (COP), Technology Executive Committee (TEC), Climate Technology Center and Network (CTCN), and Technical Assistance (TA).

The National Strategic Project team plans to implement international technology cooperation on carbon mineralization on the basis of the technology mechanism. After a feasibility study on application of carbon mineralization technology, demonstration will be operated for scale-up of the transferred technology. A bilateral network of cooperation among governments, led by the NDEs, facilitates successive measures (e.g., large-scale project and commercialization).

The MSIT also addressed the scientific agenda on climate technology after identifying prominent climate technology in the Republic of Korea. With this support from the MSIT, the importance of carbon mineralization technology and science diplomacy started to be emphasized from the 33rd meeting of the Presidential Advisory Council on Science and Technology (PACST), held in April 2016. The strategy for the development and early commercialization of carbon resource utilization technology was reported to mitigate GHGs and produce valuable chemical products.

Meanwhile, the MSIT has prepared inter-ministerial measures to strengthen science diplomacy with support from other ministries in that meeting of the PACST. These measures aim to lead international cooperative research, enhance the basis of customized and packaged partnerships, and expand the global science network. Integrating Korean official development assistance (ODA) and the Climate Technology Centre and Network (CTCN) of the UNFCCC into this science diplomacy was also proposed. After the meeting, a follow-up meeting was held in May 2016 to discuss governmental support for the development, demonstration, and commercialization of carbon utilization. The role of national research institutes in early commercialization was also emphasized.

3.2. The Intermediate Phase

Global technology cooperation was launched by the MSIT by identifying 30 pilot projects and initiating a bilateral discussion among countries. The MSIT analyzed the Technology Needs Assessments (TNAs) of potential counterpart countries and then provided the analysis results to 25 national research institutes and 5 science-specializing universities. Then, potentially cooperative items were investigated and selected on the basis of the current technologies of these organizations. Twenty organizations submitted 110 possible and available items, and then 15 prominent cooperative business models were determined. Particularly, two carbon mineralization technologies (i.e., inorganic waste recycling and abandoned mine reclamation with green cement) were proposed for Vietnam. This proposal was considered in the official meeting of the MSIT on the development strategy for carbon utilization technology (May 2016) after the 33rd meeting of the PACST. Then, the MSIT would support the establishment of domestic and international partnerships, with financing from international organizations (e.g., the Green Climate Fund (GCF) and multilateral development banks (MDBs)) in order to develop a successful model for GHG mitigation.

The Climate Technology Roadmap (CTR), initiated by the inter-ministerial body in March 2016, was completed in June 2016 to support the development of climate technology and implement the Paris Agreement from the perspective of national research and development. Ten major climate technologies in three categories were selected as follows: carbon mitigation (solar cells, fuel cells, biofuel, secondary cells, electricity information technology, and CCS), carbon utilization (byproduct gas conversion, CO_2 conversion, and CO_2 mineralization), and climate change adaptation (common platform). The analysis of the environments (market, technology, policy, and demand) of CO_2 utilization technology was provided in the CTR. In this context, the development of a technological package for the utilization of low-concentration CO_2 emissions and residual waste from power plants and factories, and a complementing demonstration project for CO_2 sequestration were also indicated.

Based on the results of the previous CTR, PACST, and discussions on global climate technology cooperation, nine National Strategic Projects were finally determined in the second meeting of the Presidential Strategy Council on Science and Technology (PSCST) in August 2016 (Figure 5). This determination characterized the top-down approach to developing a new growth engine and to improve life quality based on science and technology. The carbon mineralization technology was included in the carbon utilization category.

Growth engine of economy		Improvement of life quality	
Automatic driving car	Lightweight material	Precision medicine	Carbon utilization
Smart city	Artificial intelligence	New medicine	Particulate matter
Virtual/augmented reality			

Figure 5. The list of the National Strategic Projects in the Republic of Korea.

3.3. The Final Phase

In the final phase, the official meeting of climate technology in the New Climate Regime was held in November 2016 to discuss the growth of climate industries and facilitation of the technology mechanism of the UNFCCC for the export of technology. In particular, facilitating the technology mechanism by technical assistance of the CTCN received attention. These contexts were reflected in the Masterplan on Climate Change Response in December 2016 and concerned GHG mitigation from international cooperation. This masterplan established a basis for climate technology cooperation, particularly supporting the utilization of a climate change decision-making process by the comprehensive analysis of climate technologies and related activities. Finally, with the consideration of these discussions and this masterplan, the demonstration roadmap of the National Strategic Project was completed in December 2016. The demonstration was planned to be conducted by the following steps (Figure 5): simulation of processes (until 2016); mini-pilot demonstration (2017–2018); pilot demonstration (2018–2022); and large-scale demonstration with commercialization (from 2023) (Figure 6). Carbon mineralization is expected to sequester 10 Mt CO_2 yr^{-1} and generate an economic benefit of 15 billion dollars yr^{-1} by 2030 according to this roadmap. As already introduced in Section 2, the project consortium was organized and initiated during May–August 2017.

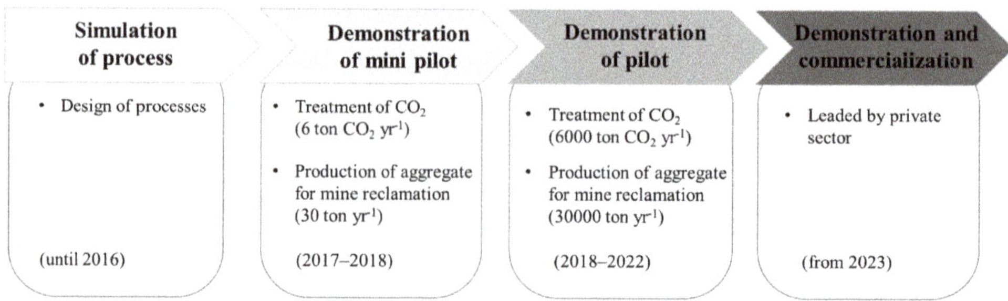

Figure 6. Process of demonstration for the National Strategic Project on carbon mineralization.

4. Discussion

The National Strategic Project on carbon mineralization was developed to facilitate the effects of carbon mineralization technology on the environment and society. Several future social and environmental benefits are expected from this research project. CO_2 sequestration is possible due to the utilization of captured CO_2 gas from plants and the substitution of original GHG-emission-intensive materials for abandoned mine reclamation [12,16]. The manufactured aggregate and cement could contribute to the establishment of a new climate industry in the context of the New Climate Regime after the Paris Agreement. Meanwhile, as the aggregate is manufactured by recycling coal ash and industrial residue from power plants and factories, environmental problems from these wastes can be solved [12]. These wastes contain metal oxide, which can chemically react with CO_2 by carbon mineralization technology [17,18]. With the consideration of environmental problems from coal ash and industrial wastes, the framework of this technology is highly beneficial. Recycling wastes and energy savings can support the establishment of circular economy [6]. Furthermore, the adverse effects of abandoned mines would be prevented after reclamation by the package technology of carbon mineralization in this research project.

However, the benefits of the carbon mineralization technology would not be improved without the successive implementation of top-down national processes. This implies that the magnitude of the effect of prominent climate technology could depend on the level of governmental support. The governmental support comprises the bottom-up approach from the identification of prominent technology by the MSIT and the top-down approach of the presidential organizations (Figure 7). First, the MSIT identified prominent climate technology (e.g., carbon mineralization). In addition, the

MSIT identified the technology needs of counterpart countries and established networks with them. The top-down approach of the presidential bodies was possible because of the support of the MSIT and inter-ministerial bodies, including addressing the scientific agenda and implementing a national system of climate technology. Finally, the cooperation and export of technology could be facilitated by the supports mentioned above.

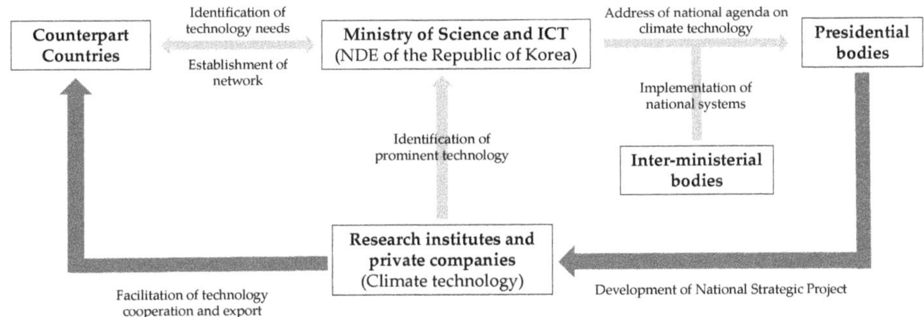

Figure 7. A framework of the Republic of Korea's governmental strategy for international climate technology cooperation. The light-gray and dark-gray arrows represent processes of inter-governmental support, and the facilitation of technology by the national top-down approach, respectively.

The MSIT of the Republic of Korea demonstrated a new model of NDE in climate technology development and cooperation. The designation of the MSIT as NDE, compared to most other countries which designated ministries related to the environment, implies a strong will of the Republic of Korea to lead international climate technology cooperation on the basis of the technological mechanism of the UNFCCC. In addition to the role of a path toward international technology transfer as NDE, the MSIT identified prominent climate technologies. Carbon mineralization technology was selected as one of the prominent climate technologies because of its CO_2 utilization efficacy, consistency with environmental problems, and the need of counterpart countries (e.g., coal ash from thermal power plants). Several technologies were then addressed in the agenda by the MSIT at the presidential meetings, in order to amplify the effectiveness and scale of these technologies in the context of international climate technology cooperation.

The inter-ministerial activities also contributed to the implementation framework of international climate technology cooperation. An inter-ministerial approach is essential because technology cooperation entails the development of technology, standardization and environmental assessment, procurement of materials, participation of corporations, and other subsidiary implementation systems. For this reason, the CTR and the Masterplan for Climate Change Response were prepared by the inter-ministerial bodies. The MSIT, Ministry of Environment, and Ministry of Trade, Industry, and Energy were also involved in the development of the National Strategic Project on carbon mineralization. Therefore, the inter-ministerial approach is essential to the facilitation of climate technology cooperation, rather than being solely enacted by the NDE.

Finally, the national top-down approach was determined by the high-level hierarchy of the government. The meetings of the PACST and PSCST thoroughly discussed the facilitation of climate technology development and cooperation, with considerations of the technology agenda from the MSIT and the inter-ministerial implementation framework. After the determination of the development of the National Strategic Projects, the demonstration roadmap for these projects was also developed. Accordingly, systemic processes and cooperation with the inclusion of various hierarchical levels of governmental bodies are recommended to facilitate prominent climate technology.

5. Conclusions

This project report suggests a new strategic framework of climate technology development. The top-down approach to developing carbon mineralization and international cooperation, involving ministerial to presidential levels, comprises the identification of technology, the establishment of an international network and implementation framework, and the development of a national project. The establishment of an international network with other governments could facilitate the impact of this technology on a global scale. This project was expected to be a pilot demonstration, sequestering 6000 tons of CO_2 and manufacturing 30,000 tons of carbonate. It was also expected to sequester 10 Mt CO_2 yr^{-1}, and provide an economic benefit of 15 billion dollars yr^{-1} by 2030 on the basis of carbon mineralization technology. Accordingly, a strategic and holistic approach of climate technology development and international cooperation from a higher organizational level would contribute to global GHG mitigation and the achievement of SDGs.

Author Contributions: J.L. analyzed the data, participated in the discussion, and wrote the project report; C.J. and K.N.S. collected the data and participated in the discussion; J.W.A. collected the data and supervised the preparation of the project report.

Funding: This research was supported by National Strategic Project (2017M3D8A2085293) and Green Technology Center Grant (C19233).

Conflicts of Interest: The authors declare no conflict of interest.

References

1. United Nations (UN). Paris Agreement. 2015. Available online: https://unfccc.int/process-and-meetings/the-paris-agreement/the-paris-agreement (accessed on 16 April 2019).
2. Giusti, L. A review of waste management practices and their impact on human health. *Waste Manag.* **2009**, *29*, 2227–2239. [CrossRef] [PubMed]
3. Thriveni, T.; Ngoc, N.T.M.; Tuan, L.Q.; Son, T.H.; Hieu, H.H.; Thuy, D.T.N.; Thao, N.T.T.; Tam, D.T.T.; Huyen, D.T.N.; Van, T.T.; et al. Technological solutions for recycling ash slag from the Cao Ngan coal power plant in Vietnam. *Energies* **2018**, *11*, 2018.
4. Verma, C.; Madan, S.; Hussain, A. Heavy metal contamination of groundwater due to fly ash disposal of coal-fired thermal power plant, Parichha, Jhansi, India. *Cogent Eng.* **2016**, *3*, 1179243. [CrossRef]
5. Lemly, A.D. Environmental hazard assessment of coal ash disposal at the proposed Rampal power plant. *Hum. Ecol. Risk Assess.* **2018**, *24*, 627–641. [CrossRef]
6. Geissdoerfer, M.; Savaget, P.; Bocken, N.M.P.; Hultink, E.J. The Circular Economy—A new sustainability paradigm? *J. Clean. Prod.* **2017**, *143*, 757–768. [CrossRef]
7. Anderson, S.T. Cost implications of uncertainty in CO_2 storage resource estimates: A review. *Nat. Resour. Res.* **2017**, *26*, 137–159. [CrossRef]
8. Smite, P.; Davis, S.J.; Creutzig, F.; Fuss, S.; Minx, J.; Gabrielle, B.; Kato, E. Biophysical and economic limits to negative CO_2 emissions. *Nat. Clim. Chang.* **2015**, *6*, 42–50. [CrossRef]
9. Shim, M.J.; Choi, B.Y.; Lee, G.; Hwang, Y.H.; Yang, J.-S.; O'Loughlin, E.J.; Kwon, M.J. Water quality changes in acid mine drainage streams in Gangneung, Korea, 10 years after treatment with limestone. *J. Geochem. Explor.* **2015**, *159*, 234–242. [CrossRef]
10. Kwon, J.C.; Nejad, Z.D.; Jung, M.C. Arsenic and heavy metals in paddy soil and polished rice contaminated by mining activities in Korea. *Catena* **2017**, *148*, 92–100. [CrossRef]
11. Kim, S.-M.; Suh, J.; Oh, S.; Son, J.; Hyun, C.-U.; Park, H.-D.; Shin, S.-H.; Choi, Y. Assessing and prioritizing environmental hazards associated with abandoned mines in Gangwon-do, South Korea: The Total Mine Hazards Index. *Environ. Earth Sci.* **2016**, *75*, 369. [CrossRef]
12. Naqi, A.; Jang, J.G. Recent progress in green cement technology utilizing low-carbon emission fuels and raw materials: A review. *Sustainability* **2019**, *11*, 537. [CrossRef]
13. Naims, H. Economics of carbon dioxide capture and utilization—A supply and demand perspective. *Environ. Sci. Pollut. Res.* **2016**, *23*, 22226–22241. [CrossRef] [PubMed]

14. Innovation for Cool Earth Forum (ICEF). Global Roadmap for Implementing CO_2 Utilization. 2016. Available online: www.globalco2initiative.org (accessed on 16 April 2019).
15. Cuéllar-Franca, R.M.; Azapagic, A. Carbon capture, storage and utilization technologies: A critical analysis and comparison of their life cycle environmental impacts. *J. CO_2 Util.* **2015**, *9*, 82–102. [CrossRef]
16. Jeon, E.-D.; Lee, K.-U.; Lee, C.-K. Development of new clean development mechanism methodology for the quantification of greenhouse gas in calcium sulfoaluminate cement. *Sustainability* **2019**, *11*, 1482. [CrossRef]
17. Dinid, A.; Quang, D.V.; Vega, L.F.; Nashef, E.; Abu-Zahra, M.R.M. Applications of fly ash for CO_2 capture, utilization, and storage. *J. CO_2 Util.* **2019**, *29*, 82–102. [CrossRef]
18. Norhasyima, R.S.; Mahlia, T.M.I. Advances in CO_2 utilization technology: A patent landscape review. *J. CO_2 Util.* **2018**, *26*, 323–335. [CrossRef]

Article

Waste Heat and Water Recovery System Optimization for Flue Gas in Thermal Power Plants

Syed Safeer Mehdi Shamsi [1,2], Assmelash A. Negash [1,2], Gyu Baek Cho [2] and Young Min Kim [2,*]

[1] Environment & Energy Mechanical Engineering, Korea University of Science and Technology, 217 Gajeong-ro, Yuseong-gu, Daejeon 34113, Korea; smshamsi@kimm.re.kr (S.S.M.S.); asmydit@kimm.re.kr (A.N.)

[2] Research Division for Environmental and Energy Systems, Korea Institute of Machinery and Materials, 156 Gajeongbuk-ro, Yuseong-gu, Daejeon 305-343, Korea; gybcho@kimm.re.kr (G.B.C.)

* Correspondence: ymkim@kimm.re.kr; Tel.: +82-42-868-7377

Received: 28 February 2019; Accepted: 26 March 2019; Published: 28 March 2019

Abstract: Fossil-fueled power plants present a problem of significant water consumption, carbon dioxide emissions, and environmental pollution. Several techniques have been developed to utilize flue gas, which can help solve these problems. Among these, the ones focusing on energy extraction beyond the dew point of the moisture present within the flue gas are quite attractive. In this study, a novel waste heat and water recovery system (WHWRS) composed of an organic Rankine cycle (ORC) and cooling cycles using singular working fluid accompanied by phase change was proposed and optimized for maximum power output. Furthermore, WHWRS configurations were analyzed for fixed water yield and fixed ambient temperature, covering possible trade-off scenarios between power loss and the number of stages as per desired yields of water recovery at ambient temperatures in a practical range. For a 600 MW power plant with 16% water vapor volume in flue gas at 150 °C, the WHWRS can produce 4–6 MWe while recovering 50% water by cooling the flue gas to 40 °C at an ambient temperature of 20 °C. Pragmatic results and design flexibility, while utilizing single working fluid, makes this proposed system a desirable candidate for practical application.

Keywords: waste heat recovery; water recovery; organic Rankine cycle; vapor compression cycle; pumped heat pipe cooling; flue gas; power plant

---

1. Introduction

Power plants that utilize fossil fuels, such as coal, pose fuel consumption problems, ultimately resulting in higher CO_2 emissions, significant water consumption and hazardous gas emissions. Utilization of waste heat recovery from the flue gas of power plants leads to its condensation and can be a simultaneous solution for the problems mentioned above.

Many studies are being conducted to find ways to utilize flue gas to reduce carbon footprints; these include employing it in the synthesis of methane and ammonia [1], as well as investigations focusing on waste heat recovery and utilization. Some studies focus on the reuse of exhaust gases for co-generation [2] and the use of this energy for pre-drying in coal power plants, which increases the thermal efficiency of the system by approximately 2% [3]. Other studies have focused on heat pump absorption technology with up to 32% energy savings within the system [4]. Zhang et al. [5] used a vapor-pump equipped gas boiler for flue gas heat recovery with system efficiency over 10%. Arietta et al. [6] investigated the Kalina cycle for waste heat recovery systems by directly converting the waste heat into electricity, which demonstrated 23% thermal efficiency. New methods of utilizing waste heat via the S-CO_2 cycle are also being conceptualized, with a cycle efficiency of approximately 18% [7]. Thermoelectric generation is also one of the promising technologies in the area of waste heat recovery where a great deal of research is being done on materials [8] as well as at a system level, but the conversion efficiencies are still below 5% [9]. The power produced by several heat recovery and renewable technologies are generally on a small scale and variable. Investigations are also being conducted to utilize the harvested power by using smart and efficient grid and electrical distribution systems which can reduce the power loss at the consumer end [10,11].

The organic Rankine cycle (ORC) is among the power cycles widely used for conversion of recovered waste heat to electrical power. Power plants based on the ORC produce power from various heat sources. These power plants range in size from 300 kW to 130 MW and have demonstrated the maturity of this technology. The cycle is well adapted for low to moderate temperature heat sources, such as waste heat from industrial plants. ORC technology is applicable to heat recovery from medium-sized gas turbines and offers significant advantages over conventional steam bottoming cycles [12]. ORCs can function in combination with low-temperature heat sources, characterized by low to moderate evaporation pressure, and still achieve better performance than steam cycles. The heating curves of ORCs can be better matched to the temperature profiles of waste heat sources, resulting in higher cycle efficiencies and higher thermal power recovery ratios than steam bottoming cycles [13]. Over the years, the ORC has been applied to waste heat recovery from different heat sources [14,15] which have also included automobile engines [16] and power plant flue gas. Koc et al. [17] analyzed a geothermal ORC from an energy and economical point of view. Liu et al. [18] on the other hand, utilized the high-grade sensible heat of a natural gas boiler present in the flue gas using the ORC, while the low-grade sensible heat and latent heat were recovered by working with fluid condensation and cooling water absorption.

Currently, fossil-fuel-based power plants consume vast quantities of water for heat rejection, fuel preparation, power augmentation, emission control, and cycle makeup purposes. At the same time, global water resources are becoming more difficult to procure as water consumption exceeds renewal [19]. Cooling systems are the most water-intensive portion of the thermoelectric generation process, presenting significant opportunities to reduce withdrawal and consumptive use of fresh water. Reuse of impaired water for cooling can reduce freshwater withdrawal and decrease water contamination as well as withdrawal-related impacts on aquatic life and the environment [20]. With the increasing trend of waste heat recovery technologies for fossil-fueled power plants, the need for water recovery from flue gas is also being seriously considered. There is a need to reduce the amount of fresh water needed for cooling purposes to enable power plant operation in more arid regions [21]. One of the methods of water recovery was investigated by Shen et al. [22], as it presented membrane technology as a potential means for water recovery from the flue gas of a lignite boiler. Similarly, Wang et al. [23] investigated a ceramic membrane suggesting that an increase in flue gas humidity can dramatically improve water and heat transfer rates and the overall heat transfer coefficient, offering a general guideline to optimize the operational parameters in low-grade heat recovery using membrane heat exchangers. Zhang et al. [24] presented the idea of reducing the flue gas temperature below the dew point of the water vapor within it to simultaneously recover latent heat and obtain clean water. The heat transfer mode of this method is a direct contact mode that utilizes the flue gas desulfurization scrubber as a flue gas water vapor condensing heat exchanger. Similarly, Bilirgen et al. [25] developed an analytical model for a flue gas condensing heat exchanger system to predict the heat transferred from the flue gas to the cooling water and the water vapor condensation rate in the flue gas. Vapor condensation from flue gas via air cooling has also been studied [26].

Other than recovering flue gas contents, heat recovery and water condensation from flue gas can help reduce emissions to the environment. In the past, researchers have focused on filterable particulate matter (FPM) because of its large emission amounts. Such active research prompted the rapid development of FPM control technology. At present, FPM is effectively controlled, and its emission concentration is extremely low. In contrast, the emission concentration of condensable particulate matter (CPM) is higher than that of FPM and requires immediate attention. Therefore, researchers are paying close attention to CPM. Nevertheless, CPM remains poorly understood [27]. Jeong et al. [21] compiled a report regarding typical acid concentrations in low-temperature condensed flue gas moisture and mercury capture efficiencies, preventing emission to the environment, but the condensation of the acidic fluids had an adverse corrosive consequence on the heat exchangers used in the system. Zukeran et al. [28] investigated the SO_2 removal rate and PM reduction for marine engines via water condensation in flue gas. As a result of water condensation, mist particles are generated which are then collected by an electrostatic precipitator. Shuangchen et al. [29] investigated the environmental influence of high humidity flue gas discharging from power plants and suggested that high humidity flue gas emissions promote the secondary transformation of air pollutants and reduce atmospheric visibility.

The objective of this study is to optimize a waste heat and water recovery system, comprised of an ORC, for power generation, with the integration of pumped heat pipe cooling and vapor compression cycle cooling, utilizing a single working fluid to perform heat extraction for power output and water recovery. Also, the characteristics of the proposed system for different yields of water and ambient temperature conditions were investigated. Condensed water and flue gas should be further treated for proper utilization as previously mentioned but are not covered in the context of this study.

2. Materials and Methods

A general schematic of the heat and water recovery of low-temperature flue gas originating from a coal-fired power plant is shown in Figure 1.

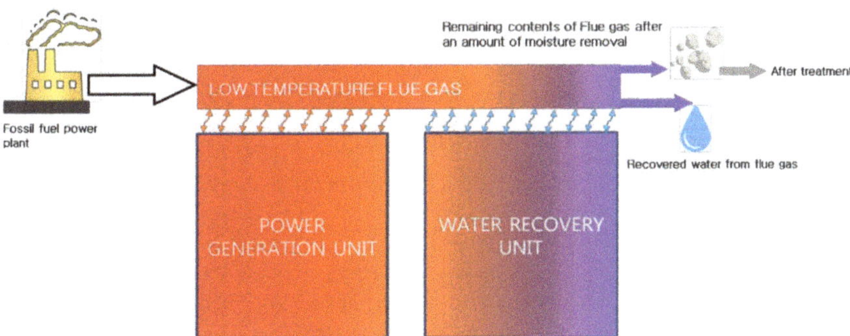

Figure 1. A general schematic of a heat and water recovery system.

The system is comprised of a power generation unit and a water recovery unit that use the same working fluid for their required purposes. The flue gas is first cooled by the heat extraction of the power unit at the high-temperature evaporator and is further cooled below the dew-point of the water that is present in the gas by the water recovery unit at the low-temperature evaporators. The design characteristics and properties of the flue gas, power generation unit, and water recovery unit are discussed further in the following text.

2.1. Flue Gas Characteristics

Table 1 shows the flue gas properties and considerations needed for the design and modeling of the heat and water recovery system.

With flue gas at 150 °C, a 600-MW coal-fired power plant was selected as the input for the heat and water recovery system. Water condensation in the flue gas begins at the dew point, i.e., when the partial pressure of the water vapor in the flue gas is equal to the saturation pressure of the water. According to Dalton's law, the partial pressure ratio of the water vapor must be equal to the partial volume ratio. For coal power plants, the pressure of the flue gas is close to atmospheric pressure (101 kPa) and 12–16% volume of water vapor corresponds to 12.1 to 16.2 kPa of water vapor pressure, causing a variation in the respective dew-point temperature from 49.4 to 55.3 °C. Figure 2 shows the variation in the dew point of the water concerning the volume percentage of water vapor.

Table 1. Flue gas properties.

Properties	Values
Fuel gas source	Coal-fired power plant (600 MW)
Flue gas mass flow rate	750 kg/s
Temperature	150 °C
Specific heat at constant pressure	1.131 kJ/kgK
Moisture content	12–16%

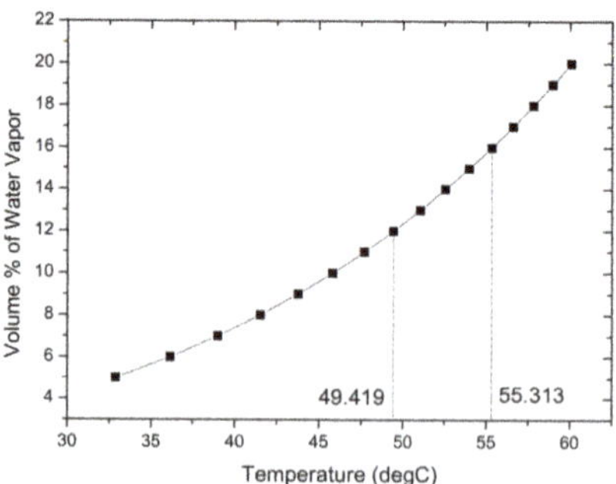

Figure 2. Variation in dew point temperature concerning the volume percentage of water vapor content in flue gas.

This study addressed a 16% volume of water vapor in the flue gas. Figure 3a relates the flue gas temperature to the saturation pressure of the water and water recovery efficiency. In this study, a target limit of 40 °C for the flue gas outlet temperature was established to obtain greater than 50% water recovery efficiency as shown in the figure. With a decrease in the flue gas temperature, the saturation pressure of the water also decreases; therefore, the water recovery efficiency increases as represented by the blue curve in Figure 3a. To recover more water from the flue gas via cooling, the amount of extracted heat from the flue gas has to increase. This is because, below the dew point temperature, latent heat comes into play because of the phase change. Thus, more energy extraction is required to decrease a particular degree of temperature as shown in Figure 3b.

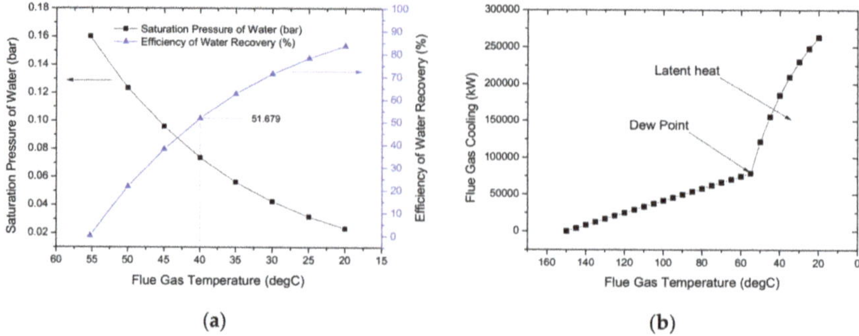

(a)   (b)

Figure 3. Characteristics of moisture content above the dew point for a 16% volume of water vapor in flue gas (**a**) variation in the partial pressure and water recovery efficiency with respect to the flue gas temperature. At 40 °C, the water recovery efficiency is greater than 50% (**b**) Cooling required for flue gas for moisture content condensation with variation in temperature.

2.2. Organic Rankine Cycle Optimization

The waste heat and water recovery system (WHWRS) is primarily comprised of two parts: an ORC that cools the flue gas and uses its heat for power generation and another part to further cool the flue gas below the dew point temperature of the moisture content present within it. Depending upon the ambient temperature, the latter can be a pumped heat pipe cooling cycle, a vapor compression cooling cycle, or a combination of both.

Table 2 shows the assumptions made for modeling of the ORC part of the system. Theoretical modeling of the WHWRS in this study was optimized for maximum power output.

Table 2. Assumptions for the ORC design.

Parameters	Values
Working fluid	R134a
Isentropic efficiency:	
Turbine	70%
Pump	70%
Pressure Decrease	3%
Pinch temperatures	
Condenser	5 °C
Evaporator	15 °C

Figure 4a shows the temperature vs. specific entropy (T-S) diagram for the ORC part of the WHWRS at a 20 °C ambient temperature. The system was modeled using the program AxCYCLED_v4.44100 with the working fluid R134a as shown in Figure 4b. Properties of the working fluid R134a were obtained from the REFPROP program.

The output parameters and efficiencies of the ORC can be calculated as follows:

Q_{eg} is the energy transfer rate at the ORC heat exchanger and is given by

$$Q_{eg} = m_{ex}C_p \left(T_{in,ex1} - T_{out,ex1}\right) \tag{1}$$

$Q_{eg,\,max}$ is the maximum allowable energy that can be transferred to the ORC part

$$Q_{eg,max} = m_{ex}C_p \left(T_{in,ex1} - T_{ambient}\right) \tag{2}$$

$P_{net,orc}$ is the net power output produced by the ORC cycle after considering the power consumption of its auxiliary equipment.

$$P_{net,orc} = P_{turbine} - P_{pump,orc} \tag{3}$$

η_{HR} is the heat recovery efficiency of the ORC cycle defined as the ratio of the recovered heat rate and the maximum allowable heat rate from the waste heat source.

$$\eta_{HR} = Q_{eg}/Q_{eg,max} \tag{4}$$

η_{cyc} is the cycle efficiency of the ORC defined as the ratio of the net power and the recovered heat rate from the waste heat source

$$\eta_{cyc} = P_{net,orc}/Q_{eg} \tag{5}$$

η_{sys} is the system efficiency of ORC defined as the ratio of the net power and the maximum allowable heat rate from the waste heat source.

$$\eta_{sys} = P_{net,orc}/Q_{eg,max} \tag{6}$$

$T_{in,ex1}$ is the temperature of the flue gas entering the ORC heat exchanger whereas $T_{out,ex1}$ is temperature of the flue gas leaving the ORC heat exchanger. m_{ex} is the mass flow rate of flue gas, C_p is the constant heating value of the flue gas, $T_{ambient}$ is the ambient temperature, $P_{turbine}$ is the work done by the turbine, $P_{pump,orc}$ is the work required by the ORC pump, $P_{net,orc}$ is the net power output produced by the ORC cycle.

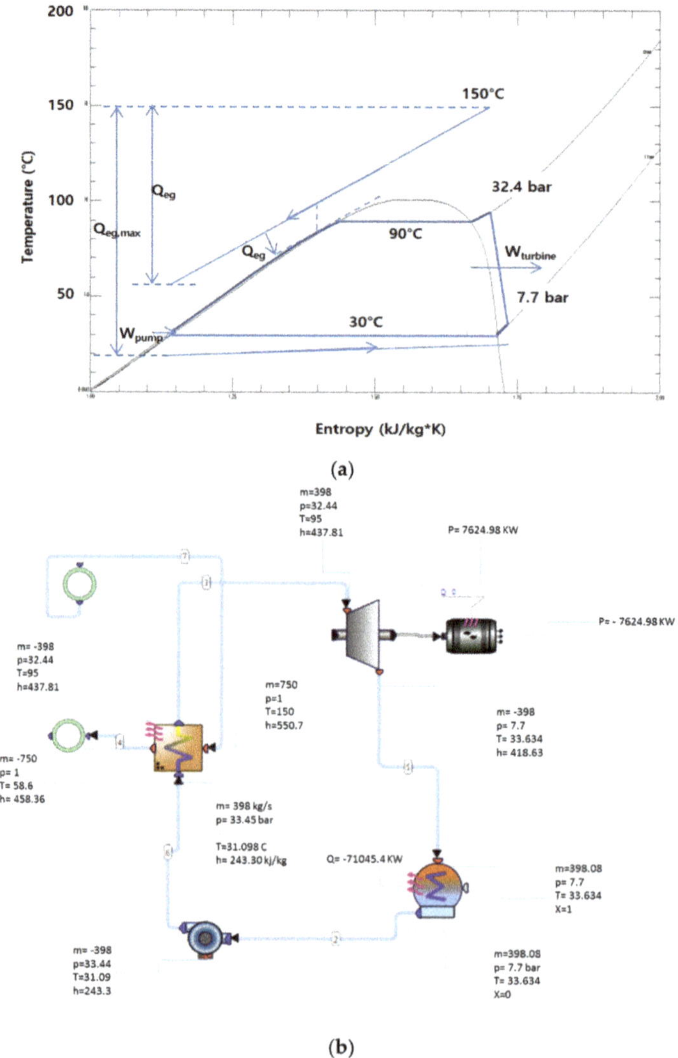

Figure 4. Values of the optimized organic Rankine cycle (ORC) showing (**a**) the temperature vs. specific entropy (T-S) diagram of the ORC for the system and (**b**) a schematic of the ORC from the AXCYCLED program.

2.2.1. High-Pressure Side Selection

The high-pressure side was mapped as the first step using the AXCYCLED program. The high-pressure side profile variation with respect to turbine inlet temperature is shown in Figure 5a. With an increase in the high-pressure side, the turbine inlet temperature increases. Although the net power increases with an increase in the turbine inlet pressure as shown in Figure 5b, For our design purposes the high-pressure side was set to 32.4 bar for ease in use of commercial heat exchangers.

Figure 5c shows the efficiency variations with the change in the high-pressure side. For the selected high-pressure side, the system thermal efficiency is approximately 6%.

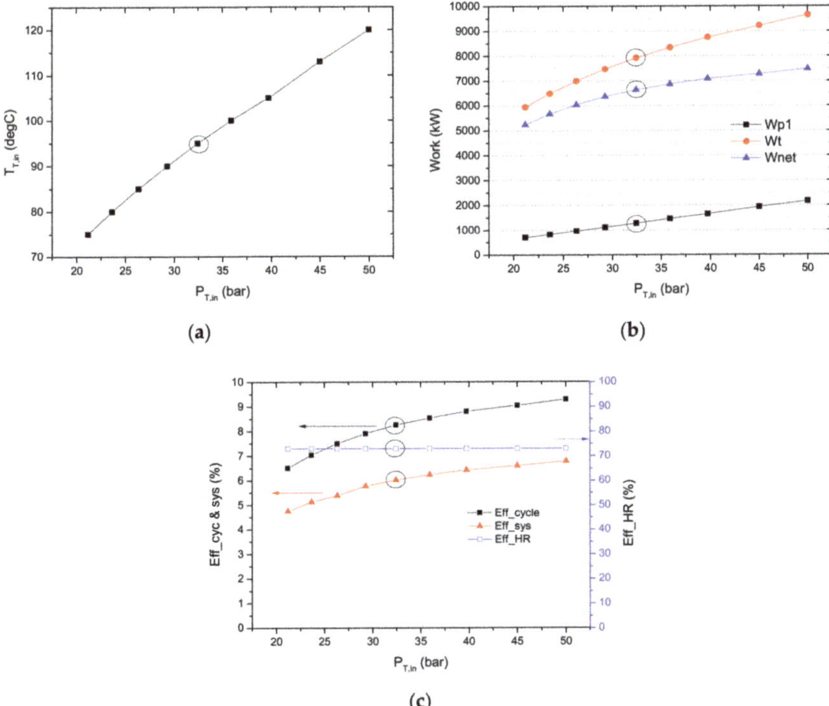

Figure 5. High-pressure side optimization for the ORC: (**a**) variation in turbine inlet with a change in the high-pressure side; (**b**) variation in turbine work, pump work, and net power output with a variation in the high-pressure side; (**c**) variation in cycle, system, and heat recovery efficiency with respect to a change in the high-pressure side.

2.2.2. Turbine Inlet Temperature Optimization

At the given high-pressure side, the turbine inlet temperature has to be increased in a superheated region up to 5 °C to maintain dryness during expansion. In this case, at the given high-pressure side of 32.4 bar, the saturation temperature of 90 °C was superheated to 95 °C. The turbine inlet temperature was optimized for maximum power output from the given waste heat source. Figure 6 shows the efficiency variation in the amount of superheating completed at the given high-pressure side. It also shows the higher heat recovery and overall system efficiency at 95 °C compared to that at higher temperatures while only the cycle efficiency is lower. At the conditions optimized for maximum power, the outlet temperature of the flue gas from the ORC part was determined at the given ambient temperature by the pinch point in the ORC evaporator.

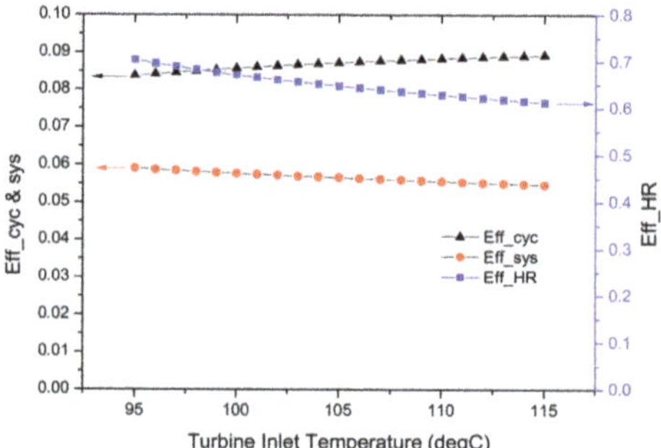

Figure 6. Variation in system, cycle, and heat recovery efficiency concerning a variation in temperature.

2.3. Water Recovery System Characterization

Following the ORC heat recovery part is a cooling system for water recovery. Figure 7 shows the five different configurations termed A, B, H, O, and X that can be implemented for the given ambient conditions. In addition to the ORC part, the water recovery system is generally comprised of a low-temperature evaporator (LT evaporator) in which heat transfer between low-temperature flue gas and the working fluid occurs for the sake of water recovery. After extracting heat at low pressure, the working fluid becomes vaporized and then passes into the common condenser to dissipate the heat.

Configuration A has an ORC system integrated into the pumped heat pipe cooling system; pumped heat pipe cooling (HPC) means only the use of a circulation pump for a phase change heat transfer. In this system, the evaporation temperature in the LT evaporator is equal to the condensation temperature of the condenser, neglecting the pressure drop. Therefore, the amount of cooling completed by the pumped heat pipe system is strictly restricted by the ambient temperature conditions. Configuration A is a two-stage system whereas configuration B is a three-stage system with the addition of a vapor compression cycle (VCC). Although a vapor compression cycle does not have any restrictions under ambient temperature conditions, it results in a loss of net power as the compressor is attached to the turbine. Configuration X had only a vapor compression cycle following the ORC system which had one large compressor. Configuration H has been considered to show the effect on power characteristics of the system if the large compressor work is distributed to various smaller compressors. In this particular case, two compressors have been used instead of one large compressor i.e. X, to signify the power consumption decrease caused by the low reversibility of two small compressors instead of a large one. Whereas configuration O has an additional HPC to reduce the power consumption further at the expense of an additional stage. The power output for the overall WHWRS (P_{net}) can be calculated as follows:

$$P_{net} = P_{net,orc} - P_{HPC} - P_{VCC} \tag{7}$$

where P_{HPC} is the power required by the pumped HPC system and P_{VCC} is the power required by the vapor compression cooling system.

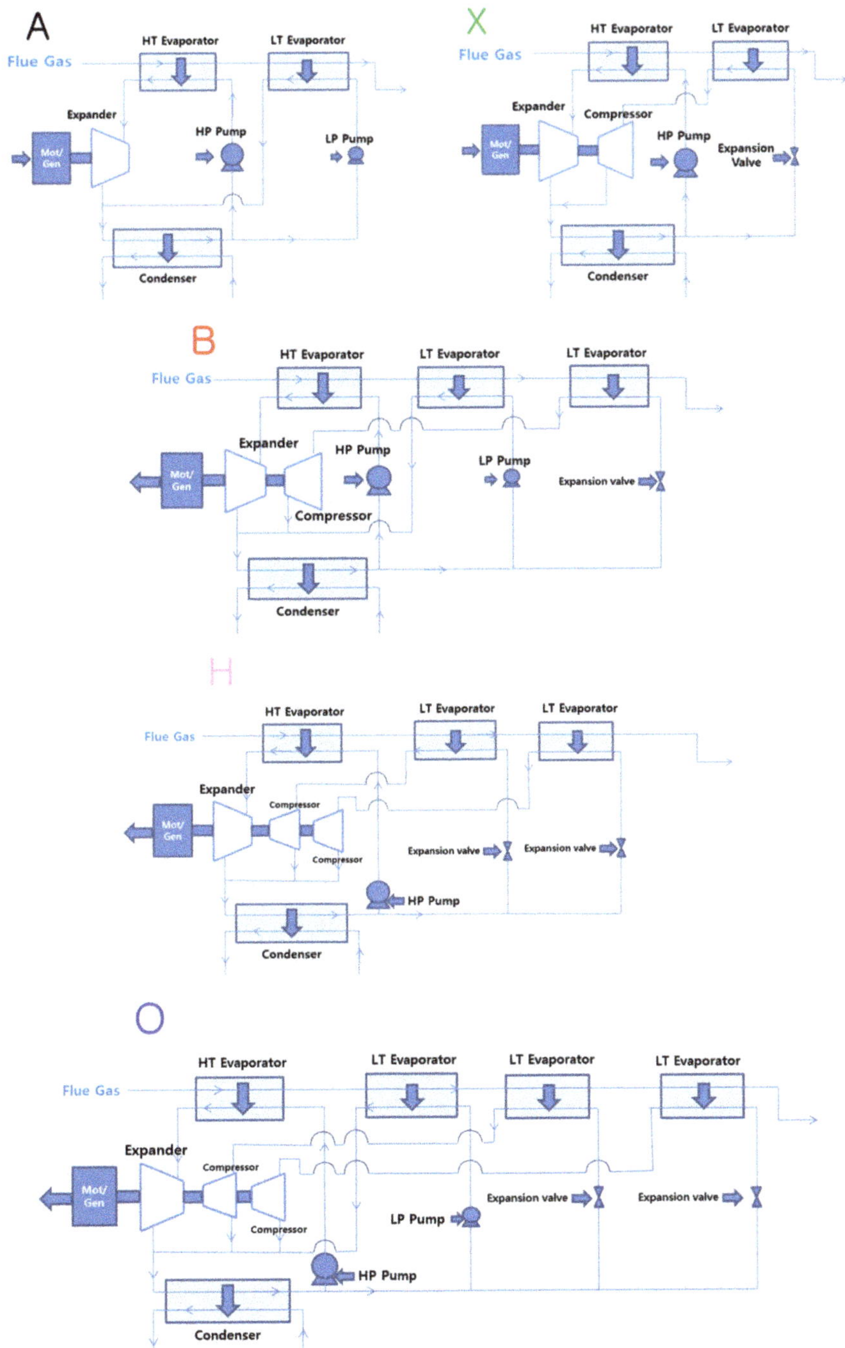

Figure 7. WHWRS configurations A, X, B, H, and O; A is composed of an ORC and heat pump cooling (HPC); X is composed of an ORC and vapor compression cooling (VCC); B is composed of ORC + HPC + VCC; H ORC + VCC + VCC, and O ORC + HPC + VCC + VCC.

3. Results

The WHWRS was analyzed in two different manners in this study. First, the characteristics of the system were analyzed for constant water recovery yield and thereafter for constant ambient temperature conditions, providing power output behaviors and the effect that the range of ambient temperature and water recovery yield had on the capital cost in terms of system stages. In both of these approaches, the five schematics mentioned above that cover every possible case of cycle configuration within these temperature ranges were used.

3.1. Constant Water Recovery Yield

This section discusses the WHWRS characteristics relative to ambient temperature conditions. The objective was to attain 40 °C for the exhaust gas outlet temperature, which in turn provided approximately 50% water recovery efficiency. Configurations A, B, H, O, and X were analyzed for net power output, water recovery, and exhaust gas outlet temperature at ambient temperatures of 40, 30, and 20 °C. Note that this study tends to suggest optimization in terms of higher power output while keeping the system stages to a minimum while attaining the desired water recovery efficiency and flue gas temperature.

3.1.1. At 40 °C Ambient Temperature

All five configurations can be seen in the graphs representing the properties of the system for the ambient condition of 40 °C as shown in Figure 8a. Figure 8a shows that three out of five configurations, i.e. X, H, and O, are those that fulfill the criteria of condensing the moisture present in the flue gas to the required temperature of 40 °C and achieving greater than 50% water recovery efficiency as shown in Figure 8a,b at 40 °C ambient temperature conditions. Configurations A and B are only able to cool the flue gas to 60 °C and 50 °C and achieve 0 and 21.6% water recovery efficiency, respectively. The figure also shows that X is two-staged, whereas H and O are three- and four-stage systems, respectively, and able to achieve the desired results. Figure 8c shows the power output distribution of all five configurations for the ambient temperature of 40 °C. Configuration X has the highest power consumption or power loss of all the configurations for this ambient temperature condition. In this case, no power output is produced by the system and the system power consumption is more than three times the power that is produced by configuration A, which is the highest power-producing configuration in the case of a 40 °C ambient temperature. O fulfills the criteria of water recovery and flue gas temperature, but despite being a four-stage system, it still consumes power for water recovery at this stage although much less than that required by X. Both B and H are three-staged systems, but only H is capable of attaining the desired results; however, its power consumption is higher than that of configuration O. Thus, at the ambient temperature of 40 °C, no configuration can obtain the desired results without the use of external power. In other words, at this point, the system can no longer be self-sufficient with the present configurations and cooling systems.

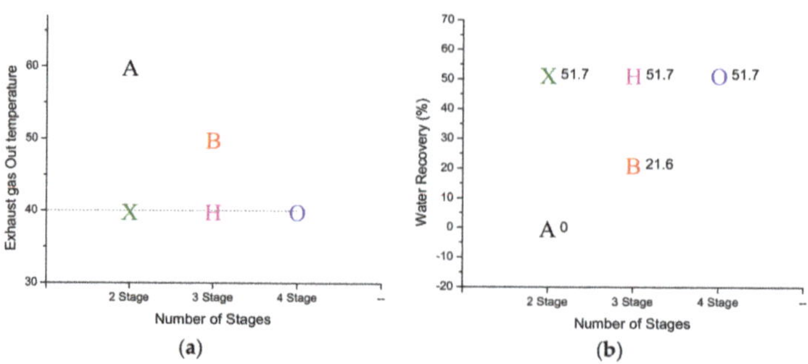

(a)   (b)

Figure 8. *Cont.*

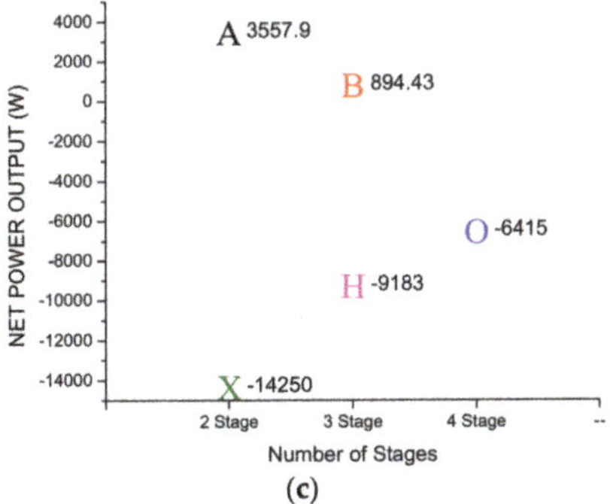

Figure 8. Properties of system configurations desirable at 40 °C ambient temperature and 50% water recovery yield: (**a**) map of configurations A, B, X, H, and O based on outlet temperature and number of stages; (**b**) map of configurations A, B, X, H, and O based on water recovery efficiency along with the number of stages; and (**c**) map of the power outputs of configurations A, B, X, H, and O with respect to number of stages.

3.1.2. At 30 °C Ambient Temperature

Figure 9 shows graphical representations of configurations feasible for an ambient temperature of 30 °C. Although A has the highest power output as compared to that of configurations B and X, it doesn't cool the flue gas up to the desired temperature of 40 °C as shown in Figure 9a; as a result it can recover only 21.6% of the water from the flue gas as shown in Figure 9b. The power output of B is higher than that of X as shown in Figure 9c. Although, X is a two-stage cycle and B a three-stage cycle, both fulfilling the criteria of greater than 50% water recovery. However, B has an advantage over X of consuming relatively less power for its vapor compression cycle as 10 °C of decreased temperature was already obtained by the pumped HPC cycle. In contrast, configuration X has an advantage over B of having one less stage as it only utilizes the vapor compression cycle.

Note that O and H are the two configurations that were not considered for this ambient temperature condition case because they have no advantage and are not comparable. H has the same number of stages as B, so it has no advantage in terms of system stages. It also uses two stages of vapor compression whereas B uses one stage of heat pump cycle and one stage of vapor compression of the same capacity, so B has dominance over H in power consumptions as well, rendering the comparison meaningless. Similarly, O is essentially B with an additional stage. As the required output can be attained without the fourth stage with less power consumption, the comparison of the configuration with an additional stage, i.e. O, becomes absurd.

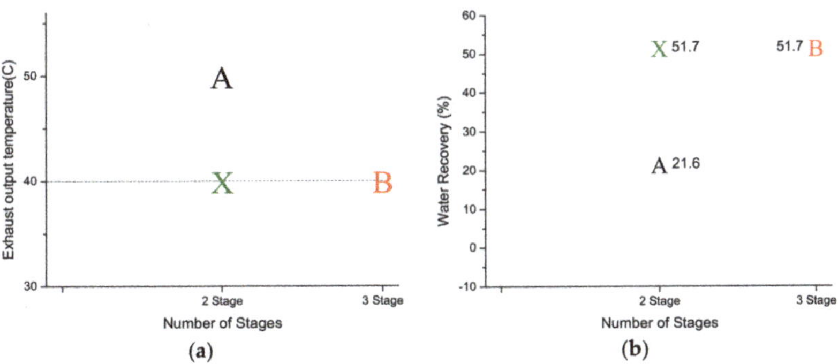

Figure 9. *Cont.*

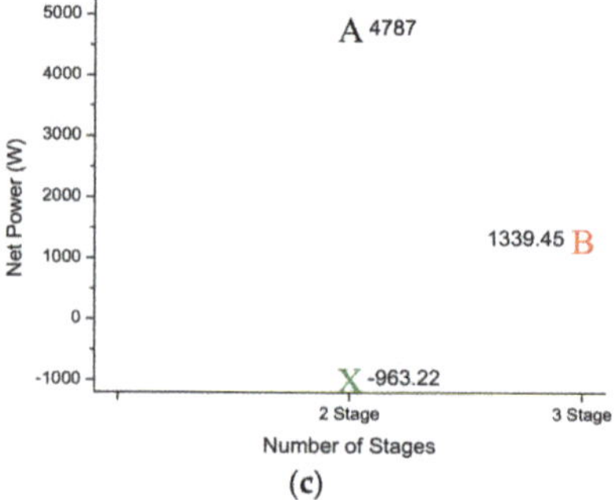

Figure 9. Properties of system configurations desirable at 30 °C ambient temperature and 50% water recovery yield: (**a**) map of configurations A, B, and X based on outlet temperature and number of stages; (**b**) map of configurations A, B, and X based on water recovery efficiency along with the number of stages; and (**c**) map of the power output of configurations A, B, and X with respect to the number of stages.

3.1.3. At 20 °C Ambient Temperature

Figure 10a–c shows that configuration A is the only candidate selected from all five configurations that can achieve the desired water recovery efficiency at a flue gas outlet temperature of 40 °C at the highest power output with the least number of stages.

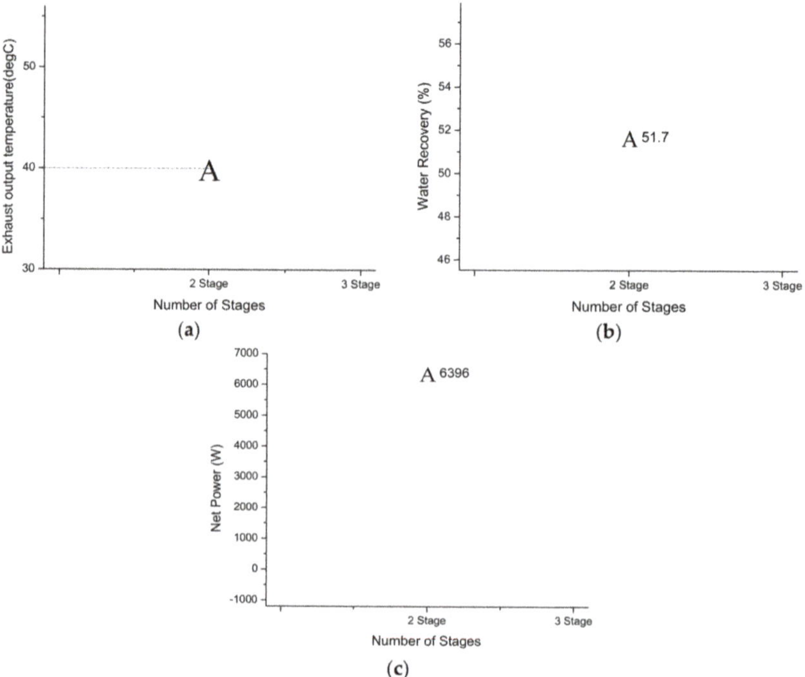

Figure 10. Properties of the system configurations for a 20 °C ambient temperature and 50% water recovery yield: (**a**) map of configuration A based on its outlet temperature and the number of stages. (**b**) Map of configuration A based on its water recovery efficiency along with the number of stages. (**c**) Map of configuration A power output with respect to the number of stages.

It eliminates all other configurations in terms of capital cost and power output desirability for the case of 20 °C. Although X is a two-stage configuration, for the fixed cooling temperature of flue gas to 40 °C, the vapor compression cycle in X results in a greater power loss than that of the pumped heat pipe cycle in A. Thus, X was not comparable to A for an ambient temperature of 20 °C in any way as it does not have any benefit over A and therefore was not considered. This is the most primary form of the WHWRS with the most benefits at hand at a generally common ambient temperature condition making A practically a desirable candidate for commercial purposes.

3.2. Constant Ambient Temperature

At the ambient condition of 20 °C, net power output, exhaust gas outlet temperature, and water recovery were analyzed relative to the increase in system stages. To avoid any confusion, only the configuration optimized for maximum power output for that particular stage was considered. For example, among all five configurations, A and X are both two-stage configurations, but instead of X, the characteristics representing a two-stage system in this section are for the A configuration because it shows a higher net power output compared to that of X. Similarly, between H and B, B will be considered for the three-stage system for the same reason, while O is the only candidate for the 4-stage configuration. Figure 11 shows a graphical representation of the exhaust gas outlet temperature, water recovery, and net power output variation concerning the system stages that were optimized for maximum power output.

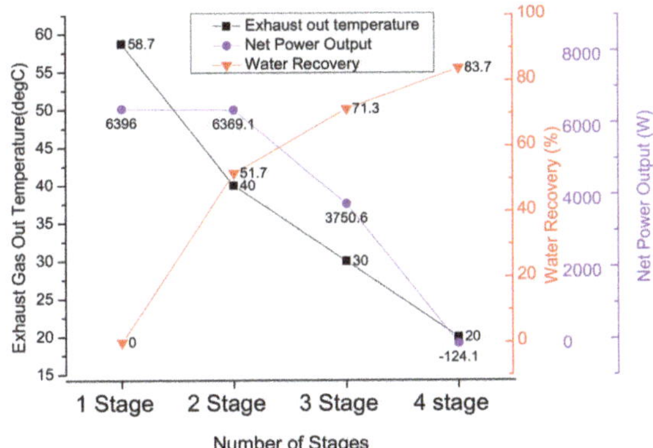

Figure 11. Variation in exhaust gas outlet temperature, water recovery efficiency, and net power output with respect to the number of system stages at a constant ambient temperature of 20 °C.

It can be seen that at the first stage there is no water recovery because at this stage only the ORC is being utilized for the 20 °C ambient temperature. The flue gas outlet temperature from the system is approximately 59 °C, which is higher than the dew point temperature of 55 °C of the moisture content present in the flue gas. However, the power output is the highest in this case relative to any other stage. The two-stage system has relatively less power output as it utilizes the pumped HPC system. Under a 20 °C ambient temperature condition, it can cool the flue gas below the dew point, down to 40 °C, increasing the water recovery efficiency from 0 to greater than 50% without a loss of a significant amount of power which it uses to operate the circulation pump of the pumped HPC system. The three-stage system represented utilizes a combination of a pumped HPC system and vapor compression cooling system because at a 20 °C ambient temperature and with our pinch point assumptions, a pumped HPC system cannot cool any further than 40 °C. Therefore, to further cool, we utilized the refrigeration cycle which as previously mentioned was restricted to create a temperature gradient of 10 °C in the flue gas to minimize the work required by the compressor. A four-stage system comprising one pumped heat pipe cycle and two vapor compression cycles collectively decreased the temperature of the flue gas to the ambient temperature of 20 °C, recovering greater than 80% of the water in the flue gas but at the cost of approximately 124 W of external power source in addition to turbine power output.

Condenser Characterization at Constant Temperature

There are multiple advantages of WHWRS in regions with inadequate water supply, including the capture of condensable particulate matter, minimizing of fresh water usage, and recovering water for potential needs within the power plant. In the case of water recovery units, whose objectives are to minimize pollution and fresh water usage, the condenser for the proposed WHWRS can be water-cooled. However, if the power plant is in a region with a low or declining water supply, the condenser should be cooled by forced air convection, either partially or totally without water cooling heat-exchangers. There is a certain amount of work input or net power loss that is required for the forced convection equipment, e.g., fan. Note that the previous calculations for all configurations have been made based on water-cooled condenser systems.

This section discusses the air-cooled condenser and compares its output with that of a water-cooled condenser, for the configurations showing maximum power output for their relative number of system stages at an ambient temperature of 20 °C. Using some design parameters for the organic Rankine cycle heat recovery power plant with a detailed analysis of the system [30], the power consumption of the air-cooled condenser of the WHWRS was briefly estimated as follows. In the reference [30], R134a was used as a working fluid, and the power consumption of the air-cooled condenser was 1224 kW to reject 58,912 kW of heat transfer rate to produce 5263 kW of net power with condensation temperature of 32.6 °C with a pinch point 6 °C at 18.8 °C ambient temperature. It was generally estimated that for fixed heat transfer rate, the lower the mass flow rate of air m_A more the rise in temperature of air T_{rise} as it passes through the condenser, i.e.,

$$m_A \propto 1/T_{rise} \tag{8}$$

The rise in air temperature for the reference case is 7.7 °C. Also, for a more mass flow rate of air, more fans, or a more powerful fan, is required, which means the mass flow rate of air is directly proportional to the power consumed by the air-cooled condenser P_{acc}.

So,

$$P_{acc} \propto 1/T_{rise} \tag{9}$$

When there is a higher heat transfer rate from the working fluid to the air-cooled condenser Q_{acc}, a higher power is required.

$$P_{acc} \propto Q_{acc} \tag{10}$$

Our case closely resembles the one mentioned here [30]. At the constant ambient condition of 20 °C, the working fluid is at 30 °C in the condenser. For the pinch point of 5 °C, the rise in temperature and the heat transfer rate of the reference case and our case can be compared to determine the power consumption of the air-cooled condenser.

Table 3 shows heat transfers and power consumption of all one to four stages of configuration, optimized for maximum power output by comparing the proportional quantities with the reference case. It can be clearly seen that for a fixed ambient temperature condition, there is a small net power difference between stage 1 and stage 2 for the water-cooled condenser system. Even though the heat rejection is same, the power required by the air-cooled condenser to compensate a large amount of latent heat from the flue gas absorbed by the working fluid makes a huge gap between stage 1 and stage 2 with an air-cooled condenser in comparison to the water-cooled condenser. Generally, the power consumption of an air-cooled condenser seems to increase with the increase in the number of stages of the system.

Table 3. Heat transfer and net power output comparison of water-cooled and air-cooled condenser stages 1–4 of WHWRS.

	Stage 1	Stage 2	Stage 3	Stage 4
Heat transfer rate Q (kW):				
Evaporator/s (in)	77,445	187,322	233,017	266,012
Condenser (out)	−71,049	−180,953	−229,267	−254,333
Power (kW):				
Turbine (out)	7625	7625	7625	7625
Pump (in)	−1228	−1255	−1255	−1255
Compressor (in)	-	-	−2618	−6493
Net power for water-cooled system	6396	6369	3751	−124
Power consumption for air-cooled condenser	959	2441	3093	3591
Net power for air-cooled system	5437	3928	658	−3715

4. Economic Analysis

Although a detailed economic analysis is beyond the scope of this work, a simple estimation of the potential cost of the system could be helpful to evaluate its usefulness. An economic model of the ORC system presented by Usman et al. [31] was used to estimate the cost of the WHWRS. In the case of a previous two-stage WHWRS represented by configuration A, applied to a 600 MW coal-fired power plant with condensation temperature of 30 °C at 20 °C ambient temperature, the component cost of the ORC unit and pumped HPC unit are presented in Tables 4 and 5, respectively. The cost of cooling fans for the air-cooled system is estimated using correlations in [32].

Table 4. Cost of ORC unit.

Component	W, Q (kW)	Cost ($)
Turbine	7625	1,946,583
Pump	1228	695,155
Evaporator	77,445	1,745,552
Condenser	71,048	1,901,733
Generator	7625	312,645
Water-cooled:		
Cooling tower	71,048	2,570,129
Total		9,171,798
Air-cooled:		
Cooling fan	959	197,859
Total		6,799,528

Table 5. Cost of pumped HPC unit.

Component	W, Q (kW)	Cost ($)
Pump	27	85,167
Evaporator	106,739	2,220,395
Condenser	106,766	2,508,585
Water-cooled:		
Cooling tower	106,766	3,860,663
Total		8,674,810
Air-cooled:		
Cooling fan	1482	275,435
Total		5,089,581

The economic benefit of water recovery was estimated as follows based on Jeong et al. [33]. Here, the assumed 16% volume of water vapor in flue gas is equivalent to 10% weight of water vapor. With 50% of water recovery efficiency, the potential annual fresh water recovery is:

- 750 kg/s × 10% × 50%/1000 m^3/kg × 3600 s/hr × 24 h × 365 days × 95% (annual availability) = 1,123,470 m^3
- Assuming a water cost of $1.69 per cubic meter [31], an annual savings of water recovery is $1,998,594

The water withdrawal in a subcritical 600 MW power plant consists of 2381 L/MWh of cooling water make-up, 400 L/MWh of FGD make-up, and 36 L/MWh of boiler make-up [34], which can be partially covered by 225 L/MWh of regenerated water as condensate out of 450 L/M of moisture in flue gas in this study.

The economic benefit of heat recovery of an ORC unit was estimated as follows, based on Walraven et al. [32], assuming the price obtained per MWh of produced electricity of $67.5/MWe, an annual availability of 95%, and the cost of operation and maintenance of 2.5% of the investment cost of the ORC and pumped HPC per year. The potential net annual income through heat recovery for the water cooled plant is:

- 6369 kW (net power) × 24 hrs × 365 days × 95% (availability) = 53,003 MWe
- 53,003 MWe × $67.5/MWe − 2.5% × ($9,171,798 + $8,674,810) = $3,131,470

Whereas for the air-cooled plant,

- 3928 kW (net power) × 24 hrs × 365 days × 95% (availability) = 32,689 MWe
- 32,689 MWe × $67.5/MWe − 2.5% × ($6,799,528 + $5,089,581) = $1,909,280

To utilize the recovered water, the condensed water usually needs further treatment such as filtration, acid removal, or demineralization [33]. The cost of water treatment could be included in a detailed economic analysis.

In addition to the economic benefits of the electric power and water recovery, the WHWRS can contribute to reduce the condensable particulate matter [27] and to decrease the humidity around the power plant, which decreases diffusion of pollutants and promotes secondary particulate matter formations. These environmental benefits were not included in the economic analysis. Furthermore, the waste heat from the condenser of the WHWRS could be utilized as a heat source for the a heat pump used for deep-cooling of the coal-fired power plant flue gas [35].

5. Discussion

This study discussed the design of a WHWRS using a single working fluid. It also analyzed various problematic scenarios of increased capital cost and power consumption, presented in practical situations by a change in the ambient temperature caused by a change in climate conditions depending on the season and location of a power plant. Thus showing that relative to a change in ambient condition, the performance of a WHWRS will vary. As this study suggests, a heat recovery ORC is to be optimized for maximum power output because the auxiliary components of a water recovery system such as a pump or compressors as well as cooling system consume power. Thus, the system must be optimized for maximum power output for it to be as self-sufficient as possible at an increased ambient temperature condition. Optimizing the high-pressure side concerning commercially available components is a reasonable approach. Superheating of the working fluid slightly above the saturation temperature is better for the longevity of the turbine.

A pumped HPC system has low power consumption as it only uses a circulation pump, but the amount of gradient it creates in temperature of the flue gas strictly depends on the ambient temperature conditions. However, for a 20 °C ambient temperature, configuration A, for a minor decrease in power output has an abrupt increase from 0% to 50% in water recovery efficiency.

Contrary to that, the vapor compression cycle has no limitation regarding ambient temperature. However, the larger the gradient it creates in the flue gas, the greater the work done by the compressor, eventually resulting in more power consumption. As shown in some cases, the power consumption to run the system exceeds the turbine power output thus external power is required. In case of 30 °C ambient temperature condition, to achieve a similar target system alters from being self-sufficient to power dependent by the difference of only one stage i.e., B having one additional stage is self-sufficient in power output whereas X is dependent on external power source although the power generated by ORC cycle is the same for both the configurations. Thus, it is always feasible to use pumped heat pipe cooling before vapor compression if the number of stages or compactness of the system is not a priority. Combination of both pumped heat pipe and vapor compression allows some of the temperature to be decreased in the flue gas by the pumped heat pipe cycle, which can reduce the work done by the compressor of the vapor compression cycle. The work of one large compressor can also be reduced by distributing the work to smaller compressors, which is beneficial in terms of power consumption because they create less irreversibility resulting into less loss as can be seen in the case of configuration H which consumes less power than X. The capital cost, however, increases as the number of stages increase because of the additional components required for additional vapor compression cycles.

The results shown by configuration A have promising potential for practical applications at regions with moderate climate conditions, i.e. 20 °C to 25 °C or lower. For regions with higher temperature climates, i.e., 40 °C through different season cycles, recovering an adequate proportion of water from the flue gas with the dew point of the moisture in the flue gas being 50 °C to 55 °C, is only possible with the presence of vapor compression cycle in the system. Therefore, all the rest of the configurations show the variation in the trade-off between power output and the number of stages, with several modifications, with ORC and VCC being the essential parts of the system. Thus, for regions with higher ambient temperatures, it depends on desirability for either higher power output or lesser number of stages. Based on the requirements, the designer can decide if either vapor compression cycle (X) is enough for the heat and water recovery according to the requirements or integration of pumped heat pipe with vapor compression cycle (B) is feasible. Distribution of a large compression work (H) into several compressors is required, or both the integration of pumped heat pipe and the distribution of the compressor work (O) fulfills his requirements of power generation while maintaining the capital cost. Also, water cooled, partially water-cooled, or air-cooled condensation has to be selected based on the availability of water. Note that, regardless of the configuration selected, the system will have better water and power yields for all ambient temperatures less than the ambient temperature at which the system is designed. As the ambient temperature becomes less and less than the design ambient temperature, the yields will increase further. Therefore, it is recommended that the system must be designed for the highest average ambient temperature condition of the region where the system is to installed.

The economic benefit may seem to tend to decrease with the increase in the number of stages, but the greater number of stages will serve the purpose of greater water recovery efficiency at lower temperatures than the design temperature. Therefore, being designed for peak temperature around the year, the system with a larger number of stages will tend to have a greater economic benefit as it will be able to recover more water and heat at the lower temperatures.

Other than the water being used for the primary purposes of cooling and condensing hazardous gases, water recovery itself reduces the incidence of white plumes and smog around the power plant, since the high humidity flue gas prevents pollutant diffusion, which promotes secondary air pollutant transformation, and contributes to smog.

In order to utilize the system for a specific site with the given objective functions and constraints, the system can be optimized by taking account of more factors related to electrical power distribution systems presented by Nguyen et al. [11] and Duong et al. [10].

6. Conclusions

Fossil fuel power plants can cause many environmental problems because of their exhaust emissions and use of significant amounts of water. Heat and water recovery from the flue gas in a thermal power plant can contribute to reducing CO_2 emissions and water requirements. Several studies utilize flue gas in different ways for heat recovery but recovering water from flue gas while recovering heat serves so many useful purposes of extra energy production, water saving, and pollution control. In this study, a WHWRS composed of an ORC and cooling cycles using the same working fluid accompanied by a phase change was proposed. The system was optimized for maximum power output by optimizing the high-pressure side and turbine inlet temperature. The design of the WHWRS depends on the desirable properties of the system. It can be either water recovery, power output, or the number of system stages (capital cost). A higher water recovery requires an increase in the number of stages and a loss of power output at a constant ambient temperature. Similarly, a higher ambient temperature requires an increase in the number of stages and an increase in power consumption to achieve the targeted water recovery efficiency. In the case of a 600 MW power plant with 16% volume of water vapor in 150 °C flue gas, the WHWRS can produce approximately 6 MWe in the case of the water-cooled condenser and 3.9 MWe in the case of the air-cooled condenser, and obtain 50% water recovery by cooling the flue gas to 40 °C at an ambient temperature of 20 °C. Other than the water being used for the primary purposes of cooling and condensing hazardous gases, water recovery itself reduces the incidence of white plumes and smog around the power plant.

Author Contributions: All authors contributed to this paper. S.S.M.S. performed the system analysis along with the preparation of the manuscript, A.N. contributed to data analysis, G.B.C. conceived the study plan and contributed to the analysis, and Y.M.K. proposed the system configurations and modeling methods and confirmed the manuscript.

Acknowledgments: This research was supported by the National Strategic Project—Cfarbon Upcycling of the National Research Foundation of Korea (NRF), funded by the Ministry of Science and ICT (MSICT), the Ministry of Environment (ME), and the Ministry of Trade, Industry and Energy (MOTIE) (NRF—2017M3D8A2085654), and the Korea Agency for Infrastructure Technology Advancement under the Ministry of Land, Infrastructure and Transport of Korean Government (Project Number: 13 Construction Research T01). The Institute of Engineering Research at Seoul National University (SNU) provided research facilities for this work. The authors are grateful for the support.

Conflicts of Interest: The authors declare no conflict of interest.

References

1. Castellani, B.; Rinaldi, S.; Morini, E.; Nastasi, B.; Rossi, F. Flue gas treatment by power-to-gas integration for methane and ammonia synthesis—Energy and environmental analysis. *Energy Convers. Manag.* **2018**, *171*, 626–634. [CrossRef]
2. García, S.G.; Montequín, V.R.; Fernández, R.L.; Fernández, F.O. Evaluation of the synergies in cogeneration with steel waste gases based on Life Cycle Assessment: A combined coke oven and steelmaking gas case study. *J. Clean. Prod.* **2019**, *217*, 576–583. [CrossRef]
3. Liu, M.; Fu, H.; Miao, G.-Y.; Wang, J.-S.; Han, X.-Q.; Yan, J.-J. Theoretical study of lignite-fired power system integrated with heat pump drying. *K. Cheng Je Wu Li Hsueh Pao/J. Eng. Thermophys.* **2016**, *37*, 929–934.
4. Xu, Z.; Mao, H.; Liu, D.; Wang, R. Waste heat recovery of power plant with large scale serial absorption heat pumps. *Energy* **2018**, *165*, 1097–1105. [CrossRef]
5. Wei, M.; Fu, L.; Zhang, S.; Zhao, X. Experimental investigation on vapor-pump equipped gas boiler for flue gas heat recovery. *Appl. Therm. Eng.* **2019**, *147*, 371–379. [CrossRef]
6. Júnior, E.P.B.; Arrieta, M.D.P.; Arrieta, F.R.P.; Silva, C.H.F. Assessment of a Kalina cycle for waste heat recovery in the cement industry. *Appl. Therm. Eng.* **2019**, *147*, 421–437. [CrossRef]
7. Liu, M.; Zhang, X.; Ma, Y.; Yan, J. Thermo-economic analyses on a new conceptual system of waste heat recovery integrated with an S-CO$_2$ cycle for coal-fired power plants. *Energy Convers. Manag.* **2018**, *161*, 243–253. [CrossRef]

8.   Liu, W.-D.; Chen, Z.-G.; Zou, J. Eco-Friendly Higher Manganese Silicide Thermoelectric Materials: Progress and Future Challenges. *Adv. Energy Mater.* **2018**, *8*, 1800056. [CrossRef]

9.   Kim, T.Y.; Kim, J. Assessment of the energy recovery potential of a thermoelectric generator system for passenger vehicles under various drive cycles. *Energy* **2018**, *143*, 363–371. [CrossRef]

10.   Duong, M.Q.; Pham, T.D.; Nguyen, T.T.; Doan, A.T.; Tran, H.V. Determination of Optimal Location and Sizing of Solar Photovoltaic Distribution Generation Units in Radial Distribution Systems. *Energies* **2019**, *12*, 174. [CrossRef]

11.   Nguyen, T.; Vu Quynh, N.; Duong, M.; Van Dai, L. Modified differential evolution algorithm: A novel approach to optimize the operation of hydrothermal power systems while considering the different constraints and valve point loading effects. *Energies* **2018**, *11*, 540. [CrossRef]

12.   Bronicki, L.Y. Organic Rankine Cycle Power Plant For Waste Heat Recovery. *Ormat Int. Inc.* **2000**, *6*, 302.

13.   Vanslambrouck, B.; Vankeirsbilck, I.; van den Broek, M.; Gusev, S.; De Paepe, M. Efficiency comparison between the steam cycle and the organic Rankine cycle for small scale power generation. In Proceedings of the Renewable Energy World Conference & Expo North America, Long Beach, CA, USA, 14–16 February 2012.

14.   Chen, C.L.; Li, P.Y.; Le, S.N.T. Organic Rankine Cycle for Waste Heat Recovery in a Refinery. *Ind. Eng. Chem. Res.* **2016**, *55*, 3262–3275. [CrossRef]

15.   Koroglu, T.; Sogut, O.S. Advanced exergoeconomic analysis of organic rankine cycle waste heat recovery system of a marine power plant. *Int. J. Thermodyn.* **2017**, *20*, 140–151. [CrossRef]

16.   Negash, A.; Kim, Y.M.; Shin, D.G.; Cho, G.B. Optimization of organic Rankine cycle used for waste heat recovery of construction equipment engine with additional waste heat of hydraulic oil cooler. *Energy* **2018**, *143*, 797–811. [CrossRef]

17.   Ergun, A.; Ozkaymak, M.; Aksoy Koc, G.; Ozkan, S.; Kaya, D. Exergoeconomic analysis of a geothermal organic Rankine cycle power plant using the SPECO method. *Environ. Prog. Sustain. Energy* **2017**, *36*, 936–942.

18.   Liu, C.; Xu, J. A novel heat recovery system for flue gas from natural gas boiler. *Huagong Xuebao/Ciesc J.* **2013**, *64*, 4223–4230. [CrossRef]

19.   Copen, J.H.; Sullivan, T.B.; Folkedahl, B.C. *Principles of Flue Gas Water Recovery System—PDF*; SIEMENS: Las Vegas, NV, USA, 2005.

20.   Pan, S.-Y.; Snyder, S.W.; Packman, A.I.; Lin, Y.J.; Chiang, P.-C. Cooling water use in thermoelectric power generation and its associated challenges for addressing water-energy nexus. *Water Energy Nexus* **2018**, *1*, 26–41. [CrossRef]

21.   Levy, E.; Bilirgen, H.; Jeong, K.; Kessen, M.; Samuelson, C.; Whitcombe, C. *Recovery of Water from Boiler Flue Gas*; Office of Research and Sponsored Programs: Bethlehem, PA, USA, 2008.

22.   Zhang, F.; Ge, Z.; Shen, Y.; Du, X.; Yang, L. Mass transfer performance of water recovery from flue gas of lignite boiler by composite membrane. *Int. J. Heat Mass Transf.* **2017**, *115*, 377–386. [CrossRef]

23.   Zhao, S.; Yan, S.; Wang, D.K.; Wei, Y.; Qi, H.; Wu, T.; Feron, P.H.M. Simultaneous heat and water recovery from flue gas by membrane condensation: Experimental investigation. *Appl. Therm. Eng.* **2017**, *113*, 843–850. [CrossRef]

24.   Wei, M.; Zhao, X.; Fu, L.; Zhang, S. Performance study and application of new coal-fired boiler flue gas heat recovery system. *Appl. Energy* **2017**, *188*, 121–129. [CrossRef]

25.   Jeong, K.; Kessen, M.J.; Bilirgen, H.; Levy, E.K. Analytical modeling of water condensation in condensing heat exchanger. *Int. J. Heat Mass Transf.* **2010**, *53*, 2361–2368.

26.   Chen, Y. Optimization of the Fin Configuration of Air-cooled Condensing Wet Electrostatic Precipitator for Water Recovery from Power Plant Flue Gas. Master's Thesis, University of Cincinnati, Cincinnati, OH, USA, 2013.

27.   Feng, Y.; Li, Y.; Cui, L. Critical review of condensable particulate matter. *Fuel* **2018**, *224*, 801–813. [CrossRef]

28.   Zukeran, A.; Ninomiya, K.; Ehara, Y.; Yasumoto, K.; Kawakami, H.; Inui, T. SOx and PM removal using electrostatic precipitator with heat exchanger for marine diesel. In Proceedings of the ESA Annual Meeting on Electrostatics, Cocoa Beach, FL, USA, 11–13 June 2013.

29.   Shuangchen, M.; Jin, C.; Kunling, J.; Lan, M.; Sijie, Z.; Kai, W. Environmental influence and countermeasures for high humidity flue gas discharging from power plants. *Renew. Sustain. Energy Rev.* **2017**, *73*, 225–235. [CrossRef]

30.   Sun, J.; Li, W. Operation optimization of an organic Rankine cycle (ORC) heat recovery power plant. *Appl. Therm. Eng.* **2011**, *31*, 2032–2041. [CrossRef]

31. Usman, M.; Imran, M.; Yang, Y.; Lee, D.H.; Park, B.-S. Thermo-economic comparison of air-cooled and cooling tower based Organic Rankine Cycle (ORC) with R245fa and R1233zde as candidate working fluids for different geographical climate conditions. *Energy* **2017**, *123*, 353–366. [CrossRef]
32. Walraven, D.; Laenen, B.; D'haeseleer, W. Economic system optimization of air-cooled organic Rankine cycles powered by low-temperature geothermal heat sources. *Energy* **2015**, *80*, 104–113. [CrossRef]
33. King, C.W. *Thermal Power Plant Cooling: Context and Engineering*; ASME Press: New York, NY, USA, 2013.
34. Carpenter, A. *Water Conservation in Coal-Fired Power Plants*; IEA Clean Coal Centre: London, UK, 2017.
35. Li, Y.; Yan, M.; Zhang, L.; Chen, G.; Cui, L.; Song, Z.; Chang, J.; Ma, C. Method of flash evaporation and condensation–heat pump for deep cooling of coal-fired power plant flue gas: Latent heat and water recovery. *Appl. Energy* **2016**, *172*, 107–117. [CrossRef]

Article

Synthesis of Nano-Calcium Oxide from Waste Eggshell by Sol-Gel Method

Lulit Habte [1], Natnael Shiferaw [2], Dure Mulatu [1], Thriveni Thenepalli [3], Ramakrishna Chilakala [4] and Ji Whan Ahn [3,*]

1   Department of Resources Recycling, University of Science & Technology, 217 Gajeong-ro, Gajeong-dong, Yuseong-gu, Daejeon 34113, Korea; luna1991@ust.ac.kr (L.H.); dure@kigam.re.kr (D.M.)
2   Korea Research Institute of Climate Change, 11, Subyeongongwon-gil, Chuncheon-si 24239, Korea; natlulit@gmail.com
3   Center for Carbon Mineralization, Mineral Resources Division, Korea Institute of Geosciences and Mineral Resources (KIGAM), 124 Gwahagno, Gajeong-dong, Yuseong-gu, Daejeon 34132, Korea; thenepallit@rediffmail.com
4   Department of Bio-based Materials, School of Agriculture and Life Science, Chungnam National University, Daejeon City 34134, Korea; chilakala_ramakrishna@rediffmail.com
*   Correspondence: ahnjw@kigam.re.kr

Received: 30 April 2019; Accepted: 3 June 2019; Published: 7 June 2019

Abstract: The sol-gel technique has many advantages over the other mechanism for synthesizing metal oxide nanoparticles such as being simple, cheap and having low temperature and pressure. Utilization of waste materials as a precursor for synthesis makes the whole process cheaper, green and sustainable. Calcium Oxide nanoparticles have been synthesized from eggshell through the sol-gel method. Raw eggshell was dissolved by HCl to form $CaCl_2$ solution, adding NaOH to the solution dropwise to agitate Ca $(OH)_2$ gel and finally drying the gel at 900 °C for 1 h. The synthesized nanoparticle was characterized by scanning electron microscope (SEM), Fourier-transform infrared spectroscopy (FTIR), X-Ray fluorescence (XRF) and X-ray diffraction (XRD). The FTIR and XRD results have clearly depicted the synthesis of calcium oxide from eggshell, which is mainly composed of calcium carbonate. The FE-SEM images of calcium oxide nanoparticles showed that the particles were almost spherical in morphology. The particle size of the nanoparticles was in the range 50 nm–198 nm. Therefore, waste eggshell can be considered as a promising resource of calcium for application of versatile fields.

Keywords: calcium oxide nanoparticle; sol-gel method; eggshell

---

1. Introduction

Ever increasing solid waste is now becoming a challenge for a sustainable world. Improper management of those wastes leads to public health and environment related problems [1]. Huge amounts of solid wastes, including municipal, industrial and hazardous wastes, have been generated worldwide. Food wastes are the major solid wastes causing problems in the environment. It was estimated that food waste would increase by 44% from 2005 to 2025 [2]. Industrialization and population growth are the major factors for the increase in solid wastes. Eggshell is a solid waste which contributes to degradation in the environment. Households, restaurants, and bakeries are the major source of eggshell [3]. The main component is pure calcium carbonate with little porosity [3]. This waste can be transformed into valuable products. Eggshells can also be used as a $CaCO_3$ source for different applications [4,5]. However, the common trend to manage waste eggshell is land filling which causes problems for the environment. Landfilling such food wastes cause unpleasant odors during biodegradation and the attached membranes attract worms and insects. Utilization of eggshells can provide benefits, not only regarding environmental concerns but also for freeing up landfill sites.

Synthesis of nanoparticles is attracting more attention because of better performance in terms of improved surface area. Metal oxide nanoparticles have many applications in diverse fields. They are considered as an active catalyst for versatile applications [6]. Metal oxide nanoparticles have been utilized as adsorbents for heavy metals removal in water and wastewater [7–11], as sorbents for CO_2 capture [12] where CO_2 capture increases with the increase in surface area of the particle [13], as heterogeneous catalysts in biodiesel production [14], as purification of exhaust gas [15] and in wall painting [16]. Calcium oxide, magnesium oxide, aluminum oxide, zinc oxide, manganese dioxide, titanium oxides, and iron oxide nanoparticles are the most commonly used nano-metal oxides. These metal oxide nanoparticles have been synthesized by several methods such as the ultrasonic-assisted method [17], the hydrogen plasma-metal reaction method [18], the biopolymer-assisted method [19] microwave-assisted method [20], facial calcination [21], co-precipitation [22], direct thermal decomposition [23], chemical co-precipitation [24], two-step process (green synthesis) [25] and two step thermal decomposition [26]. Those methods have drawbacks, such as the use of additives, high temperature, and pressure, time-consuming, expensive and complicated procedures. Sol-gel method overcomes most of the drawbacks from the above-mentioned methods. It is simple, cheap, not time consuming and no expensive equipment is required. It can also be carried out at lower temperature and with no pressure. Therefore, it can be a promising method to synthesize calcium oxide nanoparticles.

Nowadays, eggshell is being widely utilized for industrial applications [4,5,11,27]. Eggshell containing $CaCO_3$ as a major constituent can be a potential candidate for calcium oxide nanoparticle synthesis. In this study, CaO nanoparticles were synthesized by sol-gel method in which raw eggshell was dissolved by HCl to form $CaCl_2$ solution, adding NaOH to the solution dropwise to agitate $Ca(OH)_2$ gel and finally drying the gel at high temperature. This method can be considered as cheap, easy and eco-friendly. The objective of this work is to utilize waste eggshell in wastewater treatment systems.

2. Materials and Methods

2.1. Materials

Chemicals, hydrochloric acid with 35%–37% concentration and sodium hydroxide with 97% purity were purchased from Junsei Chemicals Ltd., Seoul, Korea. Waste chicken eggshells were collected from the KIGAM (Korean Institute of Geoscience and Mineral Resources) campus restaurant in Daejeon, South Korea. Raw waste eggshell is mainly composed of CaO, MgO, K_2O, P_2O_5, Na_2O and some trace compounds.

2.2. Synthesis of Calcium Oxide Nanoparticles

Synthesis of metal oxide nanoparticle through sol-gel method has four basic consecutive processes: Preparation of homogeneous solution, the formation of 'sol' by hydrolysis, the formation of 'gel' by condensation and drying of the formed gel [28]. In this experiment, a homogenous solution of metal salt, $CaCl_2$, was produced by dissolving solid $CaCO_3$ in dilute HCl as shown in Equation (1). Chicken eggshell was used as a precursor for this method. Waste eggshells were thoroughly washed with warm water and cleaned with deionized water. Then the washed sample was dried at 120 °C for 2 h, ground into a powder and sieved with 100 μm sieve size. For the preparation of calcium chloride ($CaCl_2$) solution, 12.5 gm of PES was dissolved in 250 mL of 1 M hydrochloric acid (HCl).

$$CaCO_3 \text{ (s)} + 2HCl \text{ (aq)} \rightarrow CaCl_2 \text{ (aq)} + H_2O \text{ (l)} + CO_2 \text{ (g)} \tag{1}$$

The next step was the formation of 'sol' by hydrolysis process. 'Sol' is defined as a stable dispersion of colloidal particles of precursors in a solvent due to hydrolysis reaction. During the hydrolysis process, metal hydroxide was formed. A total of 250 mL of 1 M sodium hydroxide (NaOH) was added slowly (drop by drop) to convert the homogeneous $CaCl_2$ solution formed in the previous step into 'sol' at ambient temperature (Equation (2)). Then 'gel' formation by condensation was followed. The slow addition of NaOH resulted in a low rate of nucleation and encouraged subsequent precipitation of $Ca(OH)_2$ one over another forming a highly crystalline gel. The condensation reaction resulted in small-sized particles interconnected to each other forming a rigid and highly crystalline inorganic network within the liquid. $Ca(OH)_2$ gel containing solution was aged for one night at ambient temperature.

$$CaCl_2 \text{ (aq)} + 2NaOH \text{ (aq)} \rightarrow Ca(OH)_2 \text{ (s)} + 2NaCl \text{ (aq)} \tag{2}$$

Then filtration followed where the filtrate was cleaned with distilled water in order to remove adsorbed impurities in the precipitate. The synthesizing process ended by drying the gel. In this process, the solvent (liquid phase) was removed and a significant amount of shrinkage and densification was noticed. The powder was dried at 60 °C for 24 h in an oven and calcined at 900 °C for 1 h (Equation (3)).

$$Ca(OH)_2 \text{ (s)} + \text{Heat} \rightarrow CaO \text{ (s)} + H_2O \text{ (l)} \tag{3}$$

2.3. Characterization

Characterization of powders was determined by thermogravimetric analysis (TGA), X-ray fluorescence (XRF), Fourier-transform infrared spectroscopy (FTIR), X-ray diffraction (XRD) and scanning electron microscope (SEM). Decomposition temperatures of Raw eggshell and $Ca(OH)_2$ gel were analyzed by thermogravimetric analysis (TGA) (Shimadzu DTG-60H) in a platinum crucible at a heating rate of 10 °C/min from ambient temperature to 1000 °C. The chemical analysis of raw eggshell and synthesized CaO nanoparticle were performed using X-ray fluorescence spectrometer (Shimadzu, Japan). The crystal structural analysis was analyzed by X-ray diffraction (XRD) with diffraction angles 2θ from ranging 10° to 90°and with Cu Kα radiation (λ = 1.5406 Å) as the radiation source. Size, shape and surface morphology of the nanoparticle was examined through field emission scanning electron microscope (FE-SEM), (Tuscan Mira 3 LMU FEG) with a coater (Quorum Q150T ES/10 mA, 120 s Pt coating) at an accelerating voltage of 10 kV. The FTIR spectroscopy (FT-IR) (6700 FTIR, Thermo Scientific Nicolet, Massachusetts, MA, USA) was used to determine the different functional groups present in the synthesized nanoparticle.

3. Results

Thermogravimetric analysis (TGA) curves of raw eggshell and $Ca(OH)_2$ gel are illustrated in Figure 1a,b respectively. In raw eggshell, a total weight loss of 39% was noticed and at 823 °C temperature CO_2 is released to the environment. In the case of $Ca(OH)_2$ gel, three major weight losses were observed in the analysis: 32 °C–461.89 °C, 461.89 °C–691.2 °C and 691.2 °C–1000 °C with mass changes of 18.49%, 13.8% and 1.60% respectively. The losses correspond to vaporization of physically adsorbed water, decomposition of $Ca(OH)_2$ gel to CaO and decomposition of $CaCO_3$ to CaO where CO_2 will be released [23].

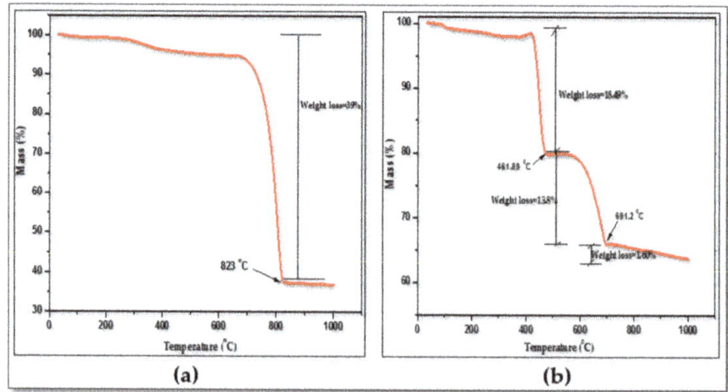

Figure 1. Thermogravimetric analysis (TGA) curve of (**a**) Raw eggshell, (**b**) Ca(OH)$_2$ gel.

The chemical composition of raw eggshell and synthesized nanoparticle is listed in Table 1. The data below shows that CaO is the major quantity in the raw eggshell, which is mainly in the form of CaCO$_3$ [29]. The High value of Ignition loss for raw eggshell represents the conversion of CaCO$_3$ to CaO and CO$_2$ during calcination [4]. After the synthesis of nanoparticle, lime was found to be the major component (86.93%).

Table 1. Chemical analysis of raw eggshell and nanoparticle.

Chemical Composition	SiO$_2$	Al$_2$O$_3$	Fe$_2$O$_3$	CaO	MgO	K$_2$O	Na$_2$O	TiO$_2$	MnO	P$_2$O$_5$	Ig. Loss
Raw eggshell (%)	<0.01	<0.01	<0.01	52.75	0.52	0.04	0.05	<0.01	<0.01	0.22	46.62
Nano-CaO (%)	0.08	0.04	0.05	86.93	1.08	0.14	1.32	<0.01	<0.01	0.43	9.3

FTIR spectrum of CaO nanoparticle, Ca(OH)$_2$ gel, raw eggshell and commercial CaO, Ca(OH)$_2$ and CaCO$_3$ are presented in Figure 2 to characterize the synthesized nanoparticle and also to compare the results with the commercial powders. The FTIR result of the raw eggshell showed broadband centering at 1415.52 cm^{-1} which is a characteristic of C–O bond showing a bond between the oxygen atom of carbonate and calcium atom [30]. In addition, there were two sharp bands at 711.62 and 875.54 cm^{-1} showing C–O bond [30]. The peaks of raw eggshell were in correspondence with commercial CaCO$_3$ except for the broadband at 2360.48 cm^{-1} which represents the N–H bond caused by the amines and amides present in the protein fiber of the eggshell membrane [31]. On the other hand, CaO nanoparticles showed peaks at 1444.42 cm^{-1}, 1064.51 cm^{-1}, and 863.95 cm^{-1} which were ascribed to C–O bond indicating the carbonation of calcium oxide nanoparticles [18,24]. The absorption peak at 3639.02 cm^{-1} has also resulted due to the O–H bond from water molecules on the surface of the nanoparticle [18,24]. The tiny peak at 2343.09 cm^{-1} might be due to atmospheric CO$_2$ [32]. This peak has also been seen in commercial CaO and Ca(OH)$_2$. The absence of a sharp absorption in the region at 1415.52 cm^{-1} indicates that the CaCO$_3$ as the basic component of the eggshell was no longer present as it was already converted to CaO. The strong band at 512 cm^{-1} identified vibration of the Ca–O band [20,24].

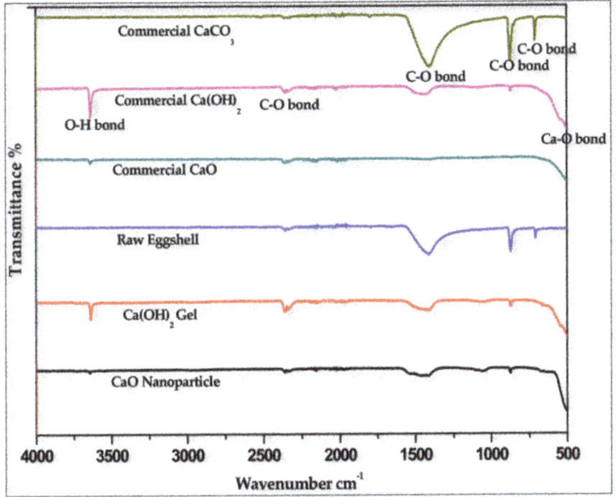

Figure 2. FTIR spectrum of CaO Nanoparticle, Ca(OH)$_2$ Gel, Raw eggshell and Commercial CaO, Ca(OH)$_2$ and CaCO$_3$.

XRD results of raw eggshell, commercial Ca(OH)$_2$, Ca(OH)$_2$ gel and synthesized calcium oxide nanoparticle are shown in Figure 3. Raw eggshell XRD pattern showed a good match with CaCO$_3$ in the calcite phase (PDF Card No. 00-081-2027). The main peak appeared at $2\theta = 29.48$. In addition, the analysis shows several peaks at 23.07, 31.52, 36.06, 39.48, 43.26, 47.64, 48.6, 56.62, 57.5, 60.92, 63.2, 64.76, 65.74, 70.29, 72.96, 76.37, 82.15, 84.9 which are assigned to (012), (006), (110), (113), (202), (018), (116), (211), (122), (214), (300), (0012), (0210), (128), (220), (114), (226) planes of calcite phase respectively. XRD patterns of Ca(OH)$_2$ gel match with portlandite phase (PDF Card No. 00-087-0673) as a major phase and also with commercial Ca(OH)$_2$ as shown the figure. The main peak of the gel appeared at $2\theta = 34.12$. In the case of synthesized calcium oxide nanoparticle (Figure 3), the XRD patterns match with calcium oxide (CaO) (PDF Card No. 99-0070). The main peak appeared at $2\theta = 37.4$. Besides, several peaks appeared at 32.22, 37.36, 53.9, 64.2, 67.4, 79.7 and 88.58 which are also assigned to (111), (200), (220), (311), (222), (400), and (331) planes of lime phase respectively. The result shows that the calcium carbonate in the raw eggshell was completely changed to calcium oxide during the synthesis.

Scherrer's Equation (Equation (4)) was used to calculate the mean crystallite size of the calcium oxide nanoparticle:

$$ d = \frac{k\lambda}{\beta \cos\theta} \tag{4} $$

where d is mean crystallite size, λ is wavelength, k is constant of Scherrer, θ is Bragg's angle and β is structural broadening. Accordingly, the mean size of crystallite was found to be 24.51 nm. Other researchers using this technique of nanoparticle synthesis (sol-gel) have also confirmed the formation of large crystallite size. It was reported that the synthesis of MgO nanoparticles with an average particle size of 27.0 nm using the sol-gel technique [33]. Other studies also reported that the synthesized calcium oxide nanoparticle had a crystallite size of 40 nm and 41 nm for two different conditions [24]. Moreover, the lattice strain of crystal was determined by Equation (5):

$$ \varepsilon = \frac{\beta}{4\tan\theta} \tag{5} $$

where θ is Bragg's angle, β is structural broadening, ε is lattice strain. Therefore, it was determined as 4.41×10^{-3}.

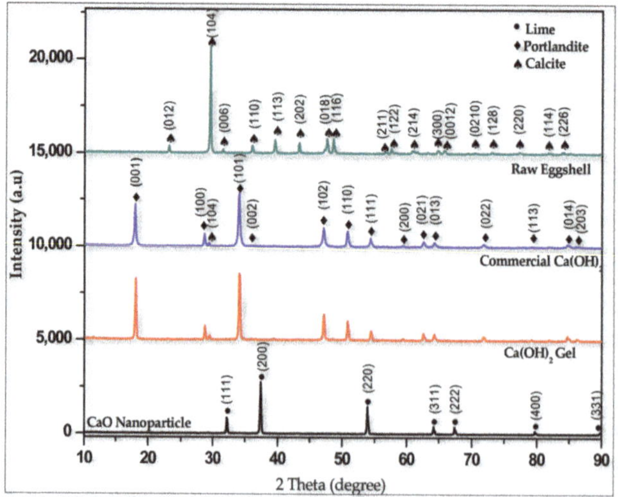

Figure 3. X-ray Diffraction of a Raw eggshell, Commercial Ca(OH)$_2$, Ca(OH)$_2$ Gel and CaO nanoparticle.

Surface morphology of Ca(OH)$_2$ gel and synthesized calcium oxide nanoparticle was studied by FE-SEM as shown in Figure 4ab. Raw eggshell had a non-porous and irregular crystal structure. In Figure 4a, the surface micro structural analysis, using SEM, of Ca(OH)$_2$ gel has been analyzed to study the transition from gel to nanoparticle obtained after drying. The result revealed that the Ca(OH)$_2$ gel has a hexagonal shape in which the size is also in nano-scale (around 450 nm). This result has also been obtained from other research of synthesizing Ca(OH)$_2$ nano-plates from oyster shells [34]. FE-SEM image shown in Figure 4b showed that the nanoparticles are approximately spherical in morphology agglomerating to each other. These agglomerates of small particles show the polycrystalline character of the nanoparticle. Other studies confirmed the spherical shape of calcium oxide nanoparticles [17,24]. Mean size of CaO nanoparticle was estimated to be 198 nm, as seen in Figure 5. Size of calcium oxide nanoparticle was reduced after drying the Ca(OH)$_2$ gel due to the release of CO$_2$ and H$_2$O. A calcination temperature of 900 °C was used, at which vaporization of absorbed water and decomposition of Ca(OH)$_2$ and CaCO$_3$ to CaO occurred. This result was also confirmed by XRD result where lime was obtained from the synthesis.

(a)   (b)

Figure 4. Scanning electron microscopy images of (a) Ca(OH)$_2$ gel, and (b) synthesized CaO nanoparticle.

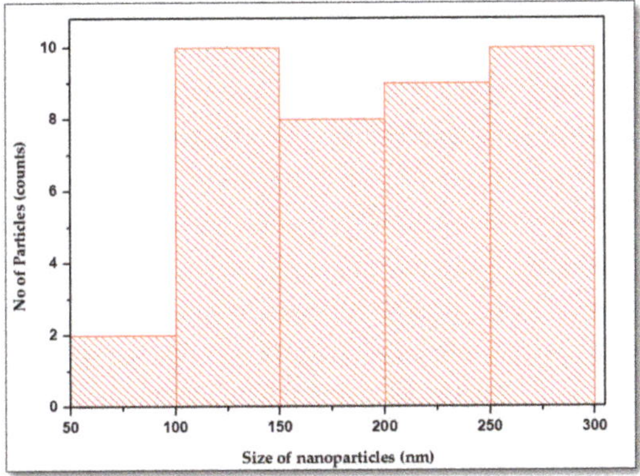

Figure 5. Particle size distribution of nano-calcium oxide.

Table 2 briefly summarizes comparison between current and other method of calcium oxide nanoparticle synthesis. Although particle size obtained in this technique is higher than other methods, it has advantages over the other methods. It is very simple, cheap, does not require expensive equipment, takes only a short time and has no additives. Moreover, in the synthesis process, ambient temperature was used, contributing to less energy consumption. Other methods used higher temperature, polymer additives, expensive equipment and took a longer time.

Table 2. Brief summary of CaO nanoparticles synthesis methods.

Method	Summary	Reference
Facial calcination	A total of 50 nm of calcium oxide nano-particles were obtained by facile thermal treating of calcite at 900 °C temperature for 5 h and then by lime hydrolysis.	Malekzadeh et al., 2012 [21]
Microwave irradiation	By microwave processing, the calcium oxide nano-particles were obtained at 160 °C temperature in the 5 min with the average particle size is 14~24 nm.	Jayanta Bhattacharya et al., 2013 [20]
Co-Precipitation	CaO nano-particles are synthesized by the co precipitation method, in presence of polyvinylpyrrolidone (PVP) reagent for control the agglomeration. The average time takes for this synthesis was 12 h at 40 °C temperature. The average size of the nano-particles are 100 nm.	Meysam Sadeghi et al., 2013 [22]
Direct thermal decomposition	Calcium oxide nano-particles were synthesized by direct thermal decomposition method at 80 °C by blowing inert argon gas with the average particle size is 91 nm~94 nm.	Fereshteh Bakhtiari et al., 2014 [23]
Chemical co-precipitation	Chemical co-precipitation was applied for the synthesis of calcium oxide nano-particles in presence of polyvinyl alcohol. The average particle size is 11 nm at 80 °C for 60 min.	Ali et al., 2015 [24]
Two step process (Green synthesis)	CaO nano-particles were synthesized from shrimp cells by two step process with the average particle size is 40 to 130 nm.	Hui-Fen Wu et al., 2015 [25]
Two step thermal decomposition	Crystallite size of calcium oxide nano-particles were obtained by 2 step thermal decomposition method.	Arul et al., 2018 [26]
Sol-gel method (present authors used)	A total of 50–198 nm of calcium oxide nanoparticles was obtained at ambient temperature, with less cost, no additives, shorter time and a calcination temperature of 900 °C for 1 h only.	Ahn et al., 2019

4. Conclusions

In this paper, calcium oxide nanoparticles were synthesized from waste eggshell through the sol-gel method. Sol-gel technique has many key advantages over other methods for synthesizing metal oxide nanoparticles, such as being simple, economic, requiring no expensive equipment, ambient temperature and having no pressure. The FTIR and XRD results have clearly depicted the synthesis of calcium oxide from eggshell, which was mainly composed of calcium carbonate. The FE-SEM images of Calcium Oxide nanoparticles showed that the particles were almost spherical in morphology. The mean particle size of the nanoparticles was 198 nm. But this result can be improved if higher temperature is used. The synthesized nanoparticle can be applied for future studies in heavy metal removal from industrial wastewater. Moreover, utilizing waste materials as a precursor for the synthesis makes the whole process cheaper, green and sustainable. Waste eggshell can further be used in future work for the synthesis of nano-calcium carbonate, which can be applied as a filler material in the automobile and paper industry.

Author Contributions: L.H., N.S., D.M., conducted the experiments and wrote the manuscript. T.T., R.C. collected the information, analyzed the data and A.J.W. corrected the final manuscript and agreed to submit this data to the sustainability journal.

Funding: This research was funded by Ministry of Science and ICT(MSIT), the Ministry of Environment (ME) and the Ministry of Trade, Industry, and Energy (MOTIE) and the Grant number [2017M3D8A2084752] and the APC National Strategic Project-Carbon Mineralization Flagship Center of the National Research Foundation of Korea (NRF).

Acknowledgments: This research was supported by the National Strategic Project-Carbon Mineralization Flagship Center of the National Research Foundation of Korea (NRF) funded by the Ministry of Science and ICT(MSIT), the Ministry of Environment (ME) and the Ministry of Trade, Industry, and Energy (MOTIE). (2017M3D8A2084752).

References

1. Ejaz, N.; Akhtar, N.; Nisar, H.; Ali Naeem, U. Environmental impacts of improper solid waste management in developing countries: A case study of Rawalpindi city. *WIT Trans. Ecol. Environ.* **2010**, *142*, 379–387. [CrossRef]
2. Mavropoulos, A.; Wilson, D.; Cooper, J.; Costas, V.; Appelqvist, B. *Globalization and Waste Management*; International Solid Waste association (ISWA): Vienna, Austria, 2012; pp. 1–55.
3. Akbar, A.; Hamideh, F. Application of eggshell wastes as valuable and utilizable products: A review. *Res. Agric. Eng.* **2018**, *64*, 104–114. [CrossRef]
4. Jose Valente, F.C.; Roberto, L.R.; Jovita, M.B.; Rosa María, G.C.; Antonio, A.P.; Gladis Judith, L.D. Sorption mechanism of Cd(II) from water solution onto chicken eggshell. *App. Sur. Sci.* **2013**, *276*, 682–690. [CrossRef]
5. Duncan, C.; Allison, R. Sustainable bio-inspired limestone eggshell powder for potential industrialized applications. *ACS Sustain. Chem. Eng.* **2015**, *3*, 941–949. [CrossRef]
6. Gerko, O. Metal oxide nanoparticles: Synthesis, characterization, and application. *J. Sol. Gel Sci. Technol.* **2006**, *37*, 161–164. [CrossRef]
7. Ming, H.; Shujuan, Z.; Bingcai, P.; Weiming, Z.; Lu, L.; Quanxing, Z. Heavy metal removal from water/wastewater by nanosized metal oxides: A review. *J. Hazard. Mater.* **2012**, *211–212*, 317–331. [CrossRef]
8. Lin, X.; Burns, R.C.; Lawrance, G.A. Heavy metals in wastewater: The effect of electrolyte composition on the precipitation of cadmium (ii) using lime and magnesia. *Water Air Soil Pollut.* **2005**, *165*, 131–152. [CrossRef]
9. El-Dafrawy, S.M.; Youssef, H.M.; Toamah, W.O.; El-Defrawy, M.M. Synthesis of nano-CaO particles and its application for the removal of Copper (II), Lead (II), Cadmium (II) and Iron (III) from aqueous solutions. *Egypt. J. Chem.* **2015**, *58*, 579–589. [CrossRef]
10. Oladoja, N.A.; Ololade, I.A.; Olaseni, S.E.; Olatujoye, V.O.; Jegede, O.S.; Agunloye, A.O. Synthesis of nano calcium oxide from a gastropod shell and the performance evaluation for Cr (VI) removal from aqua system. *Ind. Eng. Chem. Res.* **2012**, *51*, 639–648. [CrossRef]
11. Setiawan, B.D.; Oviana, R.; Fadhilah Brilianti, N.; Wasito, H. Nanoporous of waste avian eggshell to reduce heavy metal and acidity in water. *Sus. Chem. Phar.* **2018**, *10*, 163–167. [CrossRef]

12. Wenqiang, L.; An, H.; Qin, C.; Yin, J.; Wang, G.; Feng, B.; Xu, M. Performance enhancement of calcium oxide sorbents for cyclic CO_2 capture-A review. *Energy Fuels* **2012**, *26*, 2751–2767. [CrossRef]

13. Nikulshina, V.; G´alvez, M.E.; Steinfeld, A. Kinetic analysis of the carbonation reactions for the capture of CO_2 from air via the $Ca(OH)_2$–$CaCO_3$–CaO solar thermochemical cycle. *Chem. Eng. J.* **2007**, *129*, 75–83. [CrossRef]

14. Boeya, P.L.; Pragas Maniama, G.; Abd Hamid, S. Performance of calcium oxide as a heterogeneous catalyst in biodiesel production. A review. *Chem. Eng. J.* **2011**, *168*, 15–22. [CrossRef]

15. Bharathiraja, B.; Sutha, M.; Sowndarya, K.; Chandran, M.; Yuvaraj, D.; Praveen Kumar, R. *Calcium Oxide Nanoparticles as an Effective Filtration aid for Purification of Vehicle Gas Exhaust*; Srivastava, D.K., Ed.; Springer: Singapore, 2018.

16. Chelazzi, D.; Poggi, G.; Jaidar, Y.; Toccafondi, N.; Giorgi, R.; Baglioni, P. Hydroxide nanoparticles for cultural heritage: Consolidation and protection of wall paintings and carbonate materials. *J. Colloid Interface Sci.* **2013**, *392*, 42–49. [CrossRef] [PubMed]

17. Alavi, M.A.; Morsali, A. Ultrasonic-assisted synthesis of $Ca(OH)_2$ and CaO nanostructures. *J. Exp. Nano Sci.* **2010**, *5*, 93–105. [CrossRef]

18. Liu, T.; Zhu, Y.; Zhang, X.; Zhang, T.; Zhang, T.; Li, X. Synthesis and characterization of calcium hydroxide nanoparticles by hydrogen plasma-metal reaction method. *Mater. Lett.* **2010**, *64*, 2575–2577. [CrossRef]

19. Darroudia, M.; Bagherpour, M.; Ali Hosseinie, H.; Ebrahimic, M. Biopolymer-assisted green synthesis and characterization of calcium hydroxide nanoparticles. *Ceram. Int.* **2016**, *42*, 3816–3819. [CrossRef]

20. Roy, A.; Bhattacharya, J. Microwave assisted synthesis of CaO nanoparticles and use in waste water treatment. *Nano Technol.* **2011**, *3*, 565–568.

21. Roy, A.; Gauri, S.S.; Bhattacharya, M.; Bhattacharya, J. Antimicrobial Activity of CaO Nanoparticles. *J. Biomed. Nanotechnol.* **2013**, *9*, 1–8. [CrossRef]

22. Ghiasi, M.; Malekzadeh, A. Synthesis of $CaCO_3$ nanoparticles via citrate method and sequential preparation of CaO and $Ca(OH)_2$ nano particles. *Cryst. Res. Technol.* **2012**, *47*, 471–478. [CrossRef]

23. Sadeghi, M.; Husseini, M.H. A Novel Method for the Synthesis of CaO Nanoparticle for the Decomposition of Sulfurous Pollutant. *J. Appl. Chem. Res.* **2013**, *7*, 39–49.

24. Mirghiasi, Z.; Bakhtiari, F.; Darezereshki, E.; Esmaeilzadeh, E. Preparation and characterization of CaO nanoparticles from $Ca(OH)_2$ by direct thermal decomposition method. *J. Ind. Eng. Chem.* **2014**, *20*, 113–117. [CrossRef]

25. Butt, A.R.; Ejaz, S.J.; Baron, C.; Ikram, M.; Ali, S. CaO nanoparticles as a potential drug delivery agent for biomedical applications. *Dig. J. Nanomater. Biostruct.* **2015**, *10*, 799–809.

26. Gedda, G.; Pandey, S.; Lina, Y.C.; Wu, H.F. Antibacterial effect of calcium oxide nano-plates fabricated from shrimp shells. *Green Chem.* **2015**, *17*, 3276–3280. [CrossRef]

27. Arul, E.; Raja, K.; Krishnan, S.; Sivaji, K.; Jerome, D.S. Bio-Directed synthesis of calcium oxide (Cao) nanoparticles extracted from limestone using honey. *J. Nanosci. Nanotechnol.* **2018**, *18*, 5790–5793. [CrossRef] [PubMed]

28. Heung Jai, P.; Seong Wook, J.; Jae Kyu, Y.; Boo Gil, K.; Seung Mok, L. Removal of heavy metals using waste eggshell. *J. Environ. Sci.* **2007**, *19*, 1436–1441. [CrossRef]

29. Ferraz, E.; Gamelas, J.A.F.; Coroado, J.; Monteiro, C.; Rocha, F. Eggshell waste to produce building lime: Calcium oxide reactivity, industrial, environmental and economic implications. *Mater. Struct.* **2018**, *51*, 1–14. [CrossRef]

30. Ferraz, E.; Gamelas, J.A.F.; Coroado, J.; Monteiro, C.; Rocha, F. Recycling waste seashells to produce calcitic lime: Characterization and wet slaking reactivity. *Waste Biomass Valoriz.* **2018**, 1–18. [CrossRef]

31. Tizo, M.S.; Blanco, L.A.V.; Cagas AC, Q.; Dela Cruz BR, B.; Encoy, J.C.; Gunting, J.V.; Arazo, R.O.; Mabayo, V.I.F. Efficiency of calcium carbonate from eggshells as an adsorbent for cadmium removal in aqueous solution. *Sustain. Environ. Res.* **2018**, *28*, 326–332. [CrossRef]

32. Darezereshki, E. Synthesis of maghemetute nanoparticles by wet chemical method at room temperature. *Mater. Lett.* **2010**, *64*, 1471–1472. [CrossRef]

33. Aramend´ıa, M.; Borau, V.; Jiménez, C.; Marinas, J.M.; Ruiz, J.R.; Urbano, F.J. Influence of the preparation method on the structural and surface properties of various magnesium oxides and their catalytic activity in the Meerwein–Ponndorf–Verley reaction. *Appl. Catal. A Gen.* **2003**, *244*, 207–215. [CrossRef]
34. Khan, M.D.; Ahn, J.W.; Nam, G. Environmental benign synthesis, characterization and mechanism studies of green calcium hydroxide nano-plates derived from waste oyster shells. *J. Environ. Manag.* **2018**, *223*, 947–951. [CrossRef] [PubMed]

Article

Experimental Study on Hydromechanical Behavior of an Artificial Rock Joint with Controlled Roughness

Seungbeom Choi, Byungkyu Jeon, Sudeuk Lee and Seokwon Jeon *

Department of Energy Systems Engineering, Seoul National University, Gwanak-gu, Seoul 08826, Korea;
chbum092@snu.ac.kr (S.C.); snujbk1@snu.ac.kr (B.J.); ics1961@snu.ac.kr (S.L.)
* Correspondence: sjeon@snu.ac.kr; Tel.: +82-2-880-8807

Received: 22 January 2019; Accepted: 13 February 2019; Published: 15 February 2019

Abstract: Rock mass contains various discontinuities, such as faults, joints, and bedding planes. Among them, a joint is one of the most frequently encountered discontinuities in rock engineering applications. Generally, a joint exerts great influence on the mechanical and hydraulic behavior of rock mass, since it acts as a weak plane and as a fluid path in the rock mass. Therefore, an accurate understanding on joint characteristics is important in many projects. In-situ tests on joints are sometimes consumptive in terms of time and expenses so that the features are investigated by laboratory tests, providing fundamental properties for rock mass analyses. Although the behavior of a joint is affected by both mechanical and geometric conditions, the latter are often limited, since quantitative control on the conditions is quite complicated. In this study, artificial rock joints with various geometric conditions, i.e., joint roughness, were prepared in a quantitative manner and the hydromechanical characteristics were investigated by several laboratory experiments. Based on the results, a prediction model for hydraulic aperture was proposed in the form of $(e_h/e_m)^3 = \exp(-0.0462c) \times (0.8864)^{JRC}$, which was a function of the mechanical aperture, joint roughness, and contact area. Relatively good agreement between the experimental results and predicted value indicated that the model is capable of estimating the hydraulic aperture properly.

Keywords: artificial rock joint; roughness control; hydromechanical characteristics; prediction model; hydraulic aperture

---

1. Introduction

Rock mass contains various types of discontinuities in terms of size and shape, such as faults, joints, and bedding planes. Among them, a joint is a planar discontinuity that has little tensile strength showing smaller size, but more frequency than a fault. Generally, a joint exerts great influence on the mechanical behavior of rock mass, since it acts as a weak plane. At the same time, a joint has several orders higher hydraulic conductivity than rock matrix so that the majority of fluid flow in rock mass takes place through the joint [1–3]. Therefore, it is necessary to take hydromechanical characteristics of joint into account in many rock engineering applications.

For instance, in many rock engineering applications, including radioactive waste disposal [4,5], geothermal energy extraction [5,6], geological CO_2 sequestration, and deep underground tunneling, it is essential to accurately understand the hydromechanical behavior of rock joints so as to secure the stability and the efficiency of the projects. In-situ experiments can measure hydromechanical characteristics of rock mass, which consider the effects of multiple joints or joint sets. However, they are sometimes consumptive in terms of time and expenses so that the features are investigated by laboratory experiments on a joint, providing fundamental properties for rock mass analyses.

In many cases, joints are originated and gone through several geological processes under various geological conditions. A fracture void, which is a space between corresponding joint surfaces, is characterized by several joint properties, such as roughness, aperture, stiffness, contact area, etc. [7,8]. They exert complex influence on the hydromechanical characteristics of the joint, interacting with each other. Mechanical aperture is the distance between two corresponding joint surfaces in the direction perpendicular to the reference plane of the joint [7], while hydraulic aperture is calculated value based on the results of flow test. The aperture acts as a flow path in a joint, thus, several researches investigated relationships between the aperture and other joint properties [9–11]. Roughness has drawn attentions for the relationship, since it is a widely used engineering parameter and relatively easy to be measured. The roughness is usually defined by its height distribution or shape of which the quantification can be made in several different roughness parameters.

An artificial joint specimen is often used to conduct the laboratory experiments. The joint specimen is bisected mechanically by using a splitter or V block, which is similar to Brazilian jaws [3,12,13]. Although it is a simple method to create the joint specimens, it requires excessive efforts if a specific roughness is required. Therefore, some experimental works are derived from limited test conditions in terms of the joint roughness, which are mainly due to the difficulties in specimen preparation [12–16]. Reproducibility of this approach is hardly guaranteed. Therefore, casting method, which usually uses cement mortar or epoxy, is adopted as an alternative to make multiple joint specimens having similar value of roughness [17,18]. In many cases, however, the casting method also requires some joints at the early stage, since it needs original joint surfaces to make molding frames. Though it can remedy the reproducibility problem to some extent, quantitative control on the geometric property of joint is still incomplete.

In this study, a series of laboratory experiments were carried out to investigate the hydromechanical characteristics of an artificial rock joint under various experimental conditions. In order to control the joint roughness in a quantitative manner, a fractal theory, specifically the random midpoint displacement method, was adopted to mathematically define the joint surfaces. By manipulating input parameters of the method, point-cloud type data of the joint surfaces were produced with various levels of roughness. Then, the data, which contained coordinates of surface points, were imported to a 3D printer. Thanks for the recent development in 3D printing technology, laborious works in specimen preparation can be avoided to some degree. Throughout this paper, a whole experiment scheme could be classified into three parts. First, geometric properties of joint, i.e., roughness, aperture, and fracture void, were investigated. Surface information of joint was measured by a 3D laser scanner and utilized in the following experimental steps. Second, mechanical characteristics of the joint were measured by normal compression tests. Under the applied normal load, deformation behavior and corresponding variation of fracture void could be investigated. Last, flow tests were conducted under normal compression. With an assumption of cubic law, the variation of hydraulic (or equivalent) aperture under various geometric and mechanical conditions was figured out. Sufficient experimental condition was applied in terms of stress and roughness so as to obtain wide applicability and representativeness from the experiments.

Hydraulic conductivity of a joint is an important property in rock mass analyses and can be expressed in terms of hydraulic aperture. Therefore, several prediction models were proposed for estimating the hydraulic aperture [19–21]. They have their own strength and shortcomings, being bound in some cases by their own definitions. In this study, a prediction model was proposed based on the experimental results. The model was derived by multivariable, nonlinear regression analyses and was established as a function of mechanical aperture, roughness, and contact area. In order to confirm the validity and applicability of the model, comparison with the referenced models was made. It was found that the model showed acceptable predictability in engineering viewpoint.

2. Experimental Setup

Several methodologies were suggested to mathematically define an artificial joint surface. Fourier transformation [22], stochastic function, such as autocorrelation function [23], and fractal Brownian function [24] can be adopted in this procedure. In 1993, Xie [24] stated that rock joints could be successfully simulated by the fractal theory in terms of roughness, aperture, pore structure, permeability, etc. Natural rock joints usually show self-affinity, which is the feature that considers different amount of scaling in transformation. The fractal Brownian functions could take the feature into account, thus, it can be applied in estimating and/or generating models of artificial joint surfaces. In this study, random midpoint displacement method was adopted to generate synthetic surfaces. It has several benefits in application. First, it is quite straightforward, and its outcomes can be directly imported to a 3D printer. Also, the manipulation of input parameters can derive satisfactory results in controlling joint roughness [25]. Figure 1 shows an example of the printed joint specimens following the method. The uppermost pictures in Figure 1 show two corresponding joint surfaces (upper and lower), which were printed by a 3D printer, and the middle pictures show joint-embedding specimen, piling up those two surfaces to be mated. The last pictures show 3D numerical models, which had the same geometry with the printed specimens.

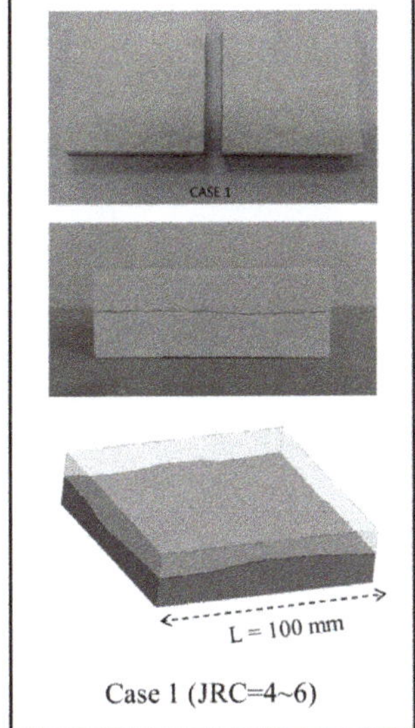

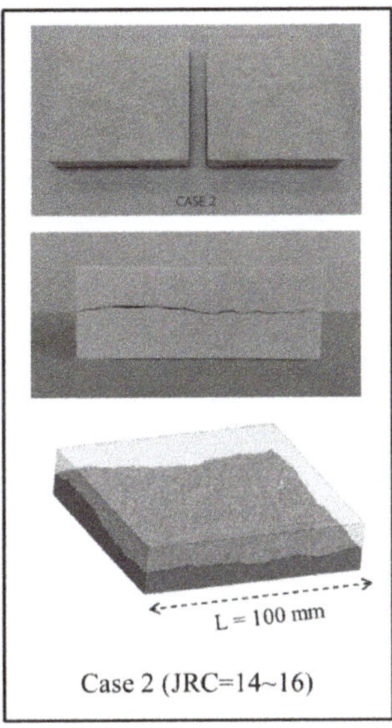

Figure 1. Printed joint specimens following the random midpoint displacement method and their 3D models (after Reference [25]).

In 2019, Choi et al. reported that the method could give satisfactory results in controlling the joint roughness, which was expressed in *JRC* (Joint Roughness Coefficient). In 1977, Barton and Choubey [26] proposed the roughness parameter by back-calculating numerous direct shear test results. Although *JRC* is not a quantitative parameter, it is widely used in engineering projects mainly due to its simplicity and practicability. Therefore, it was selected as a roughness estimator in this study. Figure 1 is an example to show how well the method controls the joint roughness. Actual laboratory experiments were conducted using cylindrical specimens, not rectangular parallelepiped as shown in Figure 1. The procedure for specimen preparation is described below.

The 3D printer used in this study was an 'additive manufacturing' type and, more precisely speaking, a binding jetting printer. It spreads a liquid state additive onto powder layer to embody predesigned objects. It is capable of making complex shape and colored objects, however, mechanical and physical properties of the printed outcome are highly dependent on printing ingredients. Unfortunately, the printer uses calcium sulfate hemihydrate powder as a main printing ingredient, which is inadequate to make a rock-like specimen in terms of strength and stiffness. Therefore, the printer produced prototype joint surfaces first. Then several aluminum casting molds, which had similar degree of roughness, were manufactured based on the prototype surfaces. Figure 2 shows the whole molding assembly including the aluminum casting mold. Joint specimens were made of cement mortar using the molding assembly, which had a cement/water ratio of 100:20 by weight with the addition of little retarder. After the whole assembly put together, cement mortar was poured into the assembly to make a half of a joint. After ~6 h of curing, the assembly was carefully dismantled. Then, the molds were reassembled incorporating the half of the joint instead of the aluminum mold. Cement mortar was poured again to make another half of the joint. Then, the assembly was taken apart. After carefully detaching the two halves of the joint, the specimens were cured more than three days to guarantee material strength under room temperature and humidity. The specimens had 54 mm of dimeter and a ratio of height to diameter was approximately 2. However, the specimen should contain some margin at the upper and lower parts so that the actual area of joint surface was 75 mm × 54 mm. In order to investigate the effects of roughness, four joint specimens having different roughness were prepared and tested throughout this study.

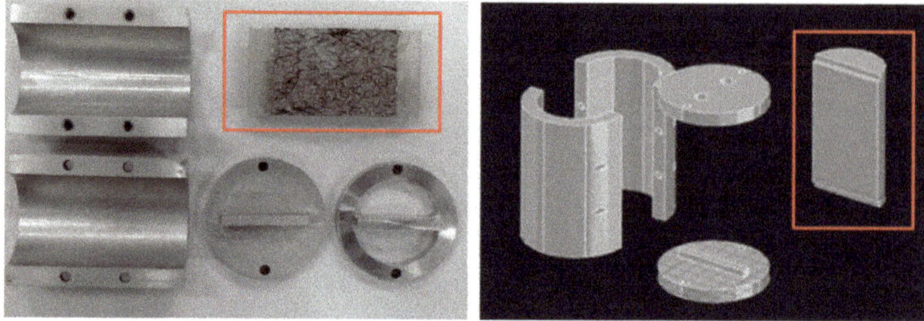

Figure 2. A picture of molding assembly (cylindrical molding frames, end plugs with top inlets, and aluminum casting molds in red box) and its computer-aided design (CAD) drawing.

In order to obtain and confirm geometric information of the joint surface, i.e., roughness, aperture, and shape of fracture void, a 3D laser scanner was used. As a non-contact measuring method, it can provide x, y, and z coordinates of surface points without any damage. Thus, many researches adopt this technique [3,12]. As mentioned above, the original surfaces were generated by the the random midpoint displacement method, however, some inevitable degradation of micro asperity occurred when manufacturing the aluminum casting mold. Therefore, there existed discrepancies between the generated surface and the cement mortar joint surface. Hence, the joint specimens were scanned again to determine the geometric properties. Joint surface of 75 mm x 54 mm was scanned and digitized with scanning interval of 0.25 mm.

Figure 3 shows the experimental setup. MTS 816 system (universal rock testing system) was used in both mechanical and hydraulic experiments. It is composed of load frame (main frame), triaxial compression unit, pore pressure intensifier, and additional accessories (Figure 3a). An actuator, which locates at the top of main frame, typically applies and controls the axial displacement or stress during experiments. In this case, axial stress of 0.1 MPa was applied to fix the specimen assembly and was servo-controlled during the whole test. The triaxial compression unit consists of triaxial compression chamber and confining pressure unit. The triaxial compression chamber, in which the in-vessel test assembly is placed, is filled with refined oil and connected to the confining pressure unit. Since the axial (vertical) stress is constrained by the actuator, the confining pressure unit applies and controls diametrical stress, which acts as normal stress on the joint surface. The pore pressure intensifier applies and controls the hydraulic pressure and it is connected to the in-vessel assembly through a pair of endcaps. Fluid injected by the intensifier (distilled water, in this case) moves into the specimen through the lower endcap and is discharged through the upper endcap. The discharged fluid is weighed by an electronic balance. Figure 3b shows the in-vessel test assembly. A pair of endcaps is located at the top and bottom parts of the joint specimen being connected to the pore pressure intensifier. Guide rods and plate are installed to prevent tilting or bending of the assembly with sufficient rigidity.

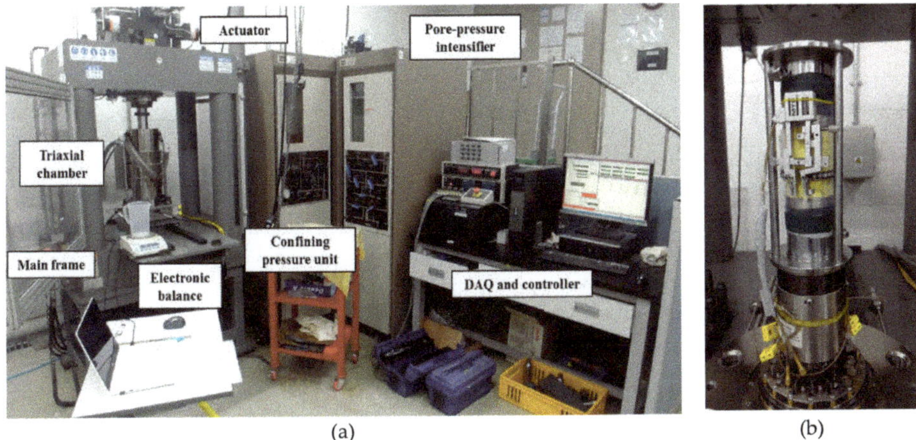

Figure 3. Experimental setup; (**a**) overall experimental setup and (**b**) in-vessel test assembly including joint specimen with the endcaps.

In normal compression tests, typical setting of uniaxial compression test was applied [27]. However, cylindrical joint specimen made by the molding assembly in Figure 2 is not suitable for normal compression, since it requires a special loading frame that is similar with Brazilian jaws, but has the exact same curvature with the specimen and may be affected by specimen arrangement. Therefore, rectangular parallelepiped specimens with the same joint characteristics and dimensions were prepared for this test. Normal stress was applied up to 10 MPa at the constant displacement loading rate of 0.005 mm/s. The stress level corresponded to the vertical in-situ stress at the depth of approximately 400 m. In addition, two loading-unloading cycles were adopted to investigate the effect of stress hysteresis. After reaching normal stress of 10 MPa under the constant loading rate, the stress was removed then the second loading cycle began. Meanwhile, a flow test was conducted in a triaxial compression chamber. In the test, a cylindrical specimen with the attachment of two endcaps was placed inside of the chamber. Normal stress on the joint surface was applied and controlled by the confining pressure unit. Applied normal stress varied from 1 MPa to 10 MPa, with the interval of 1 MPa. The pore pressure intensifier-controlled water pressure at 0.3, 0.6, and 1.0 MPa. After the target normal and water pressure were applied, fluid injection was continued until it reached constant flowrate. Two loading cycles were applied in the flow tests as well.

3. Experimental Results

3.1. Geometric Properties of Joint Surface

Artificial joint specimens with four different levels of roughness were prepared. The specimens hereinafter are referred as J1, J2, J3, and J4, respectively. They were scanned by a high-resolution laser scanner and provided with the coordinates of surface points, allowing quantitative estimation on the roughness and mechanical aperture to be made. Various studies proposed quantitative roughness parameters and their relationships with JRC. For instance, in 1979, Tse and Cruden [28] suggested roughness parameter Z_2 (Equation (1)) and the relationship with JRC (Equation (2)).

$$Z_2 = \sqrt{\frac{1}{L}\int_0^L \left(\frac{dy}{dx}\right)^2 dx} = \sqrt{\frac{1}{L}\sum_{i=1}^{n-1} \frac{(y_{i+1} - y_i)^2}{x_{i+1} - x_i}} \tag{1}$$

$$JRC = 32.2 + 32.47 \log Z_2 \tag{2}$$

where L is the length of a joint profile, dx and dy are the increments of x and y of the profile, n is the number of segments, and x_i and y_i are the x and y coordinates of i th segment, respectively.

Several quantitative parameters, including Z_2, are two-dimensional in their nature. Obviously, representativeness of a single profile is insufficient for a whole joint surface, since the joint itself is a three-dimensional structure. A few three-dimensional parameters were suggested for better representativeness and taking roughness anisotropy into consideration [29,30]. However, their practicality could not have exceeded those of two-dimensional parameters.

In this study, a compromise method was adopted in determining the joint roughness. Figure 4 shows a schematic diagram for the determination. Since all coordinates of surface points were already known, each and every profile along a certain direction could be obtained. In this study, the roughness along the long axis of a joint (x direction in Figure 4) was calculated. A joint profile would be defined by xz plane. Then, moving the xz plane along the y direction would define several profiles. Applying Equations (1) and (2) on the profiles in sequence would give a distribution of roughness so that the applicability of JRC and three-dimensional feature could be enhanced. Average of the distribution was selected as a typical value.

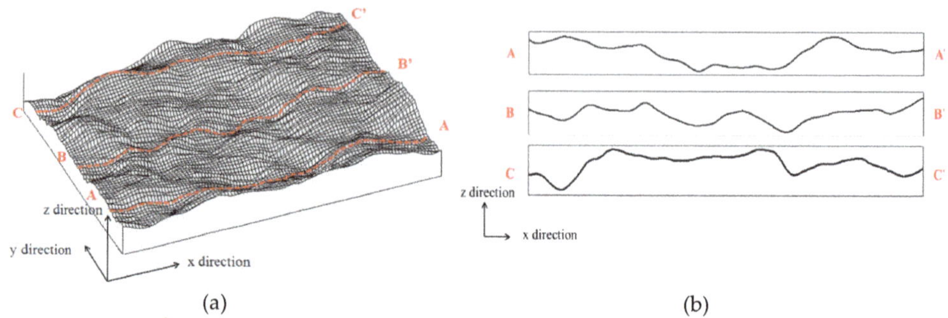

Figure 4. Schematic diagram for determining joint roughness; (**a**) three-dimensional plot of a joint and (**b**) x-z profiles for three y-values as shown in (**a**) (after Reference [25]).

The scanned coordinates could be utilized in determining the mechanical aperture as well. Several methodologies have been proposed to measure the mechanical aperture of a joint. Among them, superposition method was applied in this study. After arranging the coordinates of the upper and lower surfaces vertically, the upper surface was moved in parallel downward step-by-step until it met a certain criterion of aperture determination. When the criterion was met, the distances of all corresponding points were calculated to establish its distribution. The criterion was defined when 1% of the total surface points were in contact. It was referenced from the experimental results that the area contacted by the dead weight of the joint was approximately 1% of total surface [31]. Since only gravitational force exists, it is called initial mechanical aperture (e_{m0}). When superposing, the upper surface was translated downward by 0.001 mm at a step. Then, the distribution of aperture was calculated and all apertures in contact were set to null. Average of the distribution was selected as a typical value as well.

Five sets of specimens (a set consists of J1, J2, J3, and J4) were prepared and analyzed. Table 1 shows weighted statistics of all sets in terms of JRC and initial mechanical aperture. CV in Table 1 denotes coefficient of variation. Within applicable JRC range, which is from 0 to 20, the roughness distributed relatively equally. And judging from the CV values in the roughness, the proposed method on joint specimen preparation could give consistent results regarding joint roughness. Measured initial mechanical aperture varied in the range of 0.17~0.82 mm. CV values in the aperture were larger than those of roughness. It was attributed to the effect of micro asperity damage during molding process.

As *JRC* increased, the initial mechanical aperture increased as well so that a positive relationship could be found between them.

Table 1. Statistics of joint roughness and initial mechanical aperture.

Specimen	Roughness (*JRC*)			Initial Mechanical Aperture (mm)		
	Average	S.D.	CV (%)	Average	S.D.	CV (%)
J1	1.12	0.12	10.71	0.17	0.03	17.65
J2	7.32	0.59	8.06	0.39	0.09	23.08
J3	13.07	1.38	10.56	0.55	0.13	23.64
J4	16.47	1.46	8.86	0.82	0.17	20.73

Figure 5 shows the distribution of *JRC* and initial mechanical aperture as an example excerpted from J4. Curves in Figure 5 indicate fit normal distribution based on the results. Figure 5a shows that histogram of joint roughness followed normal distribution relatively well. However, some values were outside the applicable range of *JRC* (larger than upper limit of 20). It was because the roughness was calculated by Z_2 (Equation (1)) first, which did not have numeric limitation on the contrary. The normal distribution on joint roughness could be explained by the surface generation algorithm, where Gaussian random number is involved. Random midpoint displacement method calculates elevation of asperity following the Gaussian distribution so that the roughness may show some normality in nature.

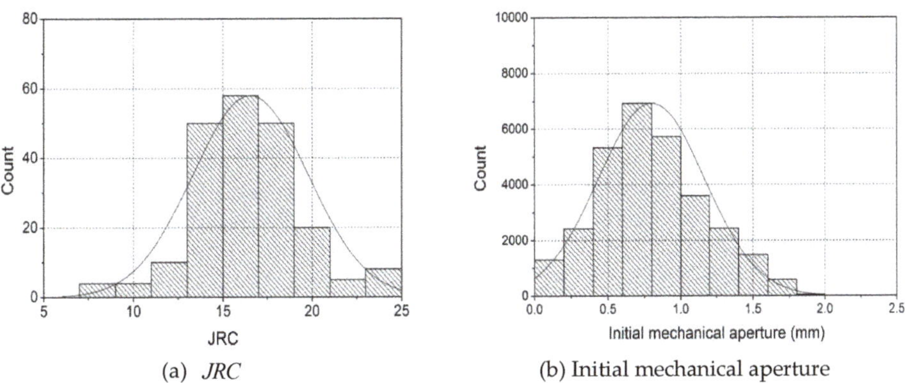

(a) *JRC*

(b) Initial mechanical aperture

Figure 5. Histogram of geometric properties; (**a**) *JRC* and (**b**) initial mechanical aperture (results of J4 specimen as an example).

Distribution of initial mechanical aperture (Figure 5b) could be presented by normal distribution as well. Aperture of natural rock joint often follows a normal or log-normal distribution [4,32,33] and the same tendency could be found in this study. Since the specimens were made of cement mortar, there must be some discrepancies from real rock joints in terms of strength and stiffness. However, from a geometric viewpoint, it could be concluded that the proposed method for generating joint specimen was capable of simulating surface geometry of the rock joint.

3.2. Mechanical Characteristics

The specimens were made of cement mortar. In order to obtain fundamental properties of the testing material, uniaxial compression tests were conducted using 10 cylindrical intact specimens. As a result, uniaxial compressive strength and Young's modulus were calculated as 48.15 MPa and 18.25 GPa on average, respectively. Therefore, the material was comparable with medium to weak strength rocks. Also, it was assumed that the material was more similar to sedimentary rocks than crystalline because of its texture of cemented grains.

In 1983, Bandis et al. [9] extensively explained behavior of joint under various stress states. Displacement of joint (or joint closure) due to normal stress can be written in Equation (3).

$$\Delta\delta_j = \Delta\delta_{net} - \Delta\delta_i \tag{3}$$

where $\Delta\delta_j$ is the deformation of joint or joint closure, $\Delta\delta_{net}$ is the net deformation, and $\Delta\delta_i$ is the deformation of intact matrix. Therefore, by subtracting the matrix component of deformation from net deformation, the joint closure can be calculated. The joint closure due to normal stress can be modelled by several equations. Goodman (1976) [34] and Bandis et al. (1983) [9] suggested hyperbolic models, while Brown and Scholz (1986) [35] proposed a logarithm model. Normal compression results of this study were applied to those three models and good agreements were achieved using each of them. Figure 6 shows an example of application to Bandis model, which is presented in Equation (4), showing good agreement especially for normal stress larger than 3 MPa.

$$\sigma'_n = \frac{\Delta\delta_j}{a - b \cdot \Delta\delta_j} \tag{4}$$

where a and b are fitting constants.

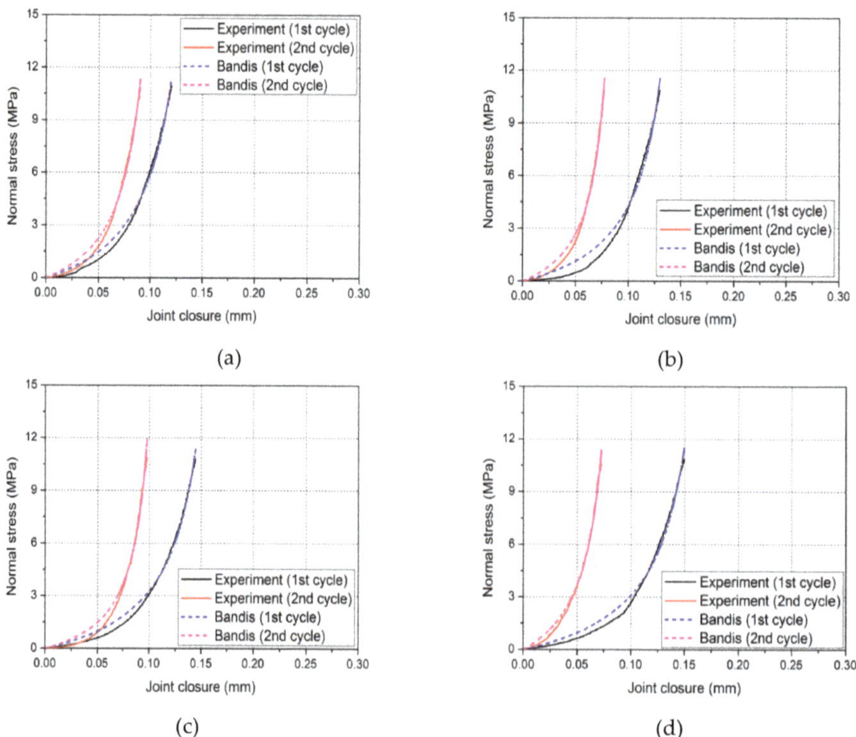

Figure 6. Application of normal compression results to model of Reference [9]; (**a**) J1 specimen, (**b**) J2 specimen, (**c**) J3 specimen, and (**d**) J4 specimen.

Overall trends were similar regardless of the roughness. Remarkable nonlinear behavior was observed at the early loading stage in both loading cycles. At the early loading region, the normal displacement greatly increased with the normal stress, then, gradually increased in a linear tendency. At the last loading stage, which corresponded with some part where the applied normal stress was close to the maximum 10 MPa, deviation from the linear tendency was not observed since the maximum applied stress (10 MPa) was not large enough compared with the material strength (48.15 MPa). In the second loading cycle, nonlinear deformation at the early loading stage was smaller than the first. It was the effect of stress hysteresis, specifically the effect of permanent closure. The magnitude of joint closure increased with the increase of roughness. It was understandable that the initial mechanical aperture and its dispersion increased with the roughness (Table 1) so that the normal displacement increased with the roughness as well because there had been more space to be deformed at the beginning.

Contact area is the area where the surfaces are in contact, transferring stresses [7]. It exerts great influence on hydromechanical behavior of a joint. In order to measure it in experiments, pressure-sensitive paper was used in this study. It is a two-sheet paper, which is composed of two polyester bases. The one is coated with a layer of micro-encapsulated color forming material and the other is coated with a layer of the color developing material. When pressure is applied to a pair of layers, some parts to which the pressure is applied, i.e., contact area, are painted. Contact area under normal compression was measured in experiments. Applied normal stresses were 0.05, 2.5, 5.0, 7.5, and 10.0 MPa, respectively. Since the paper cannot measure the contact area in real-time and cannot recover its state once color is developed, every test should be paused after reaching target stress and new paper inserted. Figure 7 shows an example of J1 specimen. Size of the paper was 75 mm × 54 mm, same as the joint specimen.

First loading cycle

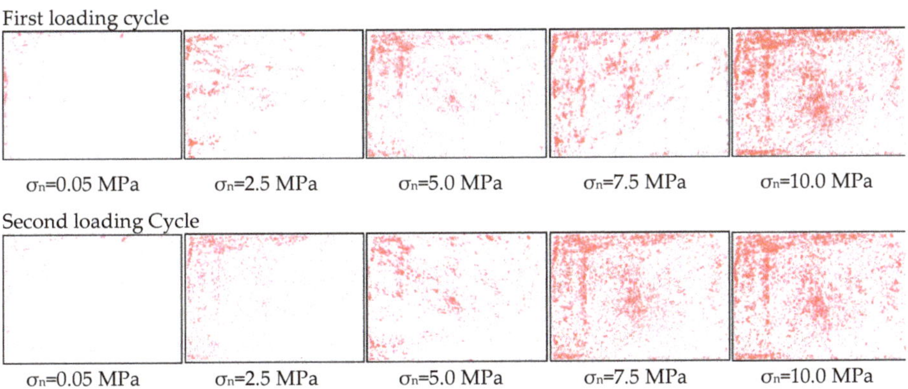

Figure 7. Variation of contact area (marked in red) of J1 specimen at different loading cycles and different normal stresses.

The contact area increased with the increase of normal stress. However, the contact occurred at specific parts even if the stress was increased to 10 MPa so that considerable area remained separated. Overall spatial distribution was similar regardless of loading cycle. However, the contact area in the second cycle was measured slightly larger than the first. In order to analyze it quantitatively, image analyses were conducted. The colored paper was carefully scanned then image files of the scanning were generated. After being imported to analyzing program, the image files were converted into binary images for convenience. Then, the contact area was calculated using the number of total pixels and the number of colored pixels. Results are shown in Figure 8.

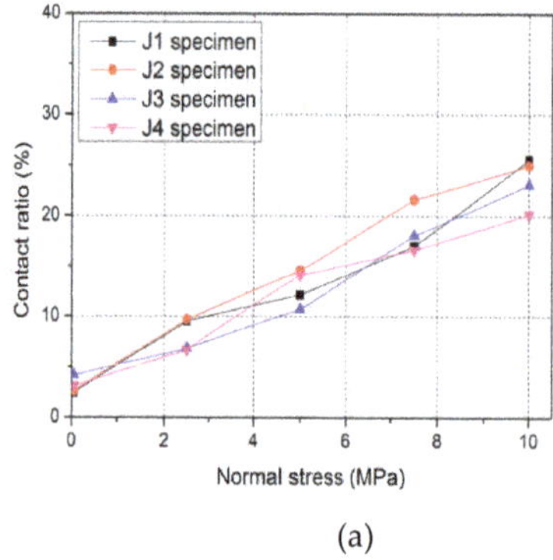
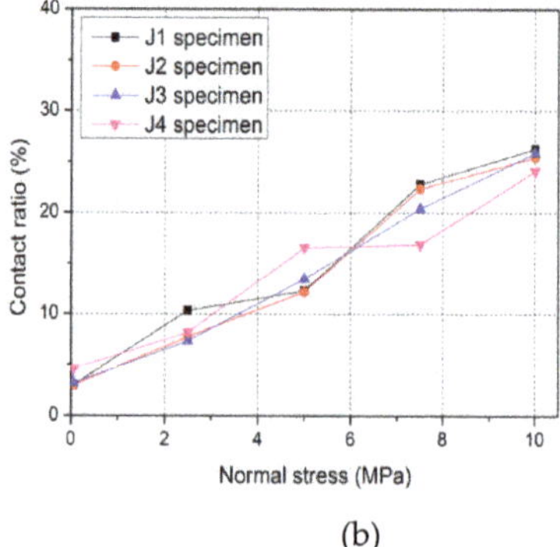

(a)   (b)

Figure 8. Results of image analyses on contact area (percentile of the number of colored pixels over total pixels; (**a**) results of the first loading cycle and (**b**) results of the second loading cycle.

As the normal stress increased, the contact area quantitatively increased as well. Under the normal stress of 10 MPa, the maximum contact area was approximately 25% (second loading cycle of J1 specimen), much more parts remained separated. It was similar to the results of Gale (1987) [36] and Pyrak-Nolte et al. (1987) [37]. In most cases, contact area in the second cycle was larger than the first, which was the effect of permanent closure and asperity deformation. The maximum amount of increase due to stress hysteresis was 7.6% (J3 specimen under the normal stress of 7.5 MPa). Under low level of normal stress (especially under 0.05 MPa), the rougher the surface, the larger the contact area was measured. However, this trend became reversed as the stress increased so that J1 specimen had the largest contact area when the stress was 10 MPa.

Mechanical aperture, as well as fracture void, varied due to applied stress. There are two methods capable of calculating the mechanical aperture under specific normal stress. One method is using results of normal compression tests and the other method is back-calculation based on the contact area. In the former method, Equation (5) can be utilized.

$$\text{First loading cycle } e_m = e_{m0} - \Delta\delta_j$$
$$\text{Second loading cycle } e_m = (e_{m0} - \delta_p) - \Delta\delta_j \tag{5}$$

where e_m is the mechanical aperture, e_{m0} is the initial mechanical aperture, and δ_p is the permanent closure. The initial mechanical aperture was measured earlier by 3D laser scanner and joint closure values were measured by MTS system in normal compression tests so that the mechanical aperture under specific stress could be calculated.

At the same time, back-calculation was possible, since the information about the initial void geometry and the contact area were measured by 3D laser scanning and pressure-sensitive paper, respectively. The back-calculation process was performed as follows. First, scanned coordinates of both upper and lower surfaces were arranged vertically. Then, the upper surface data was translated along downward step-by-step until the percentage of overlapping surface points equaled the measured contact area. When the criterion was met, not only average of the aperture but spatial distribution could be calculated. Figure 9 shows variation of aperture histogram and spatial distribution according to the normal stress, excerpted from J2.

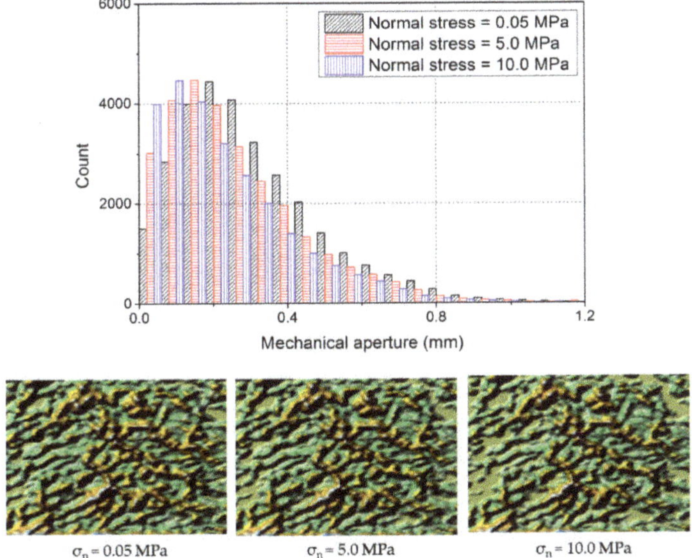

Figure 9. Variation of aperture distribution according to normal stress (results of J2 specimen as an example).

As the normal stress increased, the fracture void was compressed so that the histogram in Figure 9 was shifted toward left. Meanwhile, the spatial distribution in Figure 9 conceptually shows how the aperture distributed in the fracture void and varied by the normal stress. Yellow parts and flat parts in the bottom plot denote the large aperture and contact area, respectively. As the normal stress increased, remarkable change was observed. Flat spaces, which denote the contact area, broadened and void spaces became narrower and smaller, indicating possible channeling effect in flow experiments.

3.3. Hydraulic Characteristics

Testing assembly, which included joint specimen and two endcaps, was placed in the triaxial compression chamber. Triaxial unit controlled the normal stress, which was from 1 to 10 MPa with the interval of 1 MPa, and pore pressure unit controlled the water pressure, which was 0.3, 0.6, and 1.0 MPa. Stress hysteresis was also investigated in flow experiments. When the water or pore pressure acts on a medium, stress components of the medium can be expressed in Equation (6).

$$\sigma' = \sigma - \alpha P_w \tag{6}$$

where σ' is the effective stress, P_w is the water pressure, and α is the effective stress coefficient, sometimes referred as Biot's coefficient. Since the normal stress and water pressure were applied step-by-step, the cross-plot method for determining the effective stress coefficient of joint could be applied [38]. Procedures for applying the cross-plot method are briefly described as follows. At first, permeability, more precisely cube root of the permeability, of a joint is calculated by flow tests and is plotted as a function of water pressure at constant normal stress (Figure 10a). Horizontal lines in Figure 10a denote the cross-plotting and make intersecting points. The points are plotted again (Figure 10b) in normal stress-water pressure coordinates system so that the slope of the line in Figure 10b is the effective stress coefficient.

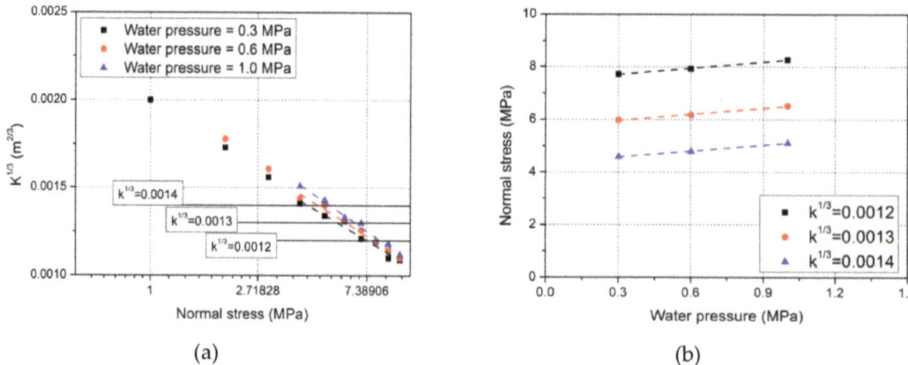

Figure 10. Cross-plot method for determining the effective stress coefficient (results of J3 specimen as an example); (**a**) relationship between normal stress-permeability, and (**b**) relationship between water pressure-normal stress (cross-plot results of (**a**))

As a result, the joint specimens, i.e., J1, J2, J3, and J4, had 0.80, 0.87, 0.75, and 0.82 of the coefficients, respectively. The coefficients were adopted when calculating the effective stress in following process.

Based on the measured flowrate and applied water pressure, flowrate per unit hydraulic head ($Q/\Delta h$) could be calculated. Figure 11 shows the effect of roughness and effective stress on the flowrate. The results were excerpted from the first loading cycle.

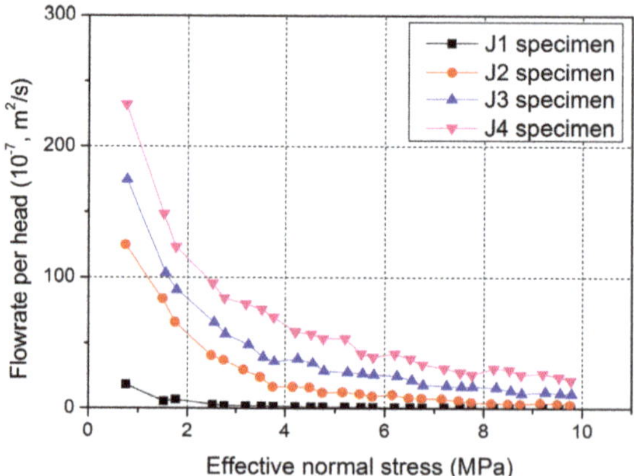

Figure 11. Variation of flowrate per head according to the roughness and effective normal stress (in the first loading cycle).

As shown in Figure 11, an increase in effective normal stress made the aperture narrower so that the flowrate decreased. An abrupt decrease at the early loading stage was observed. When comparing the flowrate under 1 MPa and 10 MPa of applied normal stress, J1 specimen showed 54 times of reduction, while J4 showed 11 times of reduction. The effect of roughness is clearly shown in Figure 11. The rougher the joint, the larger flowrate per head was measured, since the rougher joint always had larger aperture value under the same level of normal stress in this study. The same tendencies were found in the second loading cycle in terms of the effect of roughness and normal stress.

If fluid flow through a joint is steady and isothermal, the flowrate per unit hydraulic head in case of straight flow [20] can be written in Equation (7), which is commonly known as cubic law. The law is derived from smooth, parallel, and open plate model.

$$\frac{Q}{\Delta h} = Ce_h^3 = \left(\frac{W}{L}\right)\left(\frac{\rho g}{12\mu}\right)e_h^3 \tag{7}$$

where e_h is the hydraulic (equivalent) aperture, W and L are the width and length of the joint, g is the gravitational acceleration (9.81 m/s^2), ρ and μ are the density and viscosity of the fluid, respectively. Therefore, the hydraulic aperture could be calculated based on the measured flowrate. The width and length of the joint were 54 mm and 75 mm, respectively, and properties of water at 20 °C were applied (approximately 1000 kg/m^3 and 1.002 $mPa \cdot s$ for density and viscosity, respectively). Since the mechanical aperture under specific normal stress was already measured, a comparison between the mechanical and corresponding hydraulic apertures could be made, shown for J1 specimen in Figure 12.

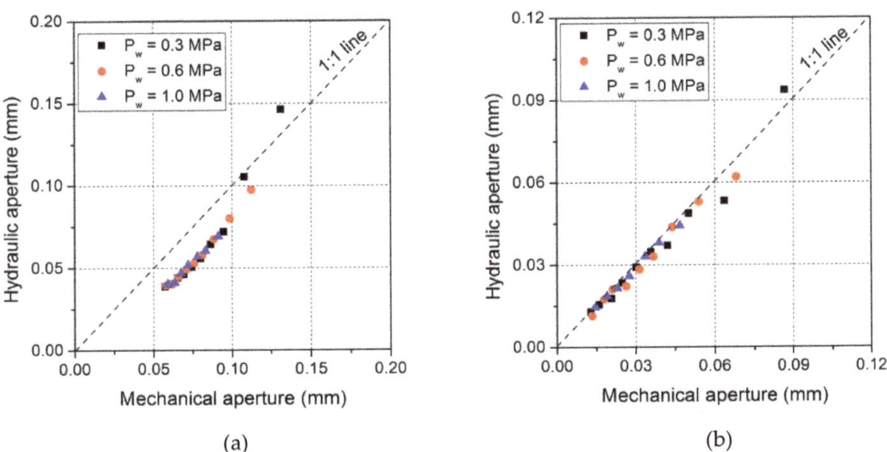

Figure 12. Comparison between mechanical and hydraulic aperture under normal stress; (**a**) results of the first loading cycle and (**b**) results of the second loading cycle (results of J1 specimen as an example).

If the two apertures are same, results lay on 1:1 line in Figure 12. It means the cubic law stands, which also indicates ideal joint conditions in Equation (7), i.e., smooth, parallel, and open condition, hold true. However, since it is not the case in real rock joint, deviations from the ideal case are inevitable. This phenomenon is affected by many factors, such as roughness, contact area, tortuosity, and so on. And the further the joint condition from the ideal one, the more deviation is measured. Even though J1 specimens had quite small roughness, deviation from 1:1 line was clearly observed in Figure 12a. A ratio of hydraulic aperture to mechanical aperture (e_h/e_m) for J1, J2, J3, and J4 specimen was measured as 0.72, 0.46, 0.39, and 0.29, respectively, on average, which meant the mechanical aperture was always larger than the hydraulic aperture. A few points in Figure 12 were located above the 1:1 line, however, it was reasonable to think that they were caused by systematic error during experiments. Therefore, the effect of roughness on the flow behavior, which impeded overall flow, could be investigated qualitatively and many references reported the same tendency [3,11]. When it comes to the stress hysteresis, Figure 12b showed that the results located much closer to 1:1 line. It could be explained that deformation of asperity due to normal stress and permanent closure made more ideal joint condition so that the experimental discrepancies were narrowed.

4. Discussion

Various studies proposed a relationship between mechanical and hydraulic apertures. By linking the apertures into an equation, predictions for hydraulic behavior of joint were conducted. Among numerous relationships, prediction models suggested by Barton et al. (1985) [19], Zimmerman and Badvarsson (1996) [20], and Matsuki et al. (2006) [39] are listed in Table 2.

$$e_h = \frac{JRC^{2.5}}{(e_m/e_h)^2} \tag{8}$$

$$\left(\frac{e_h}{e_m}\right)^3 = \left[1 - 1.5\left(\frac{\sigma_m}{e_m}\right)^2\right](1 - 2c) \tag{9}$$

$$\left(\frac{e_h}{e_m}\right)^3 = \left[1 - \frac{1.13}{1 + 0.191(2e_m/\sigma_{m0})^{1.93}}\right] \tag{10}$$

where σ_m and c in Equation (9) are the standard deviation of mechanical aperture and the contact ratio, respectively, and σ_{m0} in Equation (10) is the standard deviation of initial mechanical aperture. They have their own strength and shortcomings. Equation (8) was a rough empirical model derived from numerous data results. It is widely used, since it has relatively simple form and JRC in the equation. However, in some cases, it loses its applicability or shows estimation error especially when joints are under shear behavior. Equation (9) considers not only roughness effect, the first bracket in Equation (9), but also contact effect. Equation (10) was derived from numerical simulations.

Table 2. Comparison of referenced prediction models.

Authors	Equation
Barton et al. (1985) [19]	(8)
Zimmerman and Bodvarsson (1996) [20]	(9)
Matsuki et al. (2006) [39]	(10)

In this study, a prediction model for e_h/e_m was proposed based on the experimental results described above. It was thought that a prediction model with one parameter was not sufficient to represent the experimental results. Therefore, JRC and contact area (c) were selected as independent variables. All relevant results were arranged and calculated again, and Figure 13 shows a relationship between $(e_h/e_m)^3$ and those two variables.

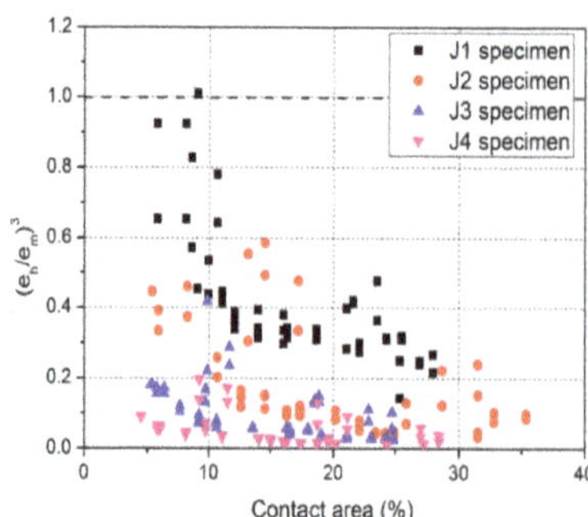
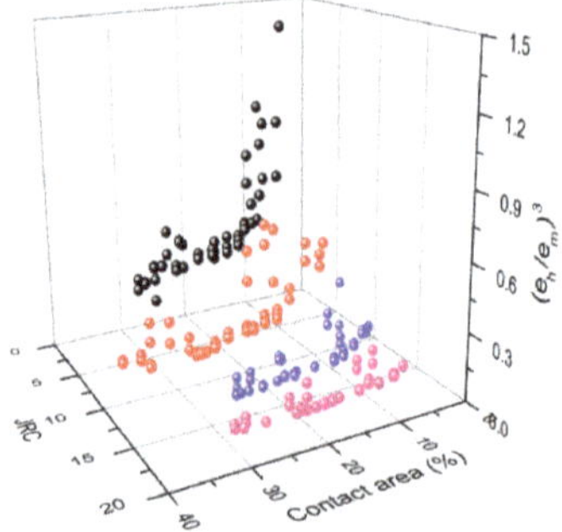

Figure 13. Relationship between $(e_h/e_m)^3$ and two independent variables.

As described in Section 3, $(e_h/e_m)^3$ decreased with the increase of JRC and c, which indicated the effect of roughness and contact area hampered the fluid flow, causing the discrepancy between those two apertures to become larger. Correlation analyses between $(e_h/e_m)^3$ and JRC, c were conducted and the results are shown in Table 3. Bivariate analyses including all three variables were performed and the results showed that both independent variables, i.e., JRC and c, had negative correlations with $(e_h/e_m)^3$. As mentioned before, $(e_h/e_m)^3$ decreased when the independent variables increased. JRC and c were found to be independent of each other, showing weak negative relationships, but statistically meaningless within 5% of significant level.

Table 3. Correlation analyses between variables.

		c	JRC	$(e_h/e_m)^3$
c	Pearson correlation	1		
	Sig. (2-tailed)	-		
JRC	Pearson correlation	−0.096	1	
	Sig. (2-tailed)	0.175	-	
$(e_h/e_m)^3$	Pearson correlation	−0.345 **	−0.693 **	1
	Sig. (2-tailed)	0.000	0.000	-

** Correlation is significant at the 0.01 level (2-tailed).

Based on the experimental results, Equation (11) was proposed. In order to consider the effects of independent variables, negative exponential functions were adopted. At the same time, coefficients of the regression model were determined considering the fact that $(e_h/e_m)^3$ should be in unity when both JRC and c equaled zero, which means flat and open joint so that the ideal cubic law condition is achieved.

$$\left(\frac{e_h}{e_m}\right)^3 = \exp(-0.0462c) \times (0.8864)^{JRC} \tag{11}$$

Figure 14 shows application of the proposed model to experimental results and its variation according to the JRC and c. As it can be seen in Figure 14a, the proposed prediction model agreed relatively with the experimental results. Adjusted r-square of three-dimensional fitting was 0.74, which is thought to be acceptable in engineering viewpoint. At the same time, root mean square error (RMSE) between the measured and predicted values was calculated as 0.1078.

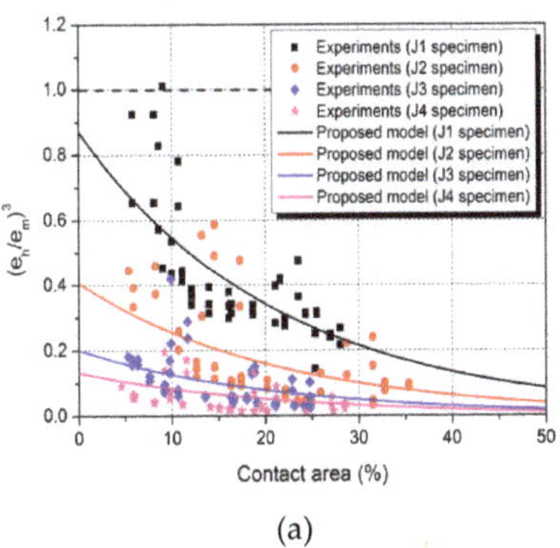

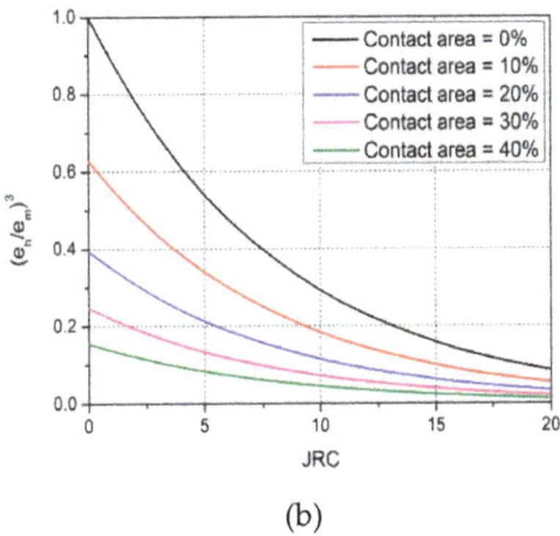

Figure 14. Evaluation of the proposed model; (**a**) application of the proposed model to experimental results and (**b**) variation of the model according to the variables.

Variations of the proposed model were investigated. Figure 14b shows the variation according to JRC with fixed contact area of 0, 10, 20, 30, and 40%. In the case of JRC equaling zero, which means a flat joint, the contact area should be 0 or 100, which corresponds with $(e_h/e_m)^3$ equaling 1 or 0. However, this case could not be modelled in this study. Based on the results of JRC=1, $(e_h/e_m)^3$ decreased to 33% and 10% when JRC increased to 10 and 20, respectively. It is noteworthy that the experimental results were obtained from a relatively wide range of JRC. On the other hand, the measured contact area in experiments had the range of 4~35%. Therefore, one needs to be cautious when a contact area larger than 35% is in consideration.

The performance of the proposed model was compared with referenced models. Equations (8)–(10) were applied to the experimental results in this study. Among several results, eight cases were excerpted and applied. Table 4 and Figure 15 show measured $(e_h/e_m)^3$ values in experiments and predicted values by Equations (8)–(10), along with predicted values by the proposed model in this study.

Table 4. Comparison between the referenced models and proposed model.

		Experiment	Equation (8)	Equation (9)	Equation (10)	Proposed
J1	Case1	0.3327	273.2437	0.2149	0.7658	0.5136
	Case2	0.3789	158.0083	−0.1617	0.5782	0.3030
J2	Case3	0.1565	0.8306	−0.1354	0.8195	0.2692
	Case4	0.0771	0.4223	−0.4362	0.7474	0.1839
J3	Case5	0.0467	0.1443	0.4792	0.8837	0.1402
	Case6	0.0920	0.0907	0.0981	0.7246	0.0731
J4	Case7	0.0289	0.0837	0.4162	0.9060	0.0802
	Case8	0.0147	0.0533	0.2914	0.8918	0.0570

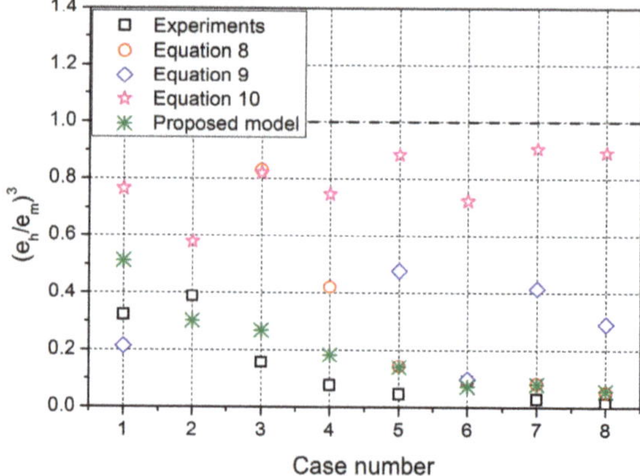

Figure 15. Comparison between the referenced models and proposed model.

In Barton model (Equation (8)), Cases 1 and 2 produced $(e_h/e_m)^3$ values that highly exceeded 1, which was not acceptable. This is because of the inherent limitation of the model, which is not capable of applying to joints with small roughness. Results of Cases 3 and 4 were less than 1, however, they still showed larger discrepancies compared with the experimental results. From Cases 5 to 8, the results became much closer to the experimental results, therefore, it is thought that Barton model is suitable for joints with relatively large roughness. In some cases, Zimmerman model (Equation (9)) derived negative $(e_h/e_m)^3$ values. This is because the first term in right side of Equation (9), which includes a ratio of standard deviation to average of mechanical aperture. By definition, the terms larger than approximately 0.8165 gave negative results. However, sometimes this could be found in experiments and corresponds to joints under high normal stress or highly weathered joints [3,39]. Equation (10) always produced $(e_h/e_m)^3$ values less than 1, however, the predicted values were relatively larger than the experiments, which was thought to be the effect of initial mechanical aperture in the equation.

For more verification, numerical simulation results by Xiong et al. (2018) [21] were referenced. They tested granite joint specimens and measured the information of fracture void using 3D laser scanner and conducted numerical simulations. Based on the simulation results, they suggested a relationship as a function of contact area and fractal dimension. Since it provided comparable results, such as mechanical aperture, roughness, and contact area, their results were excerpted and applied to Equation (11). In order to make direct comparison, the fractal dimension in Reference [21] was converted into *JRC*. Figure 16 shows the verification result.

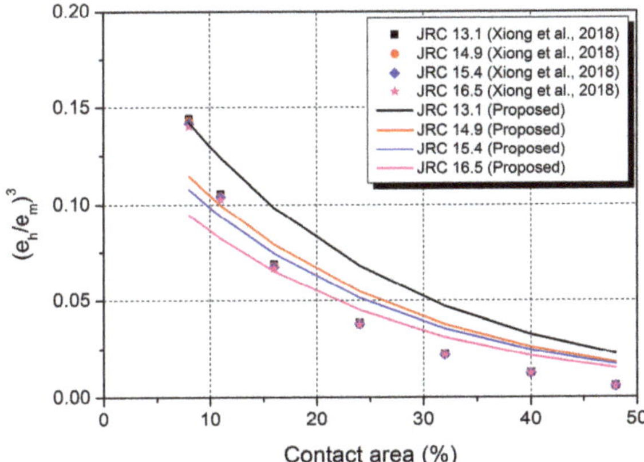

Figure 16. Verification of the proposed model by comparing with referenced data.

As shown in Figure 16, overall trends of the two models were similar to each other. $(e_h/e_m)^3$ decreased with the increase of *JRC* and with the increase of contact area. Some discrepancies were observed between them. However, the amount was quite smaller than the previous comparison. The Xiong model showed clear variation according to the contact area, but on the other hand, variation due to roughness was not significant. This was because their granite specimens had similar roughness level (*JRC* of 13~16).

Comparisons with referenced prediction models above showed that a few models could not be applied in some cases or show significant discrepancies. A couple of explanations could be deduced. First, those prediction models were derived from a specific range of geometric properties, which cannot be extended to wider cases. The matching state of the joint also affects the hydraulic features, but is not properly considered in some models. The proposed prediction model in this study considered a relatively wide range of roughness and, at the same time, consideration on the matching state was taken at some degree since the contact area was included in the model. Meanwhile, the proposed model also has a couple of limitations. First, the prediction model was constructed based on the limited data in terms of the contact area. *JRC* was distributed relatively evenly within the range of 0~20, while the contact area was 4~35%. Therefore, it could not guarantee its applicability when the input data outlays the range, especially in a contact area larger than 35%.

Analyses on hydromechanical characteristics in rock mass scale, such as discreet fracture network (DFN), should be conducted based on the accurate understanding of a joint. Several features of a joint are required in DFN, and among them, aperture is a critical parameter [40]. Distribution of mechanical aperture and its relationship with hydraulic aperture should be taken into account when building DFN models. In 2016, Luo et al. stated that most of the previous DFN studies assumed identical hydraulic and mechanical apertures in joint distribution for simplicity. In their study, remarkable discrepancies in macroscopic flowrate were observed depending on the selected relationship between apertures [41]. It indicated significant importance on prediction of hydraulic aperture so that emphasis could be made regarding understanding on characteristics of a joint.

5. Conclusions

The hydromechanical characteristics of a joint are affected by many joint parameters, such as roughness, aperture, contact area, stiffness, etc. Among them, the effects of roughness and aperture on the features are widely studied by various approaches. In many cases, artificial joint specimens are prepared to carry out laboratory experiments. However, controlling the roughness in experiments requires excessive efforts, since the specimens are often created mechanically. Therefore, some experimental works were derived from limited testing conditions in terms of the geometric properties of joint. In this study, the limitation was overcome to some extent using 3D printing technique. A fractal theory, specifically the random midpoint displacement method, was adopted to generate three-dimensional joint surfaces. By manipulating the input parameters of the method, the roughness of joint surfaces could be controlled in a quantitative manner.

Throughout the experimental works, four different joint specimens (J1, J2, J3, and J4) were prepared, which had *JRC* of 1.12, 7.32, 13.07, and 16.47, respectively. Therefore, in terms of *JRC*, the geometric condition of this study was distributed relatively evenly within the range of 0~20. Mechanical aperture and its distribution were calculated based on the scanned data as well. By adopting superposition method, initial mechanical aperture of the four joint specimens was calculated and as a result, it could be found that *JRC* and the initial mechanical aperture had positive correlation.

Normal compression tests were conducted so as to obtain joint displacement under specific stress. The maximum normal stress was 10 MPa, which corresponded with vertical in-situ stress of 400 m. As a result, the overall behavior agreed well with previous researches being simulated successfully by referenced relationships. The larger the roughness, the larger joint displacement was measured under normal compression. Variation of contact area was measured by pressure-sensitive paper. Effects of normal stress and stress hysteresis were investigated by image analyses of dyed paper, however quantitative relationship with roughness could not be found. Not only the average of aperture under normal stress, but also the spatial distribution was calculated by parallel translation of the joint surface.

Flow tests under normal compression were performed under maximum 1 MPa of water pressure. As the normal stress increased, flowrate per unit hydraulic head decreased since the aperture was compressed. Also, the rougher the surface, the larger flowrate was measured. For instance, J4 specimen showed 64 times higher flowrate than J1 specimen under 10 MPa of normal stress. Hydraulic aperture can be calculated with the assumption of cubic law. Then, comparison between the mechanical aperture and hydraulic aperture could be made. In a rough joint, deviation from the ideal condition of cubic laws are inevitable, since many factors affect fluid flow, such as roughness, contact area, channeling, etc. The mechanical and hydraulic apertures had positive correlation, however, more deviations were observed when the roughness increased.

Based on the experimental results, a prediction model for the hydraulic aperture was proposed. It was found that a model with single variable showed insufficient prediction capability so that the model was constructed as a function of roughness and contact area. As a result, $(e_h/e_m)^3$ could be expressed by those two variables having negative correlations. The effect of contact area was expressed as a negative exponential function and the effect of roughness was as an exponential function that had roughness as an exponent. The model showed an adjusted r-square of 0.74 and acceptable prediction performance comparing with several referenced models.

Author Contributions: Conceptualization, S.C.; investigation, S.C., B.J., and S.L.; writing—original draft preparation, S.C.; writing—review and editing, S.J.; supervision, S.J.

Acknowledgments: This research was supported by the National Strategic Project—Carbon Upcycling of the National Research Foundation of Korea (NRF), funded by the Ministry of Science and ICT (MSITT), the Ministry of Environment (ME), and the Ministry of Trade, Industry and Energy (MOTIE) (NRF—2017M3D8A2085654), and the Korea Agency for Infrastructure Technology Advancement under the Ministry of Land, Infrastructure and Transport of Korean Government (Project Number: 13 Construction Research T01). The institute of Engineering Research at Seoul National University (SNU) provided research facilities for this work. The authors are grateful for the support.

References

1. Makurat, A.; Barton, N.; Rad, N.S.; Bandis, S. Joint conductivity variation due to normal and shear deformation. In Proceedings of the International Symposium on Rock Joints, Leon, Norway, 4–6 June 1990; pp. 535–540.

2. Esaki, T.; Du, S.; Mitani, Y.; Ikusada, K.; Jing, L. Development of a shear-flow test apparatus and determination of coupled properties for a single rock joint. *Int. J. Rock Mech. Min. Sci.* **1999**, *36*, 641–650. [CrossRef]

3. Lee, H.S.; Cho, T.F. Hydraulic characteristics of rough fractures in linear flow under normal and shear load. *Rock Mech. Rock Eng.* **2002**, *35*, 299–318. [CrossRef]

4. Hakami, E.; Larsson, E. Aperture measurement and flow experiments on a single natural fracture. *Int. J. Rock Mech. Min. Sci.* **1996**, *33*, 395–404. [CrossRef]

5. Min, K.B.; Rutqvist, J.; Elsworth, D. Chemically and mechanically mediated influence on the transport and mechanical characteristics of rock fractures. *Int. J. Rock Mech. Min. Sci.* **2009**, *46*, 80–89. [CrossRef]

6. Brown, S. Fluid flow through rock joints: The effect of surface roughness. *J. Geophys. Res. Solid Earth* **1987**, *92*, 1337–1347. [CrossRef]

7. Hakami, E. Aperture Distribution of Rock Fractures. Ph.D. Thesis, Division of Engineering Geology, Royal Institute of Technology, Stockholm, Sweden, 1995.

8. Olsson, R.; Barton, N. An improved model for hydromechanical coupling during shearing of rock joints. *Int. J. Rock Mech. Min. Sci.* **2001**, *38*, 317–329. [CrossRef]

9. Bandis, S.; Lumsden, A.; Barton, N. Fundamentals of rock joint deformation. *Int. J. Rock Mech. Min. Sci.* **1983**, *20*, 249–268. [CrossRef]

10. Pyrak-Nolte, L.J.; Morris, J.P. Single fractures under normal stress: The relation between fracture specific stiffness and fluid flow. *Int. J. Rock Mech. Min. Sci.* **2000**, *37*, 245–262. [CrossRef]

11. Zhao, J.; Brown, E. Hydro-thermo-mechanical properties of joints in the Carnmenellis granite. *Q. J. Eng. Geol.* **1992**, *25*, 279–290. [CrossRef]

12. Park, J.; Song, J. Numerical method for the determination of contact areas of a rock joint under normal and shear loads. *Int. J. Rock Mech. Min. Sci.* **2013**, *58*, 8–22. [CrossRef]

13. Singh, K.K.; Singh, D.N.; Ranjith, P.G. Laboratory simulation of flow through single fractured granite. *Rock Mech. Rock Eng.* **2015**, *48*, 987–1000. [CrossRef]

14. Scesi, L.; Gattinoni, P. Roughness control on hydraulic conductivity in fractured rocks. *Hydrogeol. J.* **2007**, *15*, 201–211. [CrossRef]

15. Koyama, T.; Li, B.; Jiang, Y.; Jing, L. Numerical modelling of fluid flow tests in a rock fracture with a special algorithm for contact areas. *Comput. Geotech.* **2009**, *36*, 291–303. [CrossRef]

16. Xiong, X.; Li, B.; Jiang, Y.; Koyama, T.; Zhang, C. Experimental and numerical study of the geometrical and hydraulic characteristics of a single rock fracture during shear. *Int. J. Rock Mech. Min. Sci.* **2011**, *48*, 1292–1302. [CrossRef]

17. Yeo, I.; Freitas, M.; Zimmerman, R. Effect of shear displacement on the aperture and permeability of a rock facture. *Int. J. Rock Mech. Min. Sci.* **1998**, *35*, 1051–1070. [CrossRef]

18. Gentier, S.; Riss, J.; Archambault, G.; Flamand, R.; Hopkins, D. Influence of fracture geometry on shear behavior. *Int. J. Rock Mech. Min. Sci.* **2000**, *37*, 161–174. [CrossRef]

19. Barton, N.; Bandis, S.; Bakhtar, K. Strength, deformation and conductivity coupling of rock joints. *Int. J. Rock Mech. Min. Sci. Abstr.* **1985**, *22*, 121–140. [CrossRef]

20. Zimmerman, R.; Bodvarsson, G. Hydraulic conductivity of rock fractures. *Transp. Porous Media* **1996**, *23*, 1–30. [CrossRef]

21. Xiong, F.; Jiang, Q.; Chen, M. Numerical investigation on hydraulic properties of artificially-splitting granite fractures during normal and shear deformations. *Geofluid* **2018**, 1–16. [CrossRef]

22. Raja, J.; Radhakrishnan, V. Analysis and synthesis of surface profiles using Fourier series. *Int. J. Mach. Tool Des. Res.* **1977**, *17*, 245–251. [CrossRef]

23. Patir, N. A numerical procedure for random generation of rough surfaces. *Wear* **1978**, *47*, 263–277. [CrossRef]

24. Xie, H. *Fractals in Rock Mechanics*; Balkema: Rotterdam, The Netherlands, 1993; ISBN 9054101334.

25. Choi, S.; Lee, S.; Jeong, H.; Jeon, S. Development of a new method for quantitative generation of an artificial joint specimen with specific geometric properties. *Sustainability* **2019**, *11*, 373. [CrossRef]

26. Barton, N.; Choubey, V. The shear strength of rock joints in theory and practices. *Rock Mech. Rock Eng.* **1977**, *10*, 1–54. [CrossRef]

27. Ulusay, R.; Hudson, J.A. *The Complete ISRM Suggested Methods for Rock Characterization, Testing and Monitoring: 1974–2006*; ISRM Commission on Testing Method: Ankara, Turkey, 2007; ISBN 9789759367541.

28. Tse, R.; Cruden, D. Estimating joint roughness coefficients. *Int. J. Rock Mech. Min. Sci. Abstr.* **1979**, *16*, 303–307. [CrossRef]

29. Grasselli, G.; Egger, P. Constitutive law for the shear strength of rock joints based on three-dimensional surface parameters. *Int. J. Rock Mech. Min. Sci.* **2003**, *40*, 25–40. [CrossRef]

30. Park, J.; Lee, Y.; Song, J.; Choi, B. A constitutive model for shear behavior of rock joints based on three-dimensional quantification of joint roughness. *Rock Mech. Rock Eng.* **2013**, *46*, 1513–1537. [CrossRef]

31. Esaki, T.; Du, S.; Jiang, Y.; Wada, Y.; Mitani, Y. Relation between mechanical and hydraulic apertures during shear-flow coupling test. In Proceedings of the 10th Japan Symposium on Rock Mechanics, Osaka, Japan, 22–23 January 1998; pp. 91–96.

32. Iwano, M.; Einstein, H. Stochastic analysis of surface roughness, aperture and flow in a single fracture. In Proceedings of the EUROCK, Lisbon, Portugal, 21–24 June 1993; pp. 135–141.

33. Riss, J.; Gentier, S.; Sirieix, C.; Archambault, G.; Flamand, R. Degradation characteristics of sheared joint wall surface morphology. In Proceedings of the 2nd North America Rock Mechanics Symposium, Montreal, QC, Canada, 19–21 June 1996; pp. 1343–1349, ISBN 905410838X.

34. Goodman, R. *Method of Geological Engineering in Discontinuous Rocks*; West Publishing: New York, NY, USA, 1976; ISBN 0829900667.

35. Brown, S.; Scholz, C. Closure of rock joints. *J. Geophys. Res. Solid Earth* **1986**, *91*, 4939–4948. [CrossRef]

36. Gale, J. Comparison of coupled fracture deformation and fluid flow models with direct measurements of fracture pore structure and stress-flow properties. In Proceedings of the 28th US Symposium on Rock Mechanics, Tucson, AZ, USA, 29 June–1 July 1987; pp. 1213–1222.

37. Pyrak-Nolte, L.; Myer, L.; Cook, N.; Witherspoon, P. Hydraulic and mechanical properties of natural fractures in low-permeability rock. In Proceedings of the 6th ISRM Congress, Montreal, QC, Canada, 30 August–3 September 1987; pp. 225–231.

38. Walsh, J. Effect of pore pressure and confining pressure on fracture permeability. *Int. J. Rock Mech. Min. Sci. Abstr.* **1981**, *18*, 429–435. [CrossRef]

39. Matsuki, K.; Chida, Y.; Sakaguchi, K.; Glover, P. Size effect on aperture and permeability of a fracture as estimated in large synthetic fractures. *Int. J. Rock Mech. Min. Sci.* **2006**, *43*, 726–755. [CrossRef]

40. Bang, S.H.; Jeon, S.; Kwon, S. Modelling the hydraulic properties of a fractured rock mass at KAERI underground research tunnel by three-dimensional discreet fracture network. In Proceedings of the 7th Asian Rock Mechanics Symposium, Seoul, Korea, 15–19 October 2012; pp. 905–913.

41. Luo, S.; Zhao, Z.; Peng, H.; Pu, H. The role of fracture surface roughness in macroscopic fluid flow and heat transfer in fractured rocks. *Int. J. Rock Mech. Min. Sci.* **2016**, *87*, 29–38. [CrossRef]

Article

Development of a New Method for the Quantitative Generation of an Artificial Joint Specimen with Specific Geometric Properties

Seungbeom Choi, Sudeuk Lee, Hoyoung Jeong and Seokwon Jeon *

Department of Energy Systems Engineering, Seoul National University, Gwanak-gu, Seoul, Korea;
chbum092@snu.ac.kr (S.C.); ics1961@snu.ac.kr (S.L.); hyjung04@snu.ac.kr (H.J.)
* Correspondence: sjeon@snu.ac.kr; Tel.: +82-2-880-8807

Received: 10 December 2018; Accepted: 10 January 2019; Published: 12 January 2019

Abstract: A rock joint is a planar discontinuity that has significant influence on the mechanical and hydraulic characteristics of rock mass. Laboratory experiments are often conducted on a joint to investigate and provide fundamental information for rock mass analysis. Although joint roughness and mechanical aperture exert great effects on the experimental results, controlling them in quantitative manner is quite complicated and consumptive in terms of specimen preparation. A new and simple method for the quantitative generation of the joint specimen was proposed in this study. Based on random midpoint displacement method, a joint specimen with a void space inside was generated. Parametric studies for the roughness and mechanical aperture were carried out, and as a result, the two joint properties could be controlled by manipulating input parameters of random midpoint displacement method. In order to validate the proposed method, two joint specimens, which had different levels of roughness and aperture, were generated and printed. Surface coordinates of the specimens were obtained by a 3D laser scanner, and calculated to make a comparison between the target values and the estimated values. Results showed that the method was capable of generating joint specimens with satisfactory precision.

Keywords: artificial joint specimen; roughness; mechanical aperture; quantitative generation; random midpoint displacement method; 3D printer

---

1. Introduction

A rock joint is a planar discontinuity that has little tensile strength and it exerts influence on mechanical behavior of rock mass, since it acts as a weak plane. At the same time, a joint has much higher hydraulic conductivity than the rock matrix, so that the majority of fluid flow in the rock mass happens through the joint [1–3]. Therefore, it is of great importance to accurately understand the characteristics of a joint in many engineering applications, as it affects both the mechanical and hydraulic behaviors of the rock mass. For examples, in engineering projects, such as rock slope design, underground storage facilities, the repository for radioactive disposal [4,5], and geothermal energy production [5,6], etc., it is necessary to consider the effect of the joint on the stability and efficiency of the projects. Furthermore, new technologies for coping with global climate change have been developed recently, including geological storage of CO_2. There is some possibility that CO_2 may leak out and flow through the joint, which should be taken into account for guaranteeing the feasibility and stability of the storage.

In situ measurements can derive the hydraulic characteristics of rock mass, which consider the effects of multiple joints or joint sets. However, they are sometimes consumptive in terms of time and expenses, so that the features are investigated by laboratory experiments on a joint, providing fundamental properties for rock mass analyses. A joint has several relevant properties, such as roughness, aperture, stiffness, contact area, matedness, and so on [7,8]. They exert complex influences on the mechanical and hydraulic characteristics of the joint, interacting with each other. Mechanical aperture is a separation between the two corresponding joint walls. It is defined as a distance between those walls in the direction perpendicular to the reference plane at a point [7]. The aperture acts as flow path in a joint, thus, several researches have investigated the relationship between aperture and other properties [9–11]. Roughness has drawn attention for the relationship, since it is a widely-used engineering parameter, and relatively easy to be measured. The roughness is usually defined by its height distribution or shape, and there are many quantitative roughness parameters. For instance, Tse and Cruden (1979) proposed (center line average), (mean square value), (root mean square), and Z_2 (the *RMS* of the first deviation of the profile) [12]. Barton and Choubey (1977) proposed the *JRC* (joint roughness coefficient) with 10 standard joint profiles, based on numerous shear test results [13,14]. Even though *JRC* is not a quantitative parameter, it has been widely used in engineering projects, mainly due to its simplicity and practicability. The effects of aperture and roughness are of importance in the mechanical and hydraulic behaviors of a joint, so that an accurate understanding of them is quite necessary.

In many cases, artificial joint specimens are prepared, in order to carry out mechanical and/or hydraulic experiments. The joint specimens are often created mechanically by using a splitter or a V block, which applies a concentrated load at the center of rock block or cylinder similar with Brazilian jaws [15,16]. Tensile stress generated along the loading axis bisects the rock into two halves. Although it is a simple method to create a joint specimen, it requires excessive efforts if a specific joint roughness is desired. Therefore some experimental works have derived results from limited test conditions in terms of the joint roughness [3,16]. The reproducibility of this approach is hardly guaranteed so that casting method, which uses cement mortar or epoxy, is adopted as an alternative to make multiple joint specimens that have similar levels of roughness [17,18]. The casting method mitigates the reproducibility problem to some extent; however, it also has a shortcoming. The aperture, which is an important joint property along with the roughness, cannot be controlled by the method. Therefore, the geometric conditions of a joint, i.e., the roughness and aperture, could not be controlled quantitatively, and simultaneously in many cases.

In this study, a new and simple method was proposed to generate an artificial joint specimen. This method enables to control the roughness and aperture at the same time in a quantitative manner. Two corresponding joint surfaces were generated by fractal theory, specifically by the random midpoint displacement method, producing point-cloud type data of the surfaces. By using proper input parameters, the roughness as well as the aperture of a joint could be controlled and manipulated. The generated data can be directly utilized in the theoretical approach or numerical simulation, since they provide the x, y, and z coordinates of the joint surfaces. Thanks to recent developments in 3D printing technology, this method is even capable of being adopted in experimental works, overcoming the limitations that have been explained above.

2. Background Theory

Several methodologies have been proposed to make an artificial joint surface. The Fourier transformation can be used to make and evaluate the rough surfaces [19] by manipulating the number of harmonics to be calculated, and the Fourier coefficients. Stochastic functions such as the autocorrelation function can also be applied to this procedure [20] by controlling the range and standard deviation of the joint height. The fractal Brownian function is one of the viable options to make synthetic surfaces [21] with many applications being reported. Xie (1993) [21] stated that the rock joint could be successfully simulated by fractal theory in terms of roughness, aperture, pore, permeability, and so on. Natural rock joints usually show self-affinity, which is a feature that different amount of scaling should be considered in transformation. The fractal Brownian functions could take the feature into account; thus it can be applied in estimating and/or generating the artificial joint surfaces.

Among them, the random midpoint displacement method was adopted in this study to make point-cloud type data of the joint specimens. It has several benefits of application. First, it is quite straightforward, and its outcomes can be directly imported to a 3D printer. Also, the manipulation of input parameters can derive satisfactory results in controlling the joint roughness and aperture. The algorithm of the random midpoint displacement method is explained briefly in the following procedure (Figure 1).

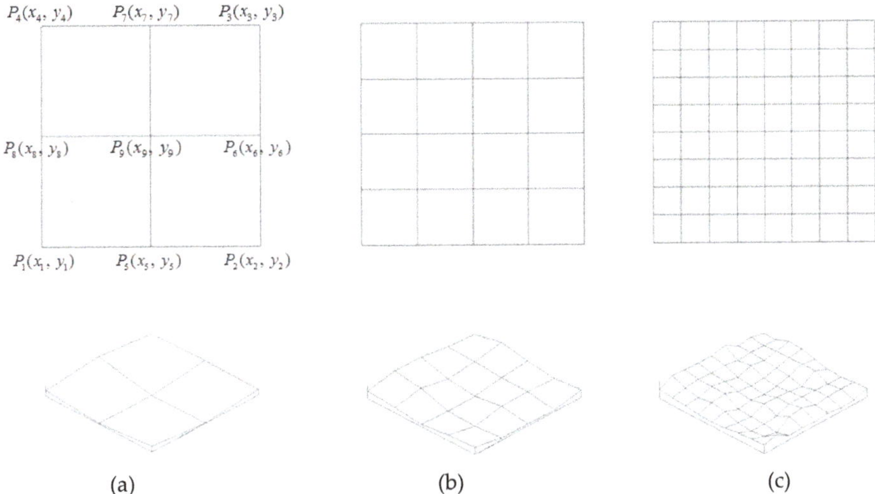

Figure 1. The generation sequence of a rough joint surface following the random midpoint displacement method: (a) Joint shape when $GL = 1$, with nine points on the joint surface; (b) Joint shape when $GL = 2$; (c) Joint shape when $GL = 3$ (After Seo and Um [22]).

For the sake of brevity, let us assume a square joint surface that has a certain side length (L). Then, the elevation of the four corner points should be determined first. If the points have the same elevation, it is called the stationary profile, and if not, it is called non-stationary. The interval between the neighboring points is calculated as Equation (1).

$$\Delta x = \Delta y = \frac{L}{2^{GL}} \tag{1}$$

where GL (generation level) is the number of generation sequences, which also defines the resolution of a joint surface.

Based on the four predetermined corner points, coordinates of midpoint ($P_9(x_9, y_9)$ in Figure 1a) are calculated by Equation (2).

$$\begin{aligned} x_9 &= \tfrac{1}{4}(x_1 + x_2 + x_3 + x_4) \\ y_9 &= \tfrac{1}{4}(y_1 + y_2 + y_3 + y_4) \\ z_9 &= \tfrac{1}{4}(z_1 + z_2 + z_3 + z_4) + D_1 \end{aligned} \tag{2}$$

where D_1 is the Gaussian random number, which follows a normal distribution that is defined by a couple of input parameters, and GL (as presented in Equation (3)).

$$N\left(0, \frac{\sigma^2(1 - 2^{2H-2})}{(2^{GL})^{2H}}\right) \tag{3}$$

where σ is a term that is related to the amplitude of point elevation, and H is the Hurst exponent. In a three-dimensional case, H has a relationship with fractal dimension (D), which is $D = 3 - H$. If the midpoint is located on a side line of the square, only two neighboring points are included in the calculation. By iterating this procedure with desired GL value, point-cloud type data of an artificial surface can be easily generated.

3. Development of a New Method and its Parametric Study

3.1. Development of a New Method for Generating Three-Dimensional Joint Specimens

As explained above, a three-dimensional joint surface could be easily generated by adopting the random midpoint displacement method. Based on this, a whole three-dimensional joint specimen, which includes two corresponding joint surfaces and a void space (collection of apertures) inside, could be generated as well. Figure 2 shows a conceptual procedure for generating a joint specimen, but it is drawn in two dimensions for clear understanding.

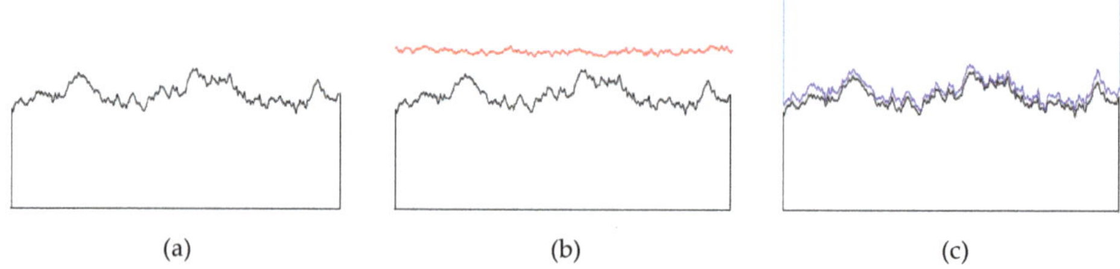

(a)   (b)   (c)

Figure 2. Conceptual procedure for generating a joint specimen that includes a void space inside.

First, a lower joint surface is generated according to the random midpoint displacement method (Figure 2a). The roughness of the lower surface can be controlled quantitatively by manipulating the input parameters. Then, a void space is generated separately by the same method (Figure 2b). The generated data can be treated as a joint surface, but at the same time, it can represent the upper bound of a void space. Thus, after generating point-cloud data of the void space then they are translated in parallel, so as to make the minimum elevation (z coordinate) of the void space zero. Each elevation of the points means the aperture of that point, which also can be controlled in a quantitative manner. At last, the lower surface and void space is summed to make an upper joint surface (Figure 2c). Figure 3 shows a detailed flowchart.

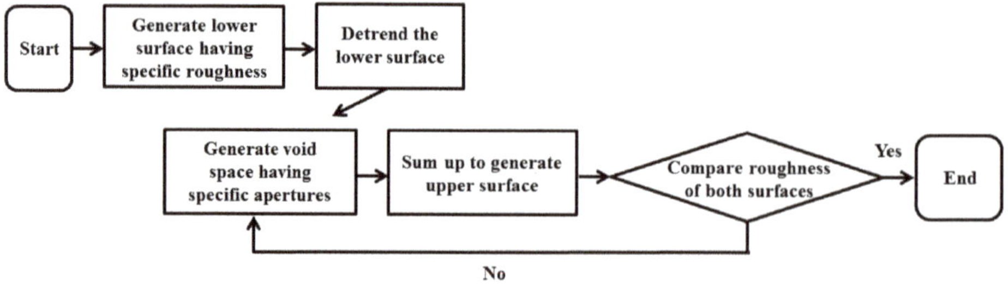

Figure 3. Flowchart for generating a joint specimen in a quantitative manner.

The generated surface may show a inclination at some degree. When it comes to evaluating the roughness quantitatively, this inclination trend should be removed. Therefore, multilinear regression was included in the flowchart so as to detrend the lower surface [3,23]. Since all coordinates of the lower surface (x_0, y_0, z_0) are known, an optimum regression plane could be found, based on the coordinates. Subtracting a trend plane (z_{trend}) from the original coordinates gives the final lower surface:

$$z = z_0 - z_{trend} \ where \ z_{trend} = n_1 x_0 + n_2 y_0 + n_3 \tag{4}$$

Meanwhile, two corresponding joint surfaces usually show a similar level of roughness, unless they are highly weathered. Thus, after generating the upper surface, a comparison of the roughness between upper and lower surfaces was included in the flowchart. If they show similar roughnesses, the procedure would end, and if not, the void space (aperture distribution) would be generated again to make another point-cloud, but with the same distribution characteristics.

3.2. Parametric Study for Generating Joint Specimen

It is known that the roughness of a joint surface generated by the random midpoint displacement method is affected by joint length (L), generation level (GL), asperity amplitude (σ), and the Hurst exponent (H) [22,24]. However, considering the testing environment, such as the size of a specimen and the resolution of a 3D printer, it is reasonable to assume that L and GL are actually determined before the surface generation. Therefore, the roughness is determined according to a combination of σ and H. The Hurst exponent, which is related to the fractal dimension, has a range of 0~1, and although there is no limit on the amplitude, it was set to 0~10 in this study.

The joint roughness can be classified into waviness and unevenness. The former defines relatively large, macroscopic scale and the latter means relatively small, microscopic scale roughness. In order to investigate the effect of input parameters, i.e., σ and H, on the roughness, FFT (Fast Fourier Transform) analyses on some joint profiles were carried out. Several joint profiles were generated with a fixed joint length of 100 mm and a generation level of 7. The amplitude (σ) varied from 1 to 10 with an interval of 1, and the Hurst exponent varied from 0 to 0.9 with the interval of 0.1. Among those profiles, six examples are presented in Figure 4.

Figure 4a–c (in the first column) shows the joint profiles and their FFT results under the conditions of varying Hurst exponents and a fixed amplitude, while Figure 4d–f (in the second column) show the results under varying amplitudes and a fixed Hurst exponent. The red column in Figure 4 means the normalized amplitude, so that its sum is 100, and the blue curve means the cumulative value. Judging from the profiles in Figure 4a–c, the joint roughness decreased with the increase of H. Also, at the same time, the graphs in Figure 4a–c show that portion of low frequency, i.e., the microscopic asperity, increased remarkably when the H increased. On the other hand, the profiles in Figure 4d–f showed that the roughness increased with the increase of σ, while the FFT results showed little difference. However, it is noteworthy that profiles in Figure 4d–f had larger scales in the y-axis than those in Figure 4a–c. Therefore, it could be concluded that the increased value of H made the overall roughness of a joint smoother, but it also made the portion of microscopic asperity larger, while σ made the overall roughness and the absolute height of a joint larger.

Several quantitative roughness parameters have been proposed; however, most of them are two-dimensional parameters. It is obvious that the representativeness of a joint profile is not enough for a whole joint surface, since the joint itself is a three-dimensional structure. In order to represent the characteristics of a joint better, a few three-dimensional roughness parameters have been proposed [25,26]. They can guarantee more representativeness, and they take joint anisotropy into consideration; however, their applicability and practicality could not have exceeded those of the two-dimensional parameters.

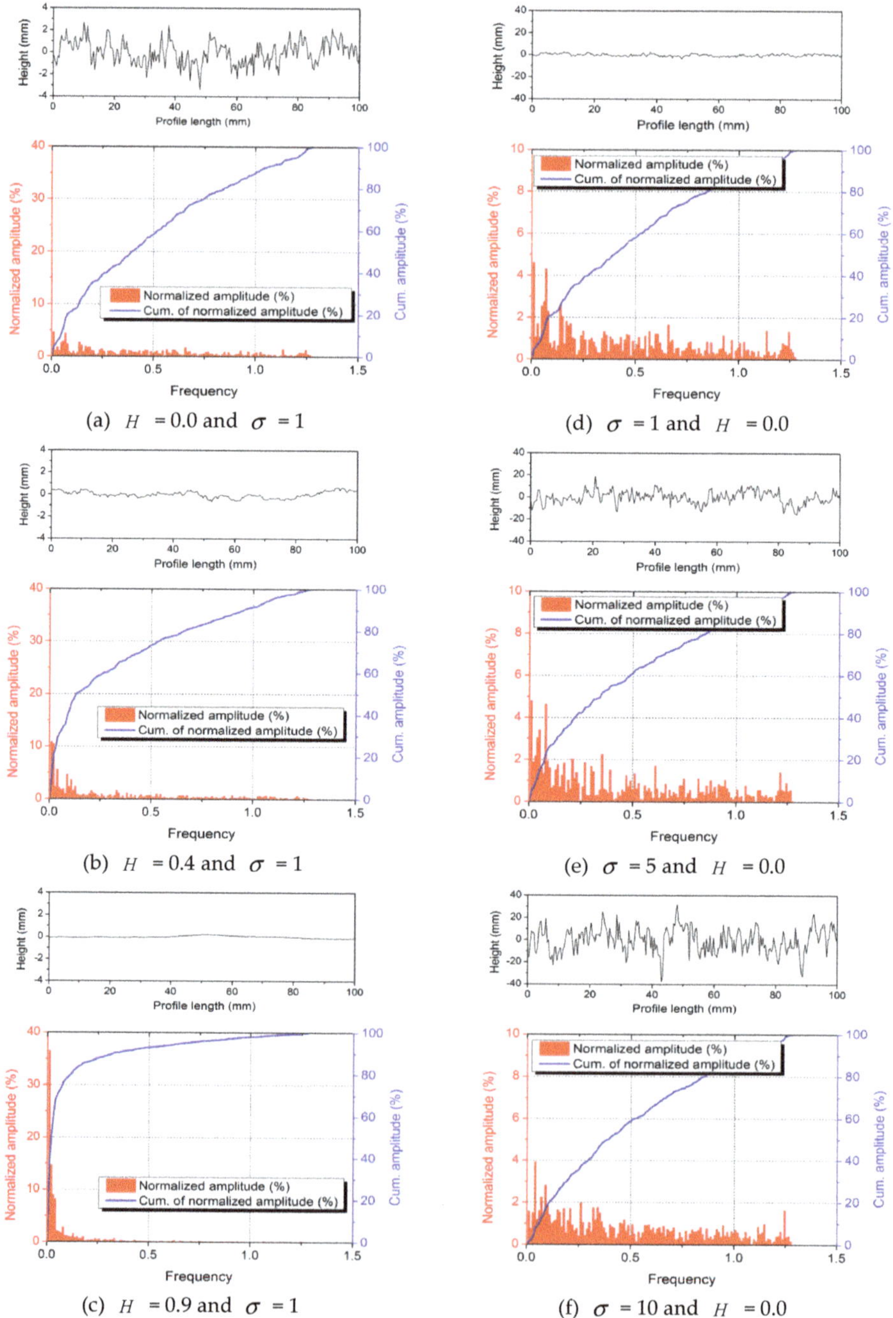

Figure 4. Joint profiles generated by a combination of input parameters (σ, H) and their Fast Fourier Transform (FFT) results. The red column denotes a normalized histogram, and the blue curve denotes the cumulative curve of the histogram.

In this study, a compromise method was adopted to determine the joint roughness. Figure 5 shows a schematic diagram for this determination. Let us assume that a long axis (the x direction in Figure 5) of the rectangle corresponds to shear direction. Then, profiles defined in the xz plane are of interest. Additionally, the moving xz plane along the y axis would define several joint profiles (Figure 5b). Each and every profile within the xz plane is obtained and calculated by following one of the quantitative two-dimensional parameters. Then, a distribution of the two-dimensional parameter could be calculated, and the average of the distribution was defined as a typical value. Tse and Cruden (1979) [12] suggested Z_2 (Equation (5)) and it was selected as an estimator in this study.

$$Z_2 = \sqrt{\frac{1}{L}\int_0^L \left(\frac{dy}{dx}\right)^2 dx} = \sqrt{\frac{1}{L}\sum_{i=1}^{n-1}\frac{(y_{i+1}-y_i)^2}{x_{i+1}-x_i}} \tag{5}$$

Several joint surfaces were generated according to the combinations of input parameters. The length of joint (L) and the generation level (GL) were set to 100 mm and 7, respectively. The amplitude (σ) varied from 1 to 10 with the interval of 1, and the Hurst exponent varied from 0 to 0.9, with an interval of 0.1. Then, another set of point-cloud data, which represented the void spaces, was generated under the same input parameters. Since the aperture values were calculated at each and every point of space, a distribution of apertures was obtained as well. The average was selected as a typical value, and the minimum elevation was set to 0. A three-dimensional plot of roughness (Z_2) and mechanical aperture are shown in Figure 6.

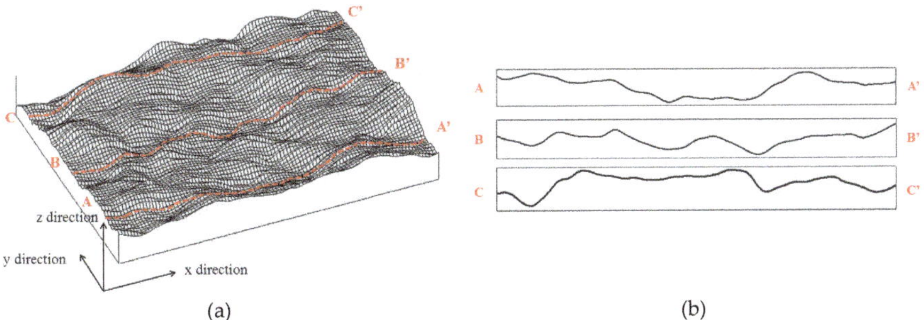

(a)   (b)

Figure 5. Schematic diagram for determining the joint roughness adopted in this study: (**a**) Three-dimensional plot of a joint; (**b**) Obtained profiles along a certain direction.

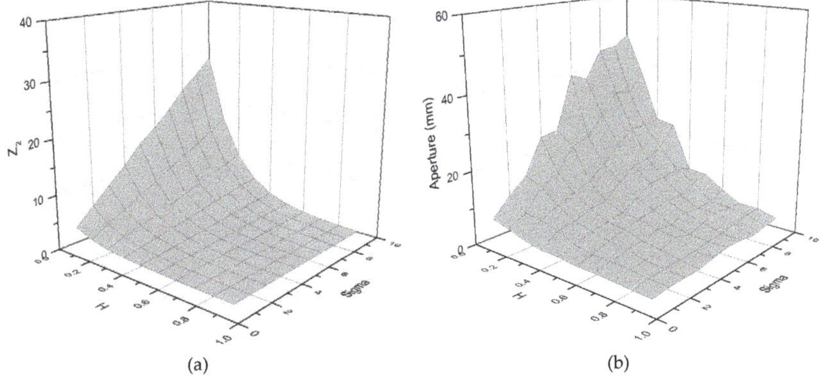

(a)   (b)

Figure 6. Three-dimensional plots from the parametric study; (**a**) results of the roughness variation (Z_2); (**b**) results of the mechanical aperture variation.

As it can be seen in Figure 6a, the roughness of a joint increased with the increase of σ and the decrease of H. This was the same tendency that was presented in Figure 4. Therefore, under the predetermined values of L and GL, the combinations of the two input parameters could be constructed, and eventually a desired joint roughness could be obtained by properly selecting a combination. Figure 6b shows a variation of mechanical aperture, according to the input parameters. The mechanical aperture increased with the increased of σ and the decrease of H. It is noteworthy that the magnitude of the mechanical aperture in Figure 6b was quite large in some cases. In those cases, an appropriate scaling should be considered, so as to obtain the desired void space.

4. Verification and Discussion

In order to validate the proposed joint-generating method, some verification works were carried out. Two joint specimens, which had specific target geometric characteristics, were generated by the proposed method, and then printed by a 3D printer. The printed specimens were scanned by 3D laser scanner, so as to obtain surface information, which was further used to measure the roughness and aperture. Then, verification was conducted by comparing the estimated results and the original target values. Table 1 shows the target geometric properties of the joints.

Table 1. Target geometric properties of the joint specimens.

	Roughness (*JRC*)	Initial Mechanical Aperture (mm)
Case 1	4~6	0.1
Case 2	14~16	0.5

As mentioned earlier, the applicability of *JRC* exceeded those of the other roughness parameters, even though it was rather qualitative. Thus, the target roughness values were expressed in *JRC*. Tse and Cruden (1979) [12] suggested a relationship that could convert Z_2 into *JRC* (Equation (6)):

$$JRC = 32.2 + 32.47 \log Z_2 \tag{6}$$

Contrary to the Z_2, *JRC* has a limited range, from 0 to 20. Therefore, the domain Z_2, should be bounded as well, when using Equation (6), which is roughly from 0.10 to 0.42. After obtaining three-dimensional plots in Figure 6 under the predetermined values of the lengths and generation level, combinations of σ and H were selected for each case. The length and generation level were set to 100 mm and 7, respectively. The lower surface of case 1 was generated under the conditions of $\sigma = 6$ and $H = 0.8$, and the void space was generated under the same input parameters, but scaled down to achieve the target aperture value. At the same time, the lower surface and the void space of case 2 were generated by the same procedure under the conditions of $\sigma = 7$ and $H = 0.7$. Figure 7 shows the printed joint specimens and their 3D models, which were generated under the same coordinates.

Then, the surfaces of the two joint specimens were scanned by a 3D laser scanner. The scanned joint area was 100 mm × 100 mm, and the scanning interval was 0.5 mm. The roughness was evaluated by the procedure explained above, and the aperture was measured by the superposition method [3,23]. The superposition method was capable of utilizing the scanned three-dimensional surface data, which were already obtained when determining the joint roughness. A brief explanation of the superposition method is as follows. First, the point data of both the lower and upper surfaces are arranged vertically by parallel and symmetric translation. Then, the upper surface is moved in parallel downward step-by-step, to make sure that it meets a certain criterion of aperture estimation. When the criterion is met, the distance of all of the corresponding points is calculated to estimate its distribution.

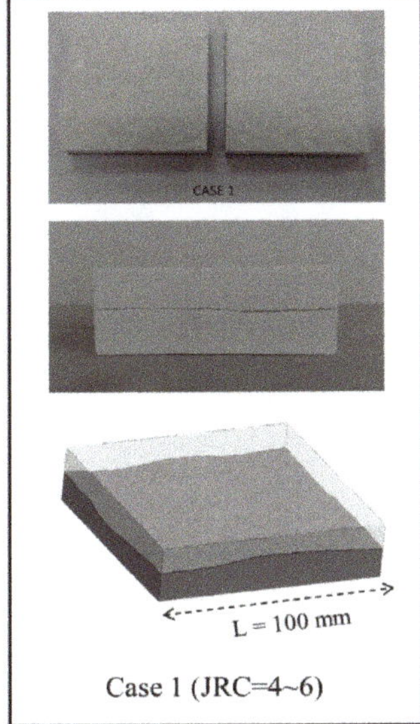
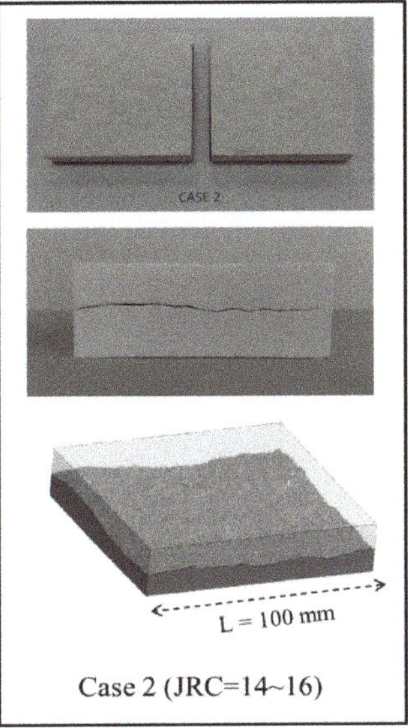

Figure 7. Picture of the printed specimens and their 3D models (side length = 100 mm).

The criterion for determining the mechanical aperture was defined when 1% of the total surface points were in superposition. It was referenced from experimental results that the area contacted by the dead weight of joint was approximately 1% of the total surface [27]. Since only gravitational force was applied to the specimen, it is called the initial mechanical aperture. Although the number of points did not directly determine the area, it was assumed that the ratio of points was equal to the ratio of the area, since a sufficient number of points was generated. When superposing, the upper surface was translated downward by 0.001 mm at a step until 1% superposition was reached. Then, the distribution of the aperture was calculated, and every aperture of the superposed points was set to zero. The calculated distributions of *JRC* and the initial mechanical aperture are shown in Figure 8, and the statistics of the distributions are presented in Table 2. The curves in Figure 8 mean a normal distribution curve that is fitted based on the estimated results.

As can be seen in Figure 8a,b, the roughness (*JRC*) roughly followed a normal distribution. Judging from the average of each case, the original target roughness was achieved quite well by the proposed method. At the same time, the estimated initial mechanical aperture also showed a good consistency with the target value. However, little discrepancy was found between them, and this was because the initial mechanical aperture was calculated when 1% of surface points were superposed. It could be mitigated with a couple of trial and error approaches, if some margin for the superposition would be included when scaling down the data of the void space. Distributions of aperture (Figure 8c,d) could be presented as the normal distribution. Natural rock joints usually followed a normal distribution or log-normal distribution [4,23,28] as well, and the same tendency could be found in this study. This result could be explained by the generation algorithm, where the Gaussian random number (D_1, in Equation (2)) was involved. Since the z coordinate was calculated following a Gaussian distribution, the roughness and aperture showed some normality in nature. Therefore, it could be concluded that the developed method was able to generate the joint surface in a quantitative manner, which could simulate similar characteristics of the natural rock joint.

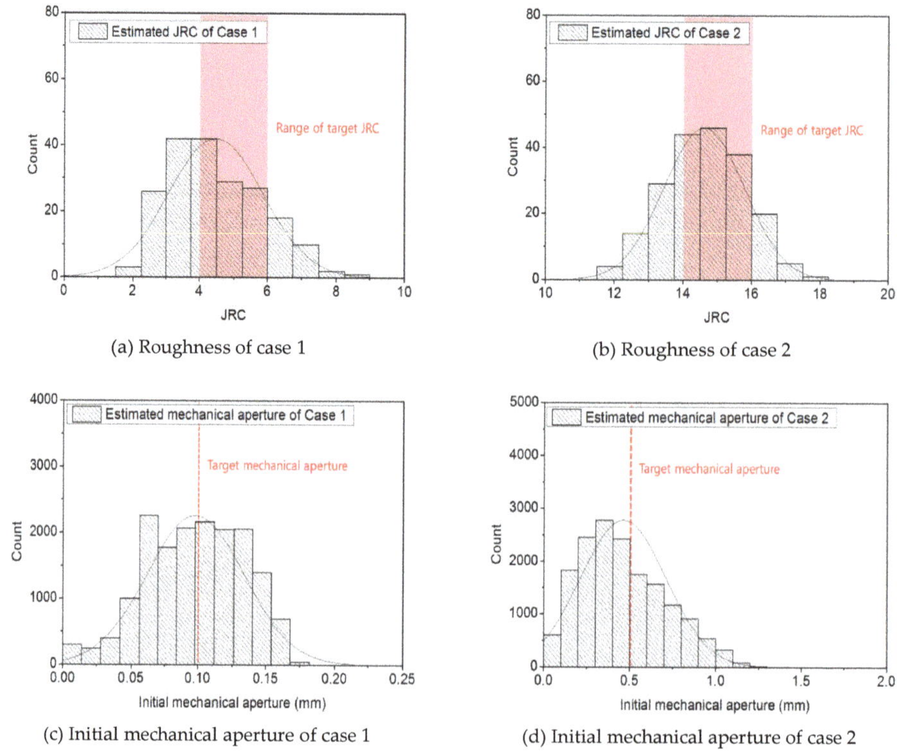

<table>
<tr><td>(a) Roughness of case 1</td><td>(b) Roughness of case 2</td></tr>
<tr><td>(c) Initial mechanical aperture of case 1</td><td>(d) Initial mechanical aperture of case 2</td></tr>
</table>

Figure 8. Estimated joint roughness (*JRC*) and initial mechanical aperture.

Table 2. Estimated results and target values of joint properties.

	Roughness (*JRC*)			Initial Mechanical Aperture (mm)		
	Target Value	**Estimated Results**		**Target Value**	**Estimated Results**	
		Ave.	**S.D.**		**Ave.**	**S.D.**
Case 1	4~6	4.47	1.39	0.1	0.0977	0.0360
Case 2	14~16	14.62	1.17	0.5	0.4611	0.2481

The method described above dealt with the generation of a single joint. However, analyses on the characteristics of the rock mass, such as DFN (Discreet Fracture Network), may require multiple joints being intersected or branched. Hakami and Stephansson (1993) [29] noted that an intersection of joints could play a significant role in the flow pattern of joint networks, even though the pore volume of intersections, are small compared to total rock mass porosity [29]. Additionally they also provided possible intersection geometries and modelling concepts. In the modelling procedure, the proposed surface-generating method in this study could be applied. Two corresponding surfaces, which are supposed to be in contact, can be generated following the method, and other than that, the rotation of the surface coordinates and the trimming points in conjunction would be required to make the intersections. Though it is out of research scope of this study, it would be promising in future studies.

5. Conclusions

A rock joint is an important feature in a rock mass that highly affects its mechanical and hydraulic properties. The joint also has several properties, such as roughness, mechanical aperture, stiffness and so on. They define the mechanical and hydraulic characteristics of the joint interacting with each other. Roughness and aperture have been drawn attention, since they exert great influence on both the mechanical and hydraulic features of the joint. However, controlling them at the laboratory scale is sometimes consumptive in terms of specimen preparation, since it requires inevitable trial and error approaches.

In this study, a new and simple method for generating an artificial joint specimen was proposed. The specimen included both the upper and lower surfaces and a void space inside. Among several joint generating algorithms, fractal theory, specifically the random midpoint displacement method, was adopted to generate the specimen. By manipulating the input parameters of the algorithm, distributions of roughness and mechanical apertures of the joint could be controlled in a quantitative manner.

A parametric study to estimate the roughness and aperture were conducted. As a result, the roughness increased with an increase in the amplitude of asperity (σ), and decreased with the increase of the Hurst exponent (H). With some predetermined testing environmental factors, such as the length of the joint and the generation level, the roughness could be controlled and manipulated quantitatively. Furthermore, it was found that H had a larger influence on the microscopic asperity, while σ had a larger influence on the macroscopic height of asperity. At the same time, the mechanical aperture, which means the separation distance between the corresponding joint surfaces, increased with the increase of σ, and decreased with the increase of H as well.

In order to validate the proposed method, two joint specimens were generated and printed by a 3D printer with the desired target joint properties. Then, the 3D laser scanning on the printed specimens was conducted to obtain surface information, and further, to calculate the roughness and mechanical aperture. By comparing the target values and estimated values, it was found that the proposed method was capable of generating joint specimens with satisfactory precision. It is anticipated that the proposed method can provide an effective and quantitative procedure for specimen preparation, which can be utilized in theoretical and numerical studies on the joint, and even in experimental works when using recent 3D printing technologies.

Author Contributions: Conceptualization, S.C.; investigation, S.C., S.L., and H.J.; supervision S.J.; writing original draft, S.C.; review and editing, S.J.

Acknowledgments: This research was supported by the National Strategic Project—Carbon Upcycling of the National Research Foundation of Korea (NRF), funded by the Ministry of Science and ICT (MSITT), the Ministry of Environment (ME), and the Ministry of Trade, Industry and Energy (MOTIE) (NRF—2017M3D8A2085654), and the Korea Agency for Infrastructure Technology Advancement under the Ministry of Land, Infrastructure and Transport of Korean Government (Project Number: 13 Construction Research T01). The institute of Engineering Research at Seoul National University (SNU) provided research facilities for this work. The authors are grateful for the support.

Conflicts of Interest: The authors declare no conflict of interest.

References

1. Makurat, A.; Barton, N.; Rad, N.S.; Bandis, S. Joint conductivity variation due to normal and shear deformation. In Proceedings of the International Symposium on Rock Joints, Leon, Norway, 4–6 June 1990; pp. 535–540, ISBN 90 6191 109 5.
2. Esaki, T.; Du, S.; Mitani, Y.; Ikusada, K.; Jing, L. Development of a shear-flow test apparatus and determination of coupled properties for a single rock joint. *Int. J. Rock Mech. Min. Sci.* **1999**, *36*, 641–650. [CrossRef]
3. Lee, H.S.; Cho, T.F. Hydraulic characteristics of rough fractures in linear flow under normal and shear load. *Rock Mech. Rock Eng.* **2002**, *35*, 299–318. [CrossRef]
4. Hakami, E.; Larsson, E. Aperture measurement and flow experiments on a single natural fracture. *Int. J. Rock Mech. Min. Sci.* **1996**, *33*, 395–404. [CrossRef]
5. Min, K.B.; Rutqvist, J.; Elsworth, D. Chemically and mechanically mediated influence on the transport and mechanical characteristics of rock fractures. *Int. J. Rock Mech. Min. Sci.* **2009**, *46*, 80–89. [CrossRef]
6. Brown, S. Fluid flow through rock joints: The effect of surface roughness. *J. Geophys. Res. Solid Earth* **1987**, *92*, 1337–1347. [CrossRef]

7.   Hakami, E. Aperture Distribution of Rock Fractures. Ph.D. Thesis, Division of Engineering Geology, Royal Institute of Technology, Stockholm, Sweden, 1995.

8.   Olsson, R.; Barton, N. An improved model for hydromechanical coupling during shearing of rock joints. *Int. J. Rock Mech. Min. Sci.* **2001**, *38*, 317–329. [CrossRef]

9.   Bandis, S.; Lumsden, A.; Barton, N. Fundamentals of rock joint deformation. *Int. J. Rock Mech. Min. Sci.* **1983**, *20*, 249–268. [CrossRef]

10. Pyrak-Nolte, L.J.; Morris, J.P. Single fractures under normal stress: The relation between fracture specific stiffness and fluid flow. *Int. J. Rock Mech. Min. Sci.* **2000**, *37*, 245–262. [CrossRef]

11. Zhao, J.; Brown, E. Hydro-thermo-mechanical properties of joints in the Carnmenellis granite. *Q. J. Eng. Geol.* **1992**, *25*, 279–290. [CrossRef]

12. Tse, R.; Cruden, D. Estimating joint roughness coefficients. *Int. J. Rock Mech. Min. Sci. Abstr.* **1979**, *16*, 303–307. [CrossRef]

13. Barton, N. Review of a new shear-strength criterion for rock joints. *Eng. Geol.* **1973**, *7*, 287–332. [CrossRef]

14. Barton, N.; Choubey, V. The shear strength of rock joints in theory and practices. *Rock Mech. Rock Eng.* **1977**, *10*, 1–54. [CrossRef]

15. Park, J.; Song, J. Numerical method for the determination of contact areas of a rock joint under normal and shear loads. *Int. J. Rock Mech. Min. Sci.* **2013**, *58*, 8–22. [CrossRef]

16. Singh, K.K.; Sing, D.N.; Ranjith, P.G. Laboratory simulation of flow through single fractured granite. *Rock Mech. Rock Eng.* **2015**, *48*, 987–1000. [CrossRef]

17. Yeo, I.; Freitas, M.; Zimmerman, R. Effect of shear displacement on the aperture and permeability of a rock facture. *Int. J. Rock Mech. Min. Sci.* **1998**, *35*, 1051–1070. [CrossRef]

18. Gentier, S.; Riss, J.; Archambault, G.; Flamand, R.; Hopkins, D. Influence of fracture geometry on shear behavior. *Int. J. Rock Mech. Min. Sci.* **2000**, *37*, 161–174. [CrossRef]

19. Raja, J.; Radhakrishnan, V. Analysis and synthesis of surface profiles using Fourier series. *Int. J. Mach. Tool Des. Res.* **1977**, *17*, 245–251. [CrossRef]

20. Patir, N. A numerical procedure for random generation of rough surfaces. *Wear* **1978**, *47*, 263–277. [CrossRef]

21. Xie, H. *Fractals in Rock Mechanics*; Balkema: Rotterdam, The Netherlands, 1993; ISBN 9054101334.

22. Seo, H.; Um, J. Generation of roughness using the random midpoint displacement method and its application to quantification of joint roughness. *Tunn. Undergr. Space* **2012**, *22*, 196–204. (In Korean) [CrossRef]

23. Iwano, M.; Einstein, H. Stochastic analysis of surface roughness, aperture and flow in a single fracture. In Proceedings of the EUROCK 1993, Lisbon, Portugal, 21–24 June 1993; pp. 135–141.

24. Choi, S.; Lee, S.; Jeon, S. Generation of a 3D artificial joint surface and characterization of its roughness. *Tunn. Undergr. Space* **2016**, *26*, 516–523. (In Korean) [CrossRef]

25. Grasselli, G.; Egger, P. Constitutive law for the shear strength of rock joints based on three-dimensional surface parameters. *Int. J. Rock Mech. Min. Sci.* **2003**, *40*, 25–40. [CrossRef]

26. Park, J.; Lee, Y.; Song, J.; Choi, B. A constitutive model for shear behavior of rock joints based on three-dimensional quantification of joint roughness. *Rock Mech. Rock Eng.* **2013**, *46*, 1513–1537. [CrossRef]

27. Esaki, T.; Du, S.; Jiang, Y.; Wada, Y.; Mitani, Y. Relation between mechanical and hydraulic apertures during shear-flow coupling test. In Proceedings of the 10th Japan Symposium on Rock Mechanics, Osaka, Japan, 22–23 January 1998; pp. 91–96.

28. Riss, J.; Gentier, S.; Sirieix, C.; Archambault, G.; Flamand, R. Degradation characteristics of sheared joint wall surface morphology. In Proceedings of the 2nd North America Rock Mechanics Symposium, Montreal, QC, Canada, 19–21 June 1996; pp. 1343–1349, ISBN 90 5410 838 X.

29. Hakami, E.; Stephansson, O. Experimental technique for aperture studies of intersecting joints. In Proceedings of the ISRM International Symposium—Eurock 93, Lisboa, Portugal, 21–24 June 1993; pp. 301–309.

Article

Analyzing the Stability of Underground Mines Using 3D Point Cloud Data and Discontinuum Numerical Analysis

Seung-Joong Lee and Sung-Oong Choi *

Department of Energy and Resources Engineering, Kangwon National University, Chuncheon 24341, Korea;
lhj3601@kangwon.ac.kr
* Correspondence: choiso@kangwon.ac.kr; Tel.: +82-33-250-6250

Received: 28 January 2019; Accepted: 11 February 2019; Published: 13 February 2019

Abstract: This study describes a precise numerical analysis process by adopting the real image of mine openings obtained by light detection and ranging (LiDAR), which can produce a point cloud data by measuring the target surface numerically. The analysis target was a section of an underground limestone mine, to which a hybrid room-and-pillar mining method that was developed to improve ore recovery was applied. It is important that the center axis and the volume of the vertical safety pillar in the lower parts match those in the upper parts. The 3D survey of the target section verified that the center axis of the vertical safety pillar in the lower parts had deviated in a north-westerly direction. In particular, the area of the lower part of the vertical safety pillar was approximately 34 m^2 lower than the designed cross-sectional area, which was 100 m^2. In order to analyze the stability of the vertical safety pillar, a discontinuum numerical analysis and safety factor analysis were conducted using 3D surveying results. The analysis verified that instability was caused by the joints distributed around the vertical safety pillar. In conclusion, investigation of the 3D survey and 3D numerical analysis techniques performed in this study are expected to provide higher reliability than the current techniques used for establishing whether mining plans require new mining methods or safety measures.

Keywords: LiDAR; pillar stability; discontinuum analysis

---

1. Introduction

The sustainable development of mineral resources globally has become an important focus area of contemporary research. In Korea, not only have overseas resources been developed, but abandoned mines have been reopened, improvements have been made to the ore recovery of currently operational mines, and technology for the selective mining of high-grade ores has been developed. Mining advancement requires not only the development of technology to increase productivity and maximize efficiency in mining work, but also the effective safety management of mines to prevent mining damage and hazards, such as subsidence.

Stability evaluation methods that may be utilized include safety management of mines, rock mass classification, safety factor analysis using empirical methods, surveying methods utilizing sensors, and numerical analysis. With the recent advancement in optical technologies, three-dimensional (3D) surveying techniques, such as 3D laser scanning and stereo photogrammetry methods, have been introduced to the mining industry and utilized actively in studies on stability evaluation.

Light detection and ranging (LiDAR), a type of 3D laser scanner, is employed in surveying equipment that emits light and measures the reflecting and returning light. LiDAR, as 3D point cloud data can be easily acquired, has been utilized rapidly and widely in various sectors, including civil engineering, construction, aviation, archeology, geology, and the game and movie industries. In the rock engineering field, LiDAR has been employed for studies on the evaluation of joint roughness [1–4], trace sampling of joints [5–7], analysis of joint orientation [8], analysis of slope stability [9,10], analysis of rockfall and toppling [11,12], analysis of blasting over-break [13,14], and geometry research of underground mines [15]. The utilization of LiDAR is expected to increase in the future in study fields that require quantitative analysis.

In this study, surveying was conducted using LiDAR to evaluate the stability of the safety pillar and stope in a mine where the hybrid room-and-pillar mining method was used, and the results were utilized directly to perform 3D stability analysis.

2. Overview of the Field

The target mine in this study, the Daesung MDI Donghae limestone mine, is located at Donghae-si, Gangwon-do, South Korea. Here, a test section of a horizontal cross-section (50 m × 50 m) around the Lv540 stope was excavated and mining work was conducted using the hybrid room-and-pillar mining method. This method is used to maximize ore recovery by collecting ores while ensuring the stability of stope using horizontal and vertical safety pillars. As shown in Figure 1, this mining method consists of a number of distinct steps: a step for gallery excavation (Figure 1a), a step for opening the upper stope (Figure 1b), a step for opening a middle stope (Figure 1c), a step for an additional opening for the upper stope (Figure 1d), a step for opening the lower stope (Figure 1e), and a step for an additional opening for the lower stope (Figure 1f). The mining method applies the opening order while considering the stability of the pillar, according to the field circumstances. To ensure the stability of a large vertical safety pillar and stope during mining work, the ores have to be recovered safely from the horizontal safety pillar. In order to do this, it is important that the center axis and the volume of the vertical safety pillar in the lower parts match those in the upper parts.

The site survey was conducted after mining ores using a general room-and-pillar mining method at the Lv520 and Lv540 stopes, as shown in Figure 2a, and then recovering ores from the horizontal safety pillar at a rate of approximately 40%. The site survey was undertaken due to an additional opening of the upper stope (Figure 1d). As verified in the planning cross-section in Figure 2b, a vertical safety pillar whose volume was 10 m × 10 m × 7 m (width, length, and height, respectively) was formed in each of the stopes. Once the 8 m thick horizontal safety pillars were all recovered, a large vertical safety pillar, 10 m × 10 m × 22 m, was left to support the upper level.

The survey results of the orientation of the joint showed that the dip direction/dip of joint set 1 distributed over the study target area was 186/71, and the dip direction/dip of joint set 2 was 272/69. Joint sets 1 and 2 were distributed as they crossed in the study target area (Figure 3). Since these two joints may act as unstable factors for the stability of the large vertical safety pillar meant to support the hanging-wall block, a comprehensive stability analysis of stopes was necessary.

This study conducted 3D surveying and discontinuum numerical analysis, using LiDAR through the study method shown in Figure 4. LiDAR was used in order to analyze the stability of the stope and the vertical safety pillar formed in the upper and lower levels of the stope at the time when the site investigation was conducted.

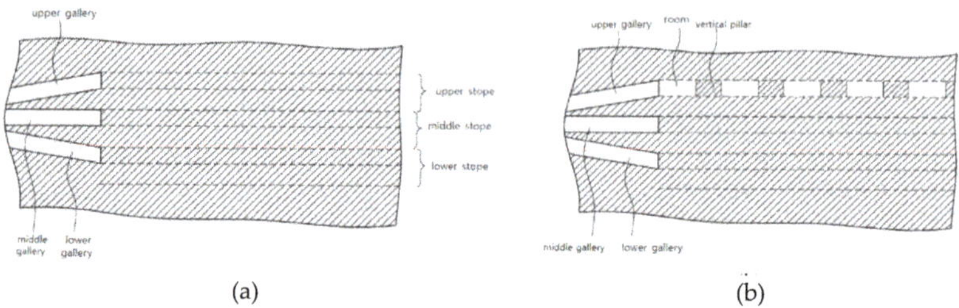

(a)   (b)

Figure 1. *Cont.*

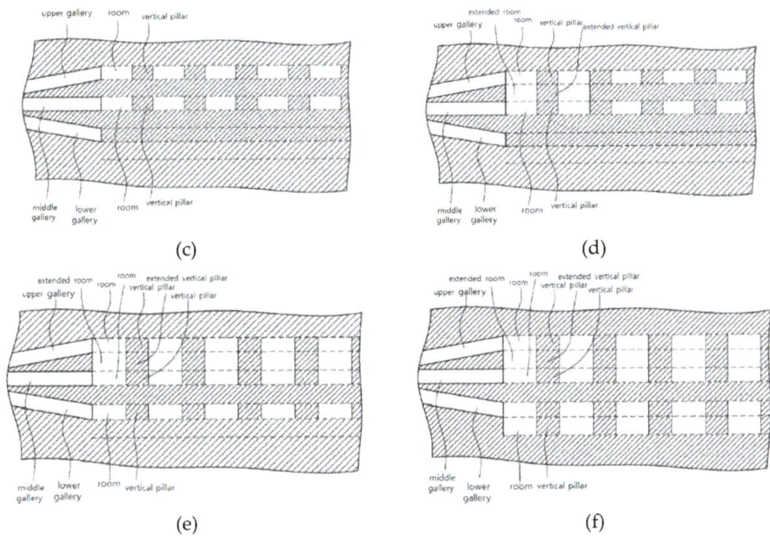

Figure 1. Sequence of hybrid room-and-pillar mining method [16]. (a) Gallery excavation; (b) opening of upper stope; (c) opening of middle stope; (d) additional opening of upper stope; (e) opening of lower stope; (f) additional opening of lower stope.

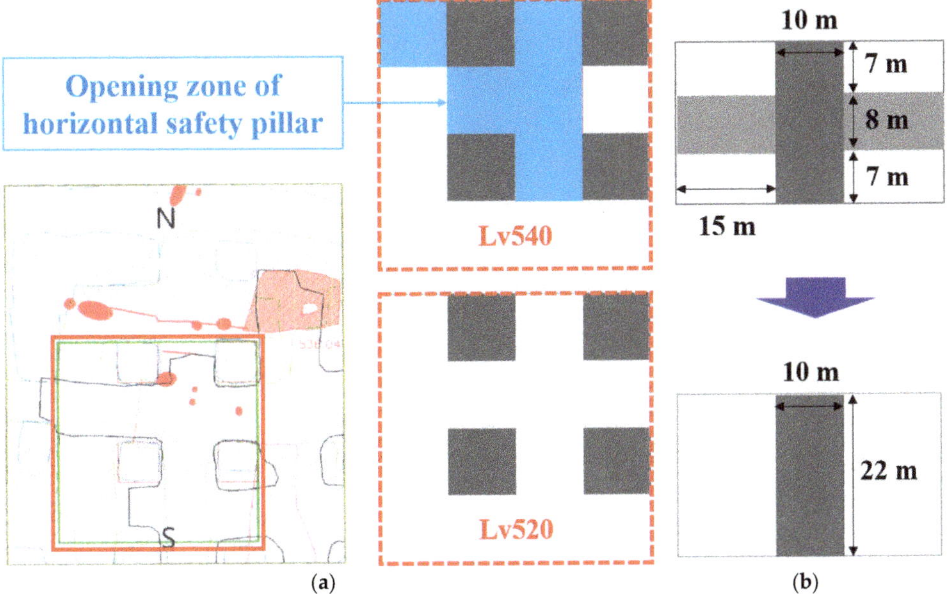

Figure 2. Plan view of excavation state, and section view of planned pillar in test bed zone. (a) Plan view of excavation state in test bed zone; (b) section view of planned pillar.

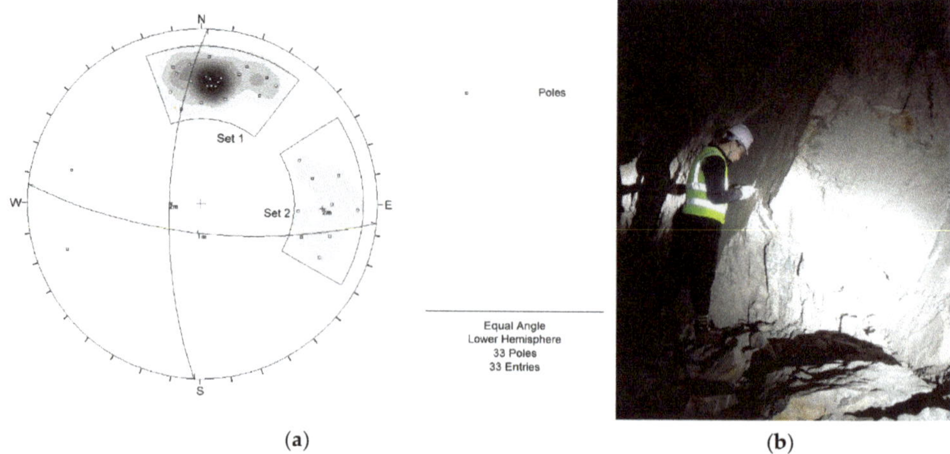

(a)   (b)

Figure 3. Results of joint distribution patterns in the site. (**a**) Orientation of joint sets in the site; (**b**) site investigation in a mine.

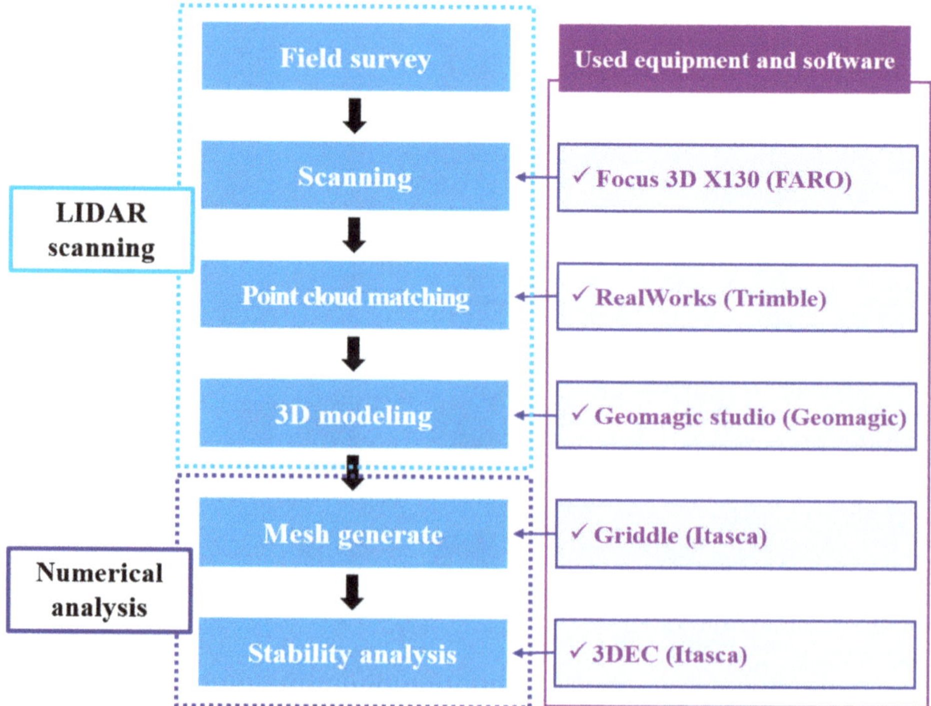

Figure 4. Process of light detection and ranging (LiDAR) scanning and numerical analysis.

3. 3D Surveying of Stope Utilizing LiDAR

Surveying work was conducted to acquire 3D point cloud data about the stope and vertical safety pillar. The continuous surface model was created using point cloud data, and analyses of cross-sections and volumes of the mining area and vertical safety pillar were conducted using this model.

3.1. Principle of LiDAR Measurement

LiDAR measures the distance something is located away from the measurement equipment by emitting light and measuring the reflecting and returning light accurately. A total station, which is surveying equipment that uses a laser, measures a distance at a single point, one point at a time, whereas LiDAR has the advantage of quickly acquiring multiple points using a mirror that rotates rapidly [8].

LiDAR measurement is divided into time of flight (ToF) and phase shift (PS) modes. The principle of the ToF mode is to measure distance by measuring the reflection time of a laser pulse signal from an object to a receiver within the measurement range. The principle of the PS mode is to calculate time and distance by measuring the phase component difference in signals reflected from the object within the measurement range, after radiating a laser beam that is continuously modulated to a specific frequency. In this study, PS mode was used in considering the volume and precision of the stope. Table 1 shows the specification of Faro Focus3D X130.

Table 1. Specification of Faro Focus3D X130.

Item	Specification
Measuring method	Phase shift
Range	0.6~130 m
Maximum measurement speed	976,000 points/sec
Ranging error	±2 mm at 25 m
Vertical field of view	305°
Horizontal field of view	360°
Vertical/Horizontal step size	0.015°

3.2. Surveying Work and Point Cloud Data Matching

Considering the site accessibility and point cloud data overlap, five points were selected in the Lv540 stope to conduct surveying work. However, surveying work was only conducted at two points in the Lv520 stope, as due to the recovery work of the horizontal safety pillar in the west area, crushed stones were piled up making this area inaccessible. The LiDAR measurement angle was set to the maximum to reduce measurement error due to obstacles or dust at each position during the measurement [17].

Figure 5 shows the point cloud data acquired at each position in the Lv540 stope. Figure 6 shows the point cloud data acquired at each position in the Lv520 stope. Various methods were available to achieve this, including coded target matching, manual matching by setting an arbitrary feature point, automatic matching through software for statistical processing of all point cloud data, and smart matching, all of which could have been employed. The smart matching technique was selected to sequentially match the point cloud data acquired at each of the measurement positions. The point cloud data acquired at SP1 to SP5 shown in Figure 5, at each position in the Lv540 stope, were matched sequentially. Later, the point cloud data acquired at SP6 and SP7 shown in Figure 6, at each position in the Lv520 stope, were matched sequentially. The matching precision in each stage was above 98%. Figure 7 shows the 3D image of the point cloud data matched sequentially from SP1 to SP7 by smart matching method.

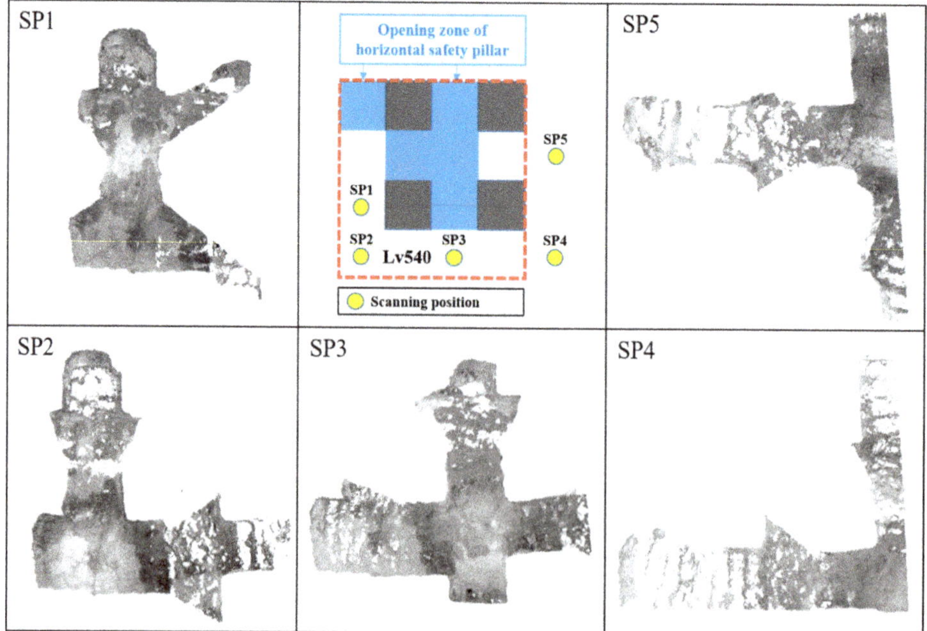

Figure 5. 3D visualization of point cloud obtained from each scanning position at Lv540.

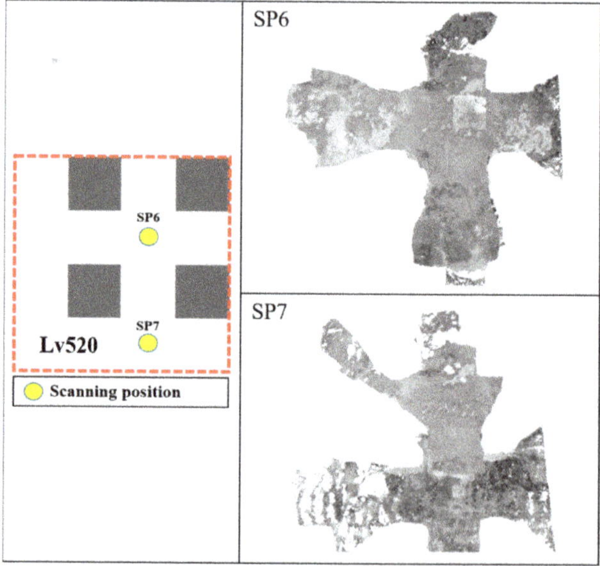

Figure 6. 3D visualization of point cloud obtained from each scanning position at Lv520.

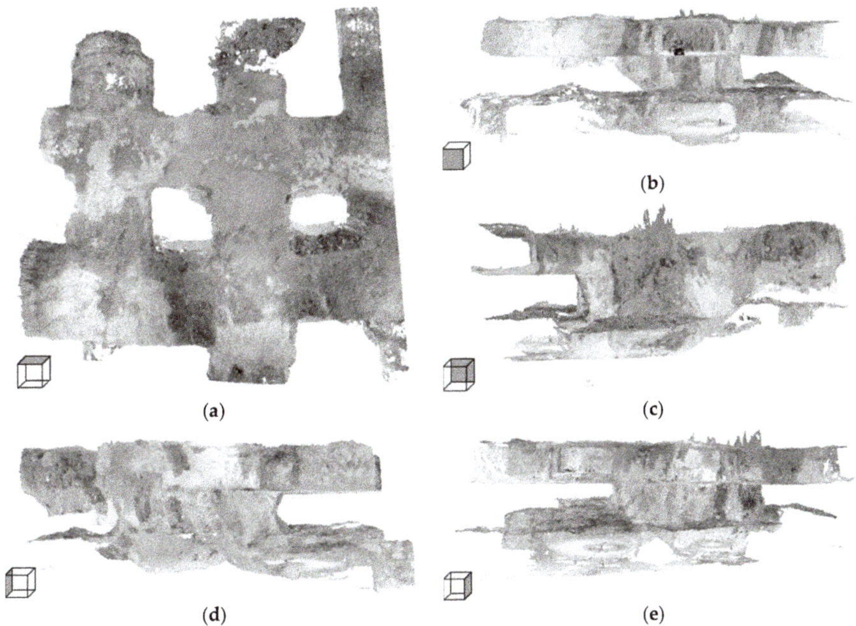

Figure 7. 3D visualization of point cloud matched by smart matching method. (**a**) Top view; (**b**) front view; (**c**) back view; (**d**) left side view; (**e**) right side view.

3.3. Analysis of Surveying Results

Figure 8 shows the model by which the matched point cloud data was converted to a 3D solid image that indicated a mining area in the stope during the site investigation. This 3D solid image was made using the automatic TIN mesh generation option of the Geomagic studio software. The volume analysis results revealed the total volume of the mining area to be 44,400 m^3. Thus, at the time of the site investigation, the mining quantity could be estimated at 119,800 tons, considering that the unit weight of limestone was 2.7 ton/m^3 (Figure 9).

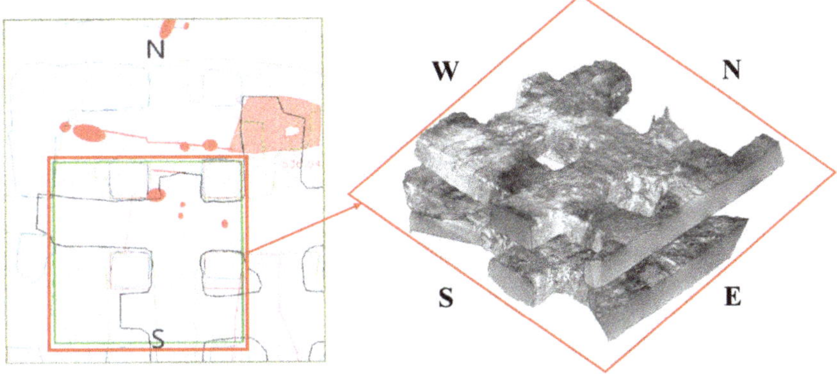

Figure 8. 3D solid image of opening zone.

Figure 9. Volume analysis of opening zone.

3.3.1. Cross-Section Analysis

The 3D solid model was employed for length surveying and a cross-section analysis of the area where a 22-m-high vertical safety pillar would later be formed. Once the mining work in the test area was complete, a total of four vertical safety pillars were left in the stope. However, the surveying work was conducted when the final excavation faces of the No. 1 vertical safety pillar (Pillar 1) and No. 2 vertical safety pillar (Pillar 2) were about 80% formed, as shown in Figure 10. Thus, the cross-section analysis was based on these two vertical safety pillars.

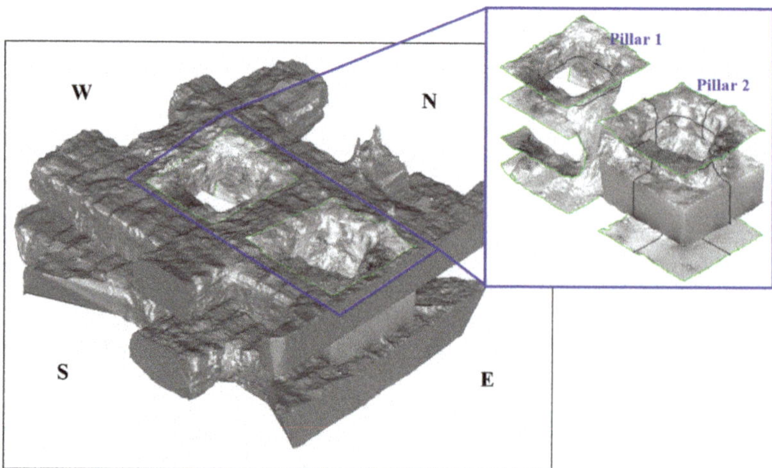

Figure 10. Formed vertical safety pillars in the test bed.

Because crushed stones were piled up in the west area of the Lv520 stope due to the recovery work on the horizontal safety pillar, it was not accessible, so accurate surveying work could not be performed in that area. Thus, for the analysis of Pillar 1, cross-section analysis was conducted at the Lv540 stope area. For the analysis of Pillar 2, both cross-section and longitudinal section analyses were conducted around the vertical safety pillars formed at the Lv520 and Lv540 stopes. The shapes of cross-section and longitudinal section were extracted using Geomagic studio software for section analysis. The measurement module of software was used for the length and area analysis of each section.

The cross-section analysis results for Pillar 2 indicated that the dimensions of the cross-section of Pillar 1 formed in the Lv540 stope were 11.71 m wide and 12.70 m long, with an area of approximately 141.61 m^2 (Figure 11). The shape of Pillar 2 formed in the Lv540 stope was more irregular than that of Pillar 1, with dimensions of 12.10 m to 15.50 m wide and 12.49 m long, and an area of approximately 161.33 m^2 (Figure 12). By contrast, the cross-section area of Pillar 2 formed in the Lv520 stope was 66.16 m^2, smaller than that of the vertical safety pillar in the Lv540 stope by 95.17 m^2.

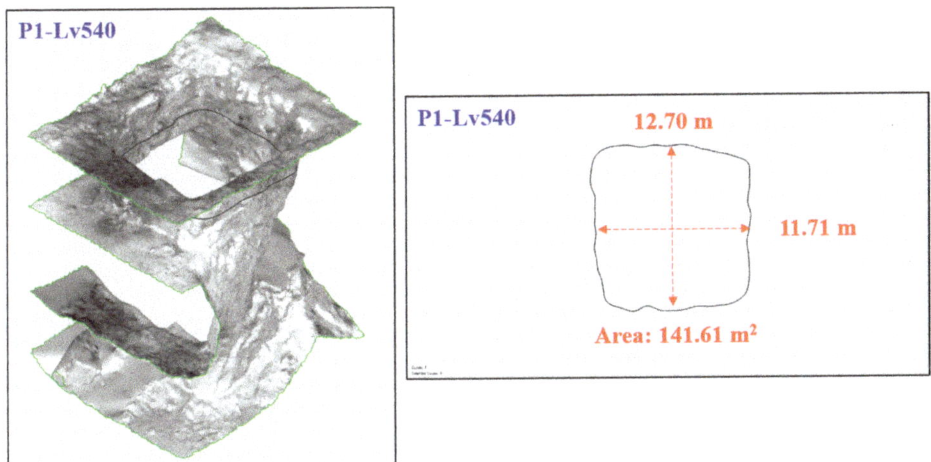

Figure 11. Cross section of Pillar 1.

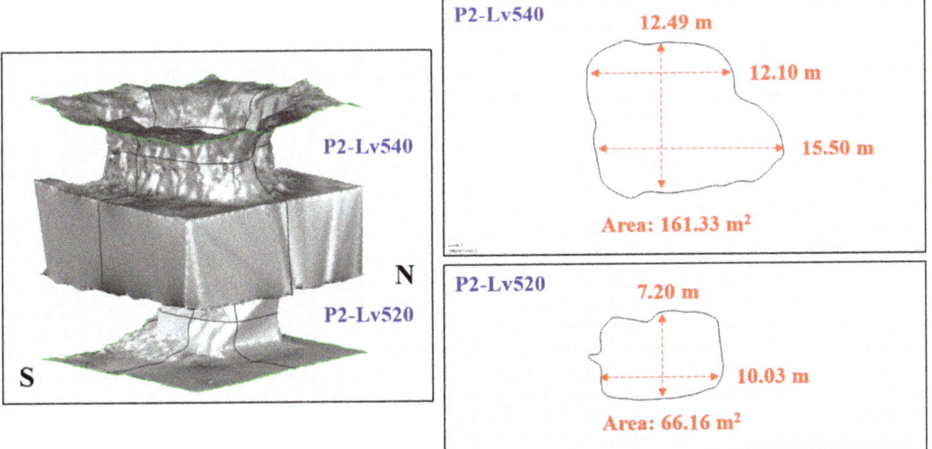

Figure 12. Cross section of Pillar 2.

The analysis results for the longitudinal section of Pillar 2 showed that the height of the vertical safety pillar was about 22.00 m in the P2-NS cross-section and approximately 23 m at the P2-EW cross-section, and its average height was approximately 22.50 m (Figure 13).

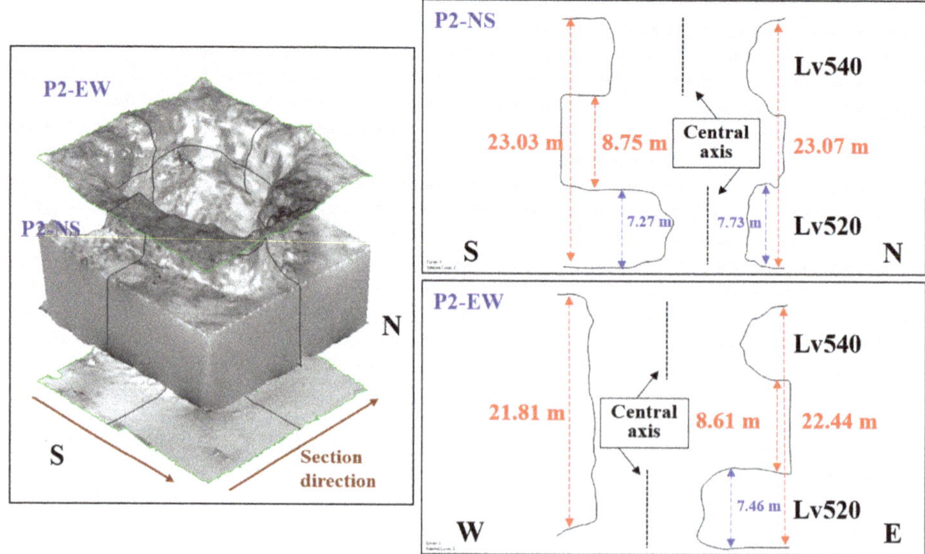

Figure 13. Longitudinal section of Pillar 2.

The above cross-section analysis results and design data of the hybrid room-and-pillar mining method were compared and analyzed. The cross-section area of the vertical safety pillar was designed to be 10 m wide and 10 m long (giving an area of 100 m^2), but the actual cross-section area of the vertical safety pillar formed in the stope was in the range of 66.16 m^2 to 161.33 m^2.

The results of the analysis verified that over-break and under-break areas were present locally in the vertical safety pillar formed in the stope. In addition, not only was the cross-section area of the vertical safety pillar formed in the Lv520 stope smaller than that of the vertical safety pillar formed in the Lv540 stope, but the center axis of the vertical safety pillar deviated to the north west (Figures 12 and 13).

This was due to the geological factor being the same as the joint distributed around the vertical safety pillar. As steep dip joints over 70° were distributed in the stope, the over-break area deviated to the south east due to separation behavior, such as the sliding of the joint, rather than the rock-crushing effects of blasting force.

3.3.2. Analysis of Orientation of the Joint

The site investigation of the joints was conducted only at the lower section of the vertical safety pillar, which was easily accessible. In contrast, the site investigation was not conducted at inaccessible areas, such as the areas where the horizontal safety pillar was recovered and where the 22 m high excavation face was formed. To overcome this site investigation challenge, a Realworks software-embedded joint orientation analysis module and a 3D point cloud were used to perform accurate and in-depth analysis of the orientation of the joints distributed around the entire stope.

Figures 14 and 15 shows the 3D point cloud image of the second vertical safety pillar in the Lv520 stope and the orientation of the joints extracted from the image. Using this method, data on 79 joints with distinctive orientation were extracted from the entire stope area.

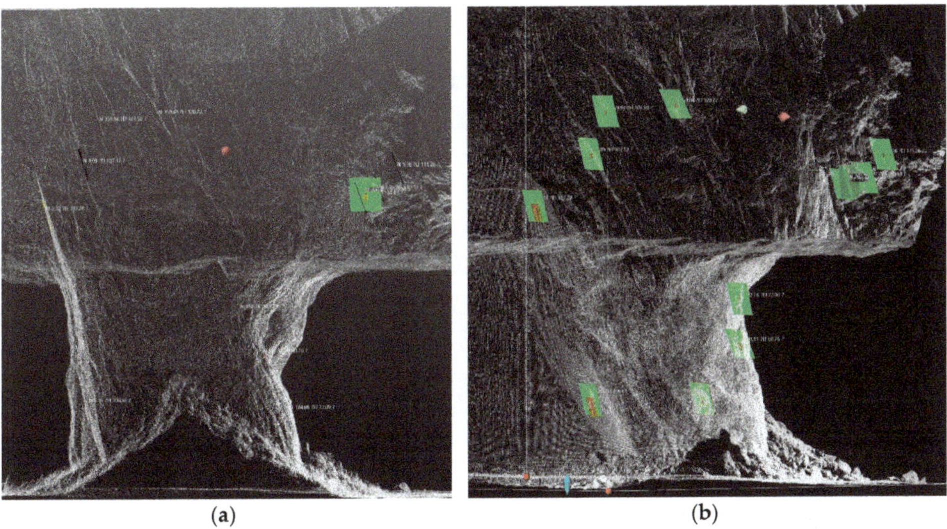

(a)   (b)

Figure 14. Obtained distribution pattern of joint set 1 from 3D point cloud. (**a**) Westerly view; (**b**) north-easterly view.

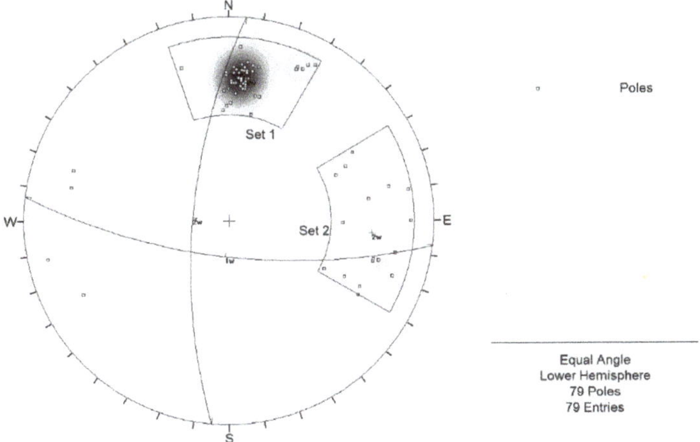

Figure 15. Analyzed orientation of joint sets from 3D point cloud.

The analysis results on the orientation of the joints showed that the dip direction/dip of joint set 1 was 185/70 and that of set 2 was 277/70.

As verified from cross-section analysis and orientation analysis results in Figure 14, the over-break occurred in the vertical safety pillar in the Lv520 stope. This over-break was concentrated in a southerly direction because the separation behavior of joint surface was caused by the effect of joint set 1 and blasting work.

Permanent support and reinforcement work are generally not conducted in the stope due to the characteristics of mine sites. The hybrid room-and-pillar mining method was pilot-operated to improve ore recovery in the stope in the test area. Thus, reinforcement work was already conducted with shotcrete and rockbolt at the vertical safety pillar and roofs to ensure the safety of workers and external visitors who visited the test area for research purposes.

Currently, no unstable factors were found during the mining phase. However, the center axes of Pillar 2 formed in the Lv540 and Lv520 stopes were not matched, and the cross-section areas differed. In addition, steep dip joints that could affect the stability of the vertical safety pillar were distributed. As such, stability analysis would be necessary for safer mining work.

4. Discontinuum Numerical Analysis for Stability Analysis in the Stope

Safety diagnosis and reinforcement measures are necessary to ensure stability in the vertical safety pillar during the mining period. This study performed 3D discontinuum numerical analysis, utilizing 3D point cloud data acquired at the site, to analyze the stability of the stope.

4.1. Numerical Analysis Model

The numerical analysis model consisted of 3D block models created in the 3DEC software package using 3D point cloud data and used the Griddle software package to reflect the actual shape of the stope. In addition, the boundary of the model was set to 100 m × 80 m × 40 m (width × length × height, respectively), considering the boundary effect of the 3D block model and mean depth of the test area, as shown in Figure 16.

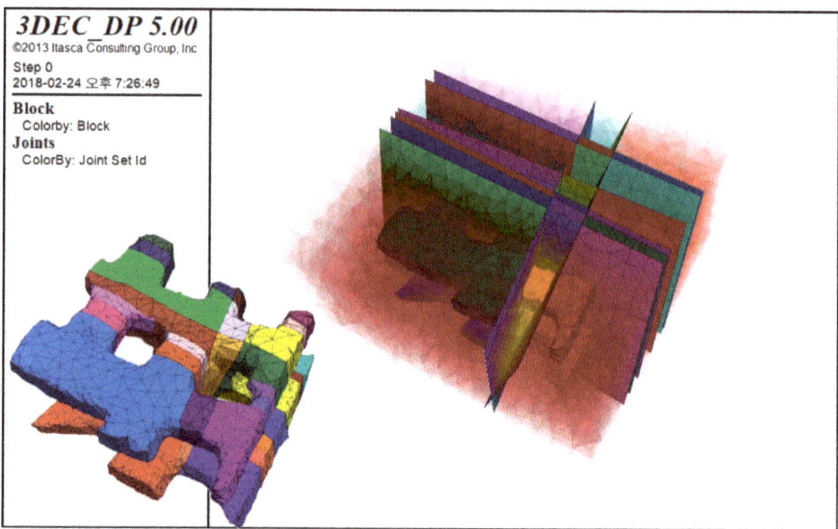

Figure 16. Generation of the numerical model for the test bed.

The joint expressed in the analysis model simulated the actual joint distributed in the vertical safety pillar. To improve the efficiency of the numerical analysis, a total of eight joints with distinctive orientation were chosen from joint sets 1 and 2. These were created on the basis of the center coordinate where the orientation analysis was complete. To generate input data for numerical analysis, a ground integer set was calculated based on the indoor rock property test, and site investigation data were utilized (Tables 2 and 3).

Table 2. Physical and mechanical properties of intact rock used in numerical analysis.

Unit Weight (ton/m^3)	Modulus of Deformation (GPa)	Poisson's Ratio	Cohesion (MPa)	Friction Angle (°)	Tensile Strength (MPa)
2.73	50.19	0.23	6.95	45.79	6.58

Table 3. Physical and mechanical properties of joint used in numerical analysis.

Normal stiffness (MPa/mm)	Shear stiffness (MPa/mm)	Cohesion (MPa)	Friction angle (°)
6.74	7.41	0.22	32.0

4.2. Numerical Analysis Results

The stability analysis was conducted based on the vertical safety pillar, and an additional stability investigation was conducted for areas that posed a risk of instability.

Figure 17 shows the distribution diagram of the displacement occurrence, displayed on the basis of the center cross-section of the vertical safety pillar. Maximum displacement occurred in the stope at approximately 7 mm. The maximum displacement that occurred in Pillar 2 was a southerly shear displacement of up to 5 mm, a dip direction for joint set 1 and, in the westerly direction, a dip direction for joint set 2 (Figure 18). However, since the displacement that occurred was relatively small, no significant stability problems would be expected currently in the stope and vertical safety pillar during the mining phase.

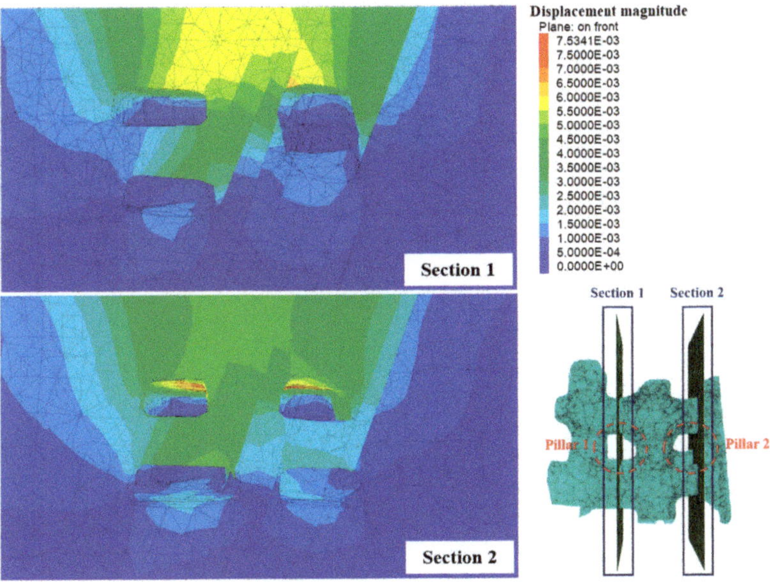

Figure 17. Displacement distribution of central section for each vertical safety pillar.

The analysis results for the maximum principal stress showed that a stress concentration section of about 1.0 MPa was present in Pillar 2 in the Lv520 stope, as verified in Section 2 of Figure 19. The stress concentration phenomenon was investigated in each direction, and the results showed that the tensile stress at a level of 1.0 MPa seemed concentrated locally in the northerly, southerly, and easterly directions of the pillar. In addition, compressive stress at a level of 4.0 MPa and tensile stress at a level of 2.0 MPa were also generated in the lower end portion in the westerly direction of the pillar (Figure 20). Furthermore, a shear failure pattern was verified due to stress concentration at the lower end portion of the vertical safety pillar, as shown in Figure 21. This was because the confining stress was released by the recovery work of the horizontal safety pillar which restrained joint set 2. As a result, the supported load of the vertical safety pillar was increased, thereby generating shear displacement along the dip direction of joint set 2.

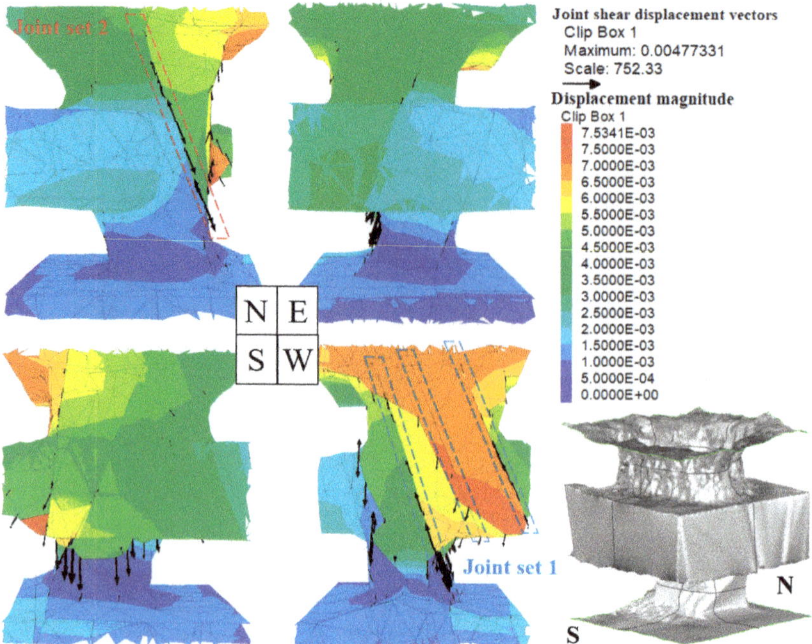

Figure 18. Displacement distribution of vertical safety pillar No. 2.

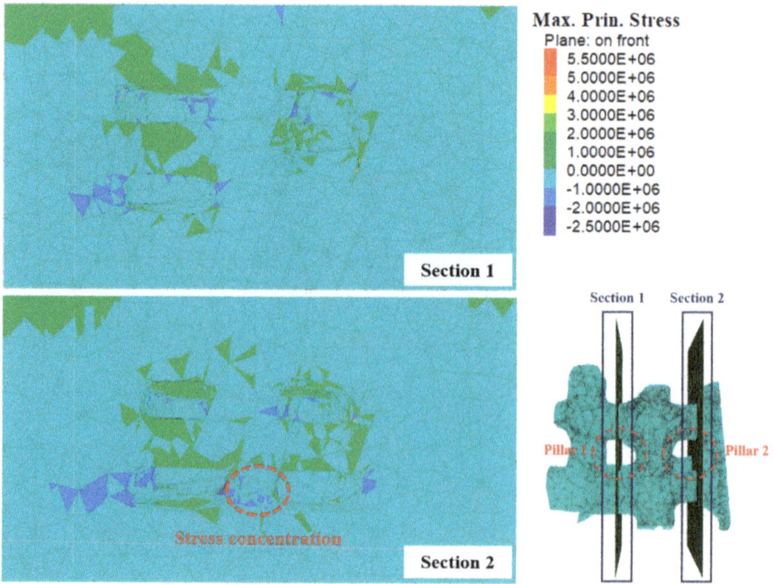

Figure 19. Maximum principal stress distribution of central section for each vertical safety pillar.

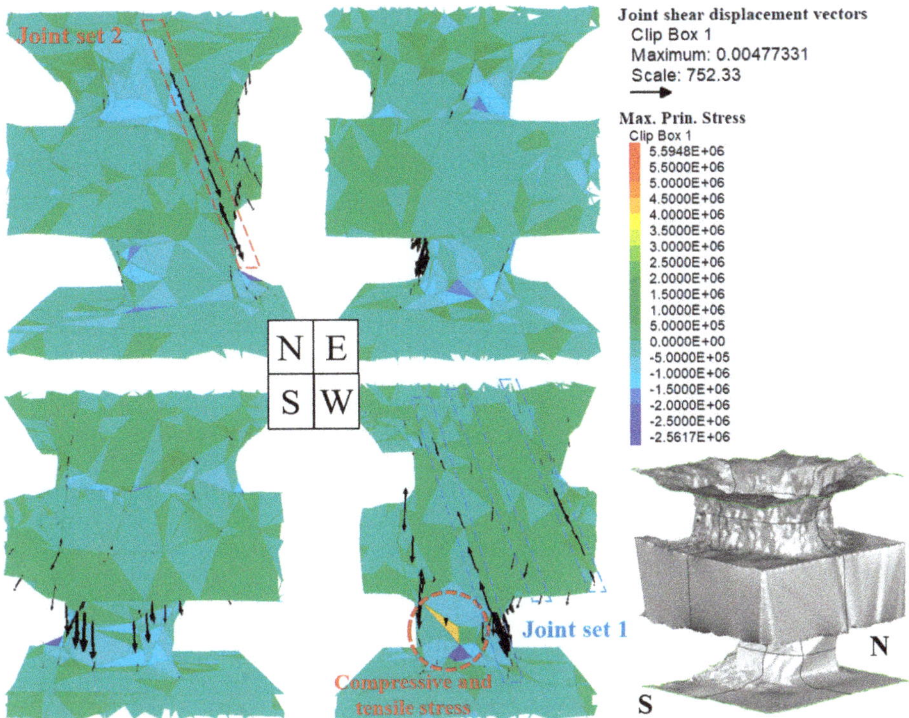

Figure 20. Maximum principal stress distribution of vertical safety pillar No. 2.

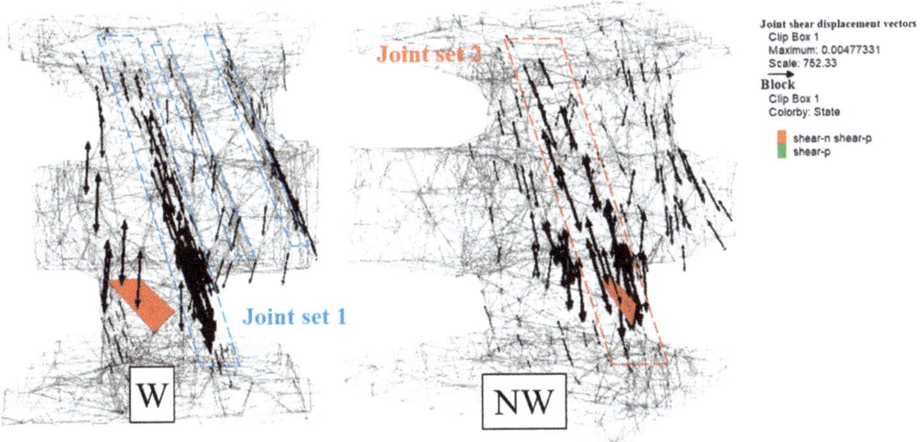

Figure 21. Failure distribution of vertical safety pillar No. 2.

Accordingly, the vertical safety pillar was unstable due to joint set 2. However, since the cross-section area of the vertical safety pillar in the Lv520 stope was smaller than that of the Lv540 stope and the center axis of the pillar deviated towards the north west, as verified in the cross-section analysis results, the effect of joint set 1 could not be overlooked if the rest of the horizontal safety pillar would be recovered later.

Currently, it seemed that the overall stability of the stope had no problems. However, since stress concentration and shear failure were present in the vertical safety pillar of the Lv520 stope, it was necessary to have a mining plan that would ensure stability during the recovery of the remaining horizontal safety pillar.

5. The Safety Factor of the Vertical Safety Pillar

The site investigation results showed that instability in the stope was not verified for the current excavation phase. However, if the horizontal safety pillar were to be recovered later, it could cause instability. Thus, taking a conservative stance, the safety factor of the vertical safety pillar was investigated.

The safety factor of Pillar 2 in the Lv520 stope was analyzed using the results derived from empirical equations proposed by previous researchers.

First, the safety factor in the empirical equations is the ratio between pillar strength and average pillar stress, which can be calculated using Equation (1). A number of empirical equations were proposed by previous researchers to calculate pillar strength. This study employed those proposed by Greenwald et al. [18] (Equation (2)) and Holland and Gaddy [19] (Equation (3)) to calculate pillar strength. The average pillar stress was calculated from the theoretical equation (Equation (4)) of the tributary area.

$$F = \frac{S_p}{\sigma_p},\tag{1}$$

$$S_p = 0.67 \times K \times \left(\frac{W_p^{0.5}}{H^{0.83}}\right),\ K = \sigma_c \times \sqrt{D},\tag{2}$$

$$S_p = K \times \left(\frac{W_p^{0.5}}{H}\right),\ K = \sigma_c \times \sqrt{D},\tag{3}$$

$$\sigma_p = \gamma \times Z \times \left(\frac{W_p + W_g}{W_p}\right)^2.\tag{4}$$

Here, F is the safety factor, S_p is the pillar strength, σ_p is the average pillar stress, K is the coefficient due to critical strength, H is the pillar height, W_p is the pillar width, σ_c is the uniaxial compressive strength, D is the diameter of the sample, γ is the unit weight, Z is the depth, and W_g is the width of the gallery.

The safety factor was calculated based on the planned cross-section at the mining planning phase, and the cross-section was measured using LiDAR at the actual stope (Figures 22 and 23, respectively). For the calculation of the safety factor, the state prior to the recovery of the horizontal safety pillar was considered to calculate the average pillar stress. In addition, an irregular cross-section shape of the pillar was assumed to be a square cross-section shape to calculate the pillar width. The pillar width, as reverse-calculated from the cross-section area, was 8.13 m. The tributary area was calculated by measuring the distance from the center of the vertical safety pillar to its neighboring pillar, as shown in Figure 23a. The average width of the gallery was calculated from the difference in the pillar width, and the reverse-calculated distance from the tributary area was 15.37 m. For the pillar height, the measured values in each of the directions was averaged and found to be 7.49 m. This value was then applied and used in the calculation of the safety factor.

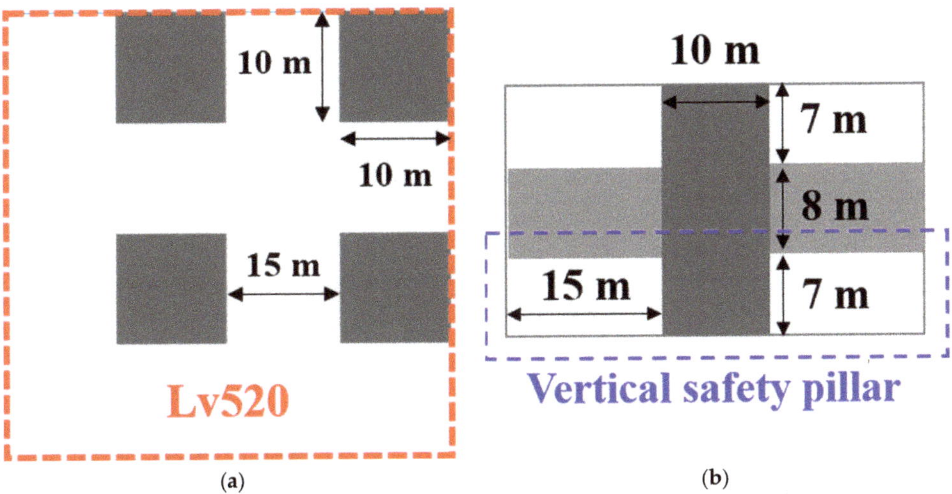

Figure 22. Planned section of Lv520 vertical pillar. (**a**) Cross section; (**b**) longitudinal section.

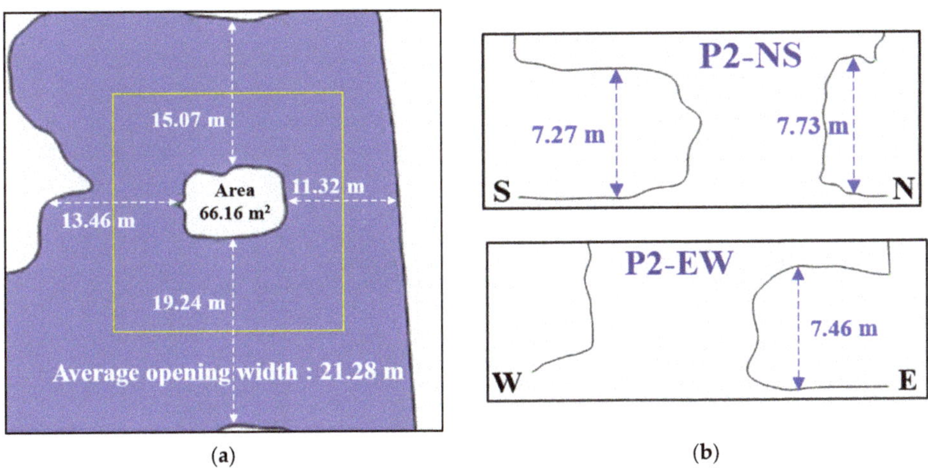

Figure 23. Measured section of Lv520 vertical pillar. (**a**) Cross section; (**b**) longitudinal section.

The safety factor that was analyzed via the previously mentioned empirical equations proposed by Greenwald et al. [18] and by Holland and Gaddy [19], obtained values of 2.0 and 2.2 at the planned cross-section, and 1.3 and 1.4 at the measured cross-section, respectively (Table 4). According to Esterhuizen et al. [20], the stability of the stope during the mining period has no problems if the safety factor is 1.5 or greater. Based on the analysis results, the safety factor of the stope at the time of the mining plan was within a safe level. However, as the cross-section area of the pillar became smaller and the width of the gallery increased due to the mining work, the safety factor was reduced to 1.5 or less.

Table 4. Calculated safety factors of pillar by each empirical equation.

	UCS	D	K	W_p	H	γ	Z	W_g	S_{pg}	S_{ph}	σ_p	F_g	F_h
Planned section	146	50	32.65	10.00	7.00	2.7	40	15.00	13.76	14.75	6.75	2.0	2.2
Measured section	146	50	32.65	8.13	7.49	2.7	40	15.37	11.73	12.43	9.02	1.3	1.4

UCS: Uniaxial Compressive Strength (MPa), D: Diameter of specimen (cm), K: Parameter by critical strength, W_p: Pillar width (m), H: Pillar height (m), γ: The unit weight of the overlying (ton/m^3), Z: Depth (m), W_g: Opening width (m), S_{pg}, S_{ph}: Pillar strength by Greenwald et al. [18], Holland & Gaddy (MPa) [19], σ_p: Pillar stress (MPa), F_g, F_h: Safety factor of pillar by Greenwald et al. [18], Holland & Gaddy [19].

As aforementioned, the safety factor analyzed with the measured cross-section was somewhat unstable. Whether or not this instability also occurred in the numerical analysis that utilized the actual measured data was checked. The safety factor was defined as the ratio of shear stress to the shear strength of the joint, as presented in Equation (5), and the safety factor of Pillar 2 was investigated using this equation:

$$\frac{\text{Shear strength}}{\text{Shear stress}} = \frac{\tau_{s,\,max}}{\tau_s}. \tag{5}$$

Here, $\tau_{s,\,max}$ refers to the maximum shear stress that can incur shear failure, assuming that the minimum effective stress applied to the element is σ_u; τ_s refers to the amount of shear stress applied to the actual element, meaning that the safety factor indicates the possibility of shear failure if it is less than 1.0.

Figure 24 shows the results, which display the ratio of the maximum shear stress with regards to the shear strength of the joint formed in Pillar 2. This figure shows that the safety factor of joint set 2 formed in the vertical safety pillar is 1.0 overall, and the safety factor of joint set 1 is greater than 2.0 overall, although falling below this in some sections. In summary, Pillar 2 was unstable due to joint set 2, which was shown to be relatively vulnerable.

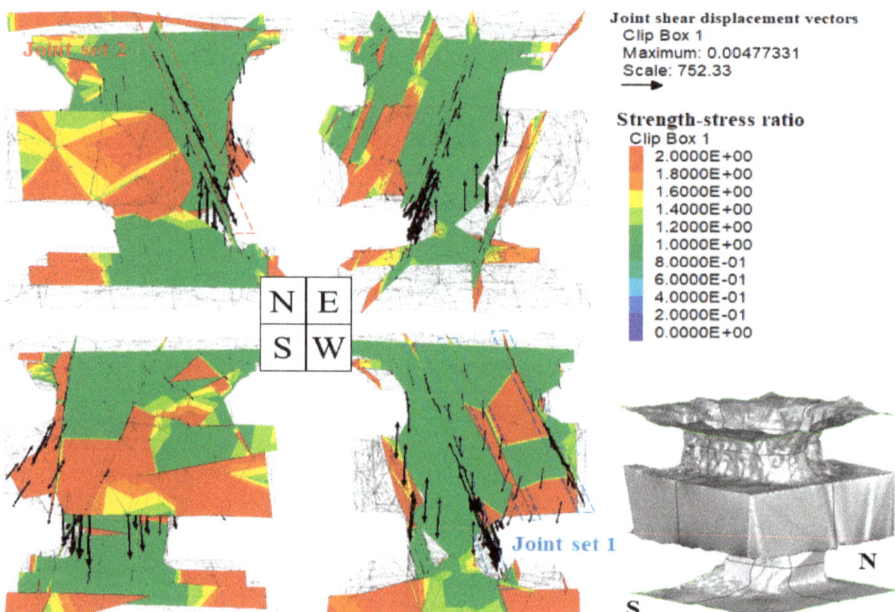

Figure 24. Ratio of shear strength and shear stress.

6. Conclusions

This study conducted 3D surveying and 3D discontinuum numerical analysis using LiDAR to analyze the stability of stopes where the hybrid room-and-pillar mining method was applied. The 3D point cloud data acquired at seven points in the Lv520 and Lv540 stopes were utilized as foundational data to construct a cross-section analysis and numerical analysis model of the vertical safety pillar.

The cross-section analyses were conducted on an area where a large-scale, 22 m high vertical safety pillar would be formed. The cross-section area of Pillar 1 located on the west side of the stope was 141.61 m^2 in the Lv540 stope, and that of Pillar 2 located on the east side was 66.16 m^2 in the Lv520 stope and 161.33 m^2 in the Lv540 stope. When the analyzed cross-section area of each of the vertical safety pillars was compared with the actual designed cross-section area of 100 m^2, over-break and under-break were present locally in the vertical safety pillar formed in the stope.

The cross-section area of Pillar 2 formed in the Lv520 stope was smaller than that formed in the Lv540 by 95.17 m^2, and the center axis of the pillar deviated to the northwest.

The results of 3D discontinuum numerical analysis showed that compressive stress at a level of 4 MPa and tensile stress at a level of 2 MPa were concentrated to the west of Pillar 2 in the Lv520 stope. This was because the confining stress was released by the recovery work of the horizontal safety pillar which restrained joint set 2. As a result, the supported load of the vertical safety pillar was increased, thereby generating shear displacement along the dip direction of joint set 2.

The safety factor of Pillar 2 in the Lv520 stope was analyzed using the results of the empirical equations proposed by Greenwald et al. [18] and Holland and Gaddy [19], as well as the results of the numerical analysis. The safety factor values calculated by via the empirical equation were 2.0 and 2.2 at the planned cross-section, and 1.3 and 1.4 at the measured cross-section, respectively. The safety factor of joint set 2, which affected the shear behavior of Pillar 2 that was calculated in the numerical analysis results, was found to be 1.0.

The actual stope shape-reflected safety factor analysis, or the safety factor analysis results using the measured cross-section and numerical analysis results, was verified as unstable based on a score of 1.5, the safety factor criterion proposed by Esterhuizen et al. [20]. If the horizontal safety pillar was to be recovered without additional reinforcement work, it would cause stability problems due to an increase in the supported load and separation behavior of joint sets 1 and 2 in Pillar 2 in the Lv520 stope.

Because reinforcement work was conducted using shotcrete and rockbolt at the actual site, the instability displayed in the numerical analysis was not verified. The results of this study are expected to be utilized as reference data in the establishment of recovery from horizontal safety pillars in the future.

As discussed above, when mining plans that require stability are established or new mining methods are applied, it is necessary to periodically evaluate the stability of stopes and pillars, using 3D surveying work, to investigate the safety factor of the surveying results and conduct 3D numerical analysis.

Author Contributions: Investigation, S.-J.L.; Methodology, S.-J.L.; Writing—original draft, S.-J.L.; Writing—review & editing, S.-O.C.

Funding: This research was supported by the National Strategic Project-Carbon Upcycling of the National Research Foundation of Korea (NRF), funded by the Ministry of Science and ICT (MSIT), the Ministry of Environment (ME), and the Ministry of Trade, Industry and Energy (MOTIE) (NRF-2017M3D8A2085342).

Conflicts of Interest: The authors declare no conflict of interest.

References

1. Abellán, A.; Oppikofer, T.; Jaboyedoff, M.; Rosser, N.J.; Lim, M.; Lato, M.J. Terrestrial laser scanning of rock slope instabilities. *Earth Surf. Process. Landf.* **2014**, *39*, 80–97. [CrossRef]

2.   Cai, M.; Kaiser, P.K.; Uno, H.; Tasaka, Y.; Minami, M. Estimation of rock mass deformation modulus and strength of jointed hard rock masses using the GSI system. *Int. J. Rock Mech. Min. Sci.* **2004**, *41*, 3–19. [CrossRef]

3.   Cai, M.; Kaiser, P.K.; Tasaka, Y.; Minami, M. Determination of residual strength parameters of jointed rock masses using GSI system. *Int. J. Rock Mech. Min. Sci.* **2007**, *44*, 247–265. [CrossRef]

4.   Lee, S.; Jeon, S. A Study on the roughness measurement for joints in rock mass using LIDAR. *Tunn. Undergr. Space* **2017**, *27*, 58–68. (In Korean) [CrossRef]

5.   Kim, C.; Kemeny, J. Measurement of joint roughness in large-scale rock fracture using LIDAR. *Tunn. Undergr. Space* **2009**, *19*, 52–63. (In Korean)

6.   Oh, S. Extraction of rock discontinuity orientation by laser scanning technique. Master's Thesis, Seoul National University, Seoul, Korea, February 2011.

7.   Park, S.; Lee, S.; Lee, B.; Kim, C. A study on reliability of joint orientation measurements in rock slope using 3d laser scanner. *Tunn. Undergr. Space* **2015**, *25*, 97–106. (In Korean) [CrossRef]

8.   Lee, S.; Jeon, S. A study on the extraction of slope surface orientation using LIDAR with respect to triangulation method and sampling on the point cloud. *Tunn. Undergr. Space* **2016**, *26*, 46–58. (In Korean) [CrossRef]

9.   Kasperski, J.; Delacourt, C.; Allemand, P.; Potherat, P.; Jaud, M.; Varrel, E. Application of a Terrestrial Laser Scanner (TLS) to the study of the Séchilienne landslide (Isère France). *Remote Sens.* **2010**, *2*, 2785–2802. [CrossRef]

10.   Oppikofer, T.; Jaboyedoff, M.; Blikra, L.; Derron, M.H.; Metzger, R. Characterization and monitoring of the Åknes rockslide using terrestrial laser scanning. *Nat. Hazards Earth Syst. Sci.* **2009**, *9*, 1003–1019. [CrossRef]

11.   Abellán, A.; Jaboyedoff, M.; Oppikofer, T.; Vilaplana, J.M. Detection of millimetric deformation using a terrestrial laser scanner: Experiment and application to a rockfall event. *Nat. Hazards Earth Syst. Sci.* **2009**, *9*, 365–372. [CrossRef]

12.   Rosser, N.J.; Petley, D.N.; Lim, M.; Dunning, S.A.; Allison, R.J. Terrestrial laser scanning for monitoring the process of hard rock coastal cliff erosion. *Q. J. Eng. Geol. Hydrogeol.* **2005**, *38*, 363–375. [CrossRef]

13.   Lee, J.C.; Moon, D.Y.; Kim, N.S.; Seo, D.J. Calculation of over cutting volume on tunnel using 3D laser scanner. In Proceedings of the KSCE Regular Conference, Gwangju, Korea, 12–13 October 2006; pp. 4608–4611.

14.   Lee, S.J.; Choi, S.O.; Lee, S.; Jeon, S.; Jin, Y.H.; Jung, M.S. Analysis of blasting overbreak using stereo photogrammetry in an underground mine. *Tunn. Undergr. Space* **2016**, *26*, 348–362. (In Korean) [CrossRef]

15.   Gawronek, P.; Makuch, M.; Mitka, B.; Bozek, P.; Klapa, P. 3D scanning of the historical underground of Benedictine Abbey in Tyniec (Poland). In Proceedings of the 17th International Multidisciplinary Scientific GeoConference (SGEM 2017), Vienna, Austria, 27–29 November 2017.

16.   Kim, Y.B.; Chung, S.K.; Jo, S.H.; Kim, C.O.; Um, W.W. Hybrid room-and-pillar mining method. KR Patent No. 1015657890000, 29 October 2015.

17.   Konič, S.; Ribičič, M.; Vulič, M. Contribution to a rock block slide examination by a model of mutual transformation of point clouds. *Atca Carsologica* **2009**, *38*, 107–116. [CrossRef]

18.   Greenwald, H.P.; Howarth, H.C.; Hartman, I. *Experiments on strength of small pillars of coal in the Pittsburgh bed*; U.S. Department of the Interior: Washington, DC, USA, 1939.

19.   Holland, C.T.; Gaddy, F.L. The strength of coal mine pillar. In Proceedings of the 6th U.S. Symposium on Rock Mechanics, Rolla, MI, USA, 28–30 October 1964; pp. 450–466.

20.   Esterhuizen, G.S.; Dolinar, D.R.; Ellenberger, J.L. Pillar strength and design methodology for stone mines. In Proceedings of the 27th International Conference on Ground Control in Mining, Morgantown, WV, USA, 29–31 July 2008; pp. 241–253.

 *sustainability*

Article

Stiffness and Cavity Test of Concrete Face Based on Non-Destructive Elastic Investigation

Jinpyo Hong [1], Seokhoon Oh [1,*] and Eunsang Im [2]

1   Department of Energy and Resources Engineering, Kangwon National University, Chuncheon 24341, Korea; hongjp0509@kangwon.ac.kr
2   Infrastructure Research Center, K-water Institute, Daejeon 34350, Korea; esim89@kwater.or.kr
*   Correspondence: gimul@kangwon.ac.kr; Tel.: +82-33-250-6258

Received: 7 November 2018; Accepted: 20 November 2018; Published: 24 November 2018

Abstract: A non-destructive testing (NDT) method was used in a concrete face rockfill dam (CFRD) to identify the condition of the concrete face slab and detect any existing cavities between the concrete face slab and the underlying support layer. The NDT for the concrete face slab was conducted using the impulse response (IR) method and the electrical resistivity tomography (ERT) method with the application of non-destructive electrodes. Information regarding the dynamic stiffness and average mobility of the concrete was obtained based on the mobility-frequency of the IR method, and cavity detection under the plate structures was analyzed using the two-dimensional (2D) electrical resistivity section of the ERT method. The results of the IR method showed that zones with low dynamic stiffness and high average mobility were expected to be found in concrete of poor quality and in cavities beneath the concrete face slab. The results of the ERT method showed that zones with high resistivity were expected to be cavities between the concrete face slab and the underlying support layer. As a result, the tendency toward low dynamic stiffness, high average mobility, and high resistivity in both methods implies unstable concrete conditions and the possible occurrence of a cavity. The results of the two methods also showed a good correlation, and it was confirmed that the NDT method was reliable in terms of cavity estimation.

Keywords: concrete face rockfill dam; face slab; cavity; impulse response method; electrical resistivity tomography

---

1. Introduction

Concrete has been applied to various types of structures worldwide, including reinforced concrete structures, dams, tunnel lining, concrete retaining walls, and pavement. However, the destruction and cracking of concrete owing to various influences of these structures as time passes has caused serious damage to human life and property. Therefore, safety inspections and evaluations of concrete structures are particularly important.

Concrete face rockfill dams (CFRDs) are relatively easy to construct and economical to manage, and have less impact from weather and earthquakes than other types of dams [1]. The concrete face slab considered herein is a concrete plate-type structure located on the upstream face of a rockfill dam body, and is used as an anti-seepage structure. Therefore, damage occurring in the cavity behind the concrete face slab may cause a deformation and collapse of the dam body. If a cavity occurs beneath the concrete slab, cracks may occur from hydraulic pressure or a dead load, which may cause problems such as a load being applied to the lower part of the plinth owing to a decrease in frictional force between the concrete face slab and the zone materials of the dam body.

Non-destructive testing (NDT) methods such as ground penetrating radar (GPR), electric, seismic, and radioactive surveys are widely used safety inspection methods of concrete face slabs. There are also computer image analysis techniques for assessing the condition of concrete structures, such as assessing the degree of cracking of the structure [2–4]. Other studies have analyzed CFRD stability problems through a numerical analysis or model experiment. Seo et al. conducted centrifuge model tests for simulating the behavior of CFRD with increasing water levels [5]. Cheng and Zhang conducted a study on the dynamic response and deformation law by changing the waveform of the input wave using a centrifuge model test [6]. Arici analyzed the state of the stress and the cracking behavior of a concrete face slab over both the short and long term during the lifecycle of a CFRD [7]. Zhang et al. detected cavities that may exist beneath a concrete face slab of a seawall using GPR [8].

The stability problems caused by a concrete face slab are as follows: The cracking behavior of a concrete face slab of CFRD under intense seismic excitation is not the opening of the vertical joints but rather the cracking of the concrete plate from crushing between the concrete face slab and the underlying support layer and an excessive settlement of the fill [9,10]. The major factors indicating the deformation characteristics after an initial impoundment are the crest and face settlements [11]. The cracking of the concrete face slab is associated with seepage [12]. Therefore, a concrete face slab that applies the core function of the CFRD should have mechanical properties preventing cracking from an external force, or tensile stress of the concrete from moment generation [13]. In other words, the vulnerability of a concrete face slab adversely affects the stability of the CFRD.

In this study, we investigated the conditions of the concrete face slab and detected the cavity between the concrete face slab and underlying support layer for a safety inspection and quality control of the CFRD. As the NDT methods, we applied impulse response (IR) and electrical resistivity tomography (ERT) methods using non-destructive electrodes. IR methods are widely used in the safety inspection of concrete cracks, cavities underneath the concrete, pavement, and other elements, and are applied to investigate the conditions of the concrete based on the dynamic stiffness and average mobility. ERT methods using non-destructive electrodes have been developed to reduce the contact resistance of the concrete and prevent the destruction of the object of investigation, and are used to investigate the conditions of the concrete through a two-dimensional (2D) electrical resistivity section. Therefore, the NDT of a concrete face slab was carried out through an analysis and correlation between the two methods.

2. Study Object

The target object was a CFRD located in the northern district of the Republic of Korea, with a dam height of 125 m, a crest length of 601 m, a maximum reservoir capacity of 2.63 billion m^3, and a ratio of upstream slope to downstream slope of 1:1.5 (Figure 1a). Step 1 of the dam was completed in 1989, and step 2 in 2006.

In general, the thickness of the concrete face slab was 0.3 m at the top, except for the case of a low dam height, and changes according to the depth of the water (H) while descending to the bottom of the concrete face slab [14]. Therefore, the thickness of the concrete face slab for step 1 was thicker than that of step 2. In addition, the steel reinforcements were designed using a 0.4–0.5% steel ratio to avoid cracks from the tensile stress that occurs through changes in temperature, and were arranged at intervals of 200–300 mm to cope effectively with cracks.

Figure 1b shows a schematic of the IR and ERT survey lines of the study object, namely, CF-9, -13, -19, and -23 of a concrete face slab, with a width of 15 m upstream, all of which are located at the center. The IR survey was carried out in steps 1 and 2, and the ERT survey was carried out at CF-13 and -19 in step 2. Both surveys were conducted from top to bottom. Through the IR survey, 21 stations were acquired for a length of 79 m in step 2, and 23 stations were acquired for a length of 80 m in step 1. The spacing of the stations was set to 2.5 m for the cavity estimation examined through a GPR survey [15], and 5 m for the other section. Through the ERT survey, 32 channel cables were used, and the electrode spacing and line length were set to 1 and 31 m, respectively, when taking into account the thickness of the concrete face slab and the steel reinforcement.

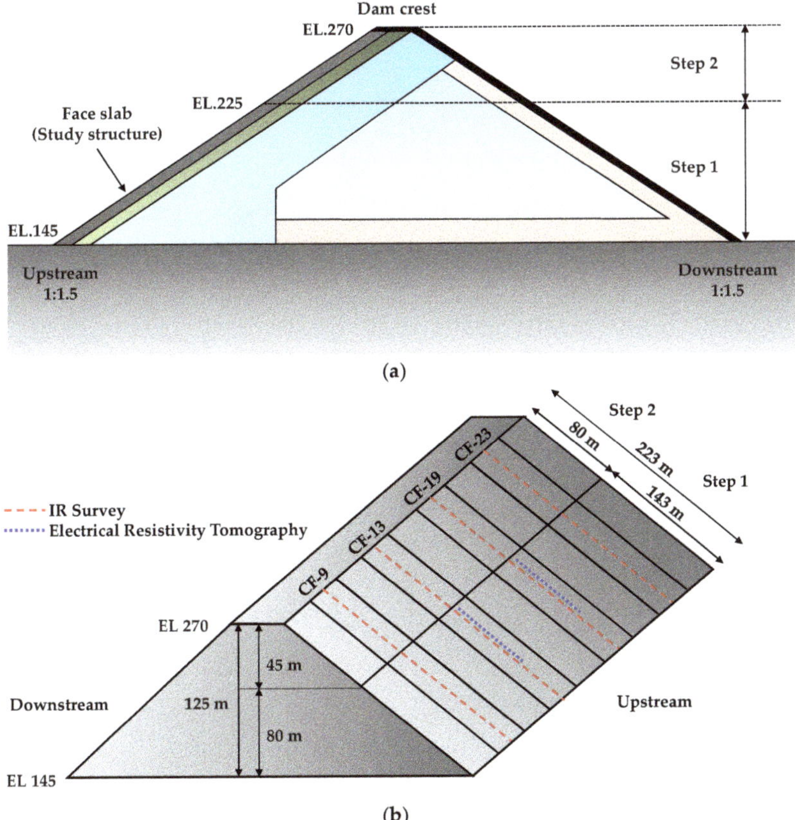

Figure 1. (**a**) Schematic of the concrete face rockfill dam (CFRD) that was used as the study object. The concrete face slab was located on the upstream slope and was divided into steps 1 and 2. The concrete construction time and concrete thickness of each step were different. (**b**) Schematic of impulse response (IR) and electrical resistivity tomography (ERT) survey lines. Each line was centrally located at CF-9, -13, -19, and -23 on the concrete face slab with a width of 15 m. The IR survey was carried out on steps 1 and 2, and the ERT survey was carried out on CF-13 and -19 of step 2.

3. Application of Non-destructive Testing to Test Bed

An IR survey is a method used to detect the internal defects of a structure by applying a mechanical impact to the concrete surface based on elastic waves. This technique can be widely applied to concrete structures such as bridges, roads, and dams, and can be used to quickly investigate large areas [16]. In addition, it has the advantage of non-destructively investigating the cracks and cavities of an applied concrete structure.

Figure 2 shows a schematic of the impact response tester developed by Chung et al. [17] and a graph of the mobility-frequency of a plate-like object such as a typical concrete face slab. Figure 2a shows the data acquisition device (DAQ), impact hammer with a load cell, and geophone applied. During an IR measurement, communication between the DAQ and a laptop PC is achieved through the wireless fidelity (Wi-Fi) ad hoc mode, and data are saved wirelessly. The stored data is transformed into the frequency-domain using a fast Fourier transform algorithm after applying bandpass filtering to the spectrum of the time-domain generated by the impact source and geophone velocity transducer. When an impact is applied to the study object using an impact hammer, the object on the plate reacts in bending mode within a frequency range of 0 to 800 Hz, and the geophone measures the velocity of the bending motion [18]. The IR method obtains the transfer function from the impact force measured in the load cell attached to the impact hammer and the velocity of the bending motion measured using the geophone. The transfer function is defined as the ratio of the velocity spectrum to the force spectrum, which is called mobility. The mobility at a frequency range of 0–800 Hz also includes the parameter information based on the conditions and integrity of the concrete study object. Figure 2b shows an example graph of the dynamic stiffness and average mobility in a plate-like object such as a concrete face slab. The dynamic stiffness is defined as the reciprocal of the slope at a frequency of less than 80 Hz where the mobility increases linearly in the mobility-frequency graph. The dynamic stiffness indicates the flexibility around the measurement point and depends on the quality of the study object, the thickness of the plate, and the supporting state of the plate. The average mobility is defined as the arithmetic mean of the mobility within the 100–800 Hz frequency range. The elastic wave caused by the applied impact source is attenuated by the intrinsic rigidity of the plate. The average mobility within the 100–800 Hz range is directly related to the plate density and thickness. The average mobility increases when the thickness of the plate becomes thinner or the distance between two layers of the specimen decreases.

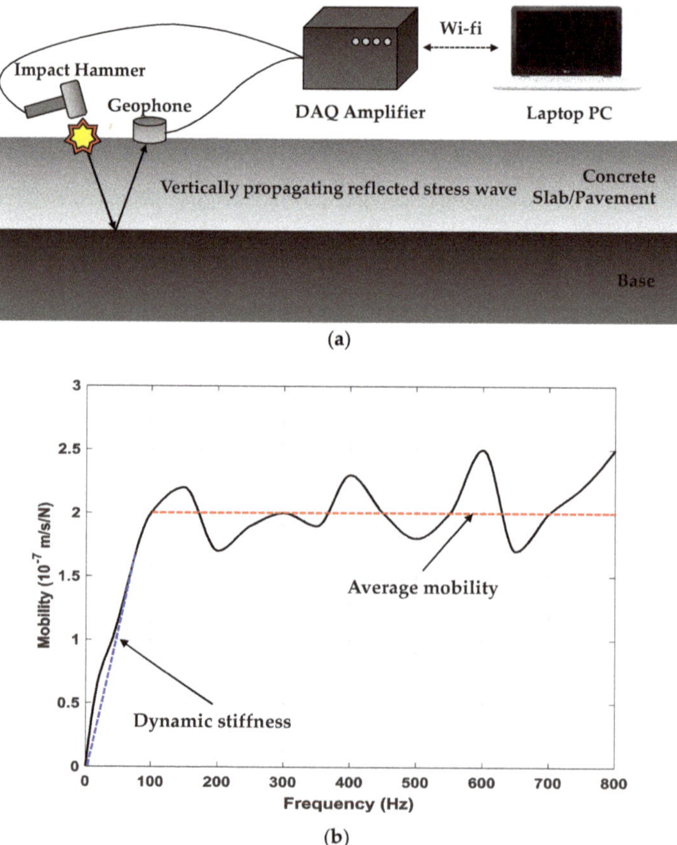

Figure 2. (**a**) Schematic of the impulse response tester. This equipment consisted of an impact hammer, a data acquisition device (DAQ) amplifier, a geophone, and a laptop PC. (**b**) The mobility plot of a typical concrete face slab. The dynamic stiffness was defined as the reciprocal of the slope at a frequency of less than 80 Hz. The average mobility was defined as the arithmetic mean of the mobility within the 100–800 Hz frequency range.

Various electrical resistivity measurement techniques for concrete have been developed, including a uniaxial method and a four-point method [19]. These techniques are mainly for quality control and a durability evaluation of concrete. In this study, however, the ERT method of a dipole-dipole array was used to obtain a 2D section of the electrical resistivity distribution. In general, an ERT survey is applied to underground exploration, and is a geophysical exploration technique for investigating the geological structure and ground condition by measuring the potential difference between electrodes caused by the electric current flowing from the current electrode. In addition, the non-destructive electrode was used because the electrodes of the stainless steel used in the ERT survey could cause destruction of the study object.

Figure 3 shows a schematic diagram of an ERT survey using a dipole-dipole array. The potential difference (ΔV) was measured by flowing a current (I) during the ERT survey, as characterized through Equations (1) and (2) below:

$$\rho_a = k \cdot \frac{\Delta V}{I}, \tag{1}$$

$$k = \pi n(n+1)(n+2)a, \tag{2}$$

where ρ_a is the apparent resistivity of an underground medium, k is the geometrical factor, a is the electrode spacing, and n is the electrode separation index. However, because the measured ρ_a is not the true resistivity, it must be calculated through an analysis technique. For this, DIPRO for Windows [20] was used.

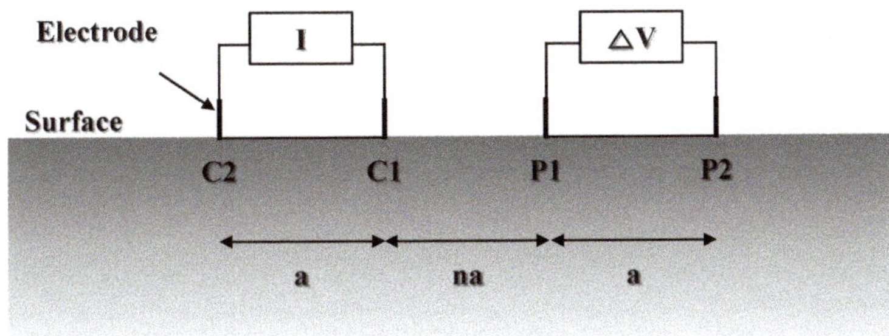

Figure 3. A dipole-dipole array of a typical electrical resistivity tomography (ERT) survey. The resistivity of the study object was calculated by measuring the potential difference between two current and two potential electrodes.

Figure 4 shows the concrete model manufactured for reviewing the applicability of an ERT survey of concrete, as well as the results of inverse and forward modeling. The concrete model was made to confirm the effects of the steel reinforcement and cavities, and was $2.4 \times 6.0 \times 1.0 \, \text{m}^3$ in size (Figure 4a). Steel reinforcements were arranged at the center of the left side of the concrete at 0.2 m intervals with a 25 mm diameter mesh. The cavity was located at the center of the bottom of the concrete and was $2.4 \times 2.0 \times 0.3 \, \text{m}^3$ in size. Sand, gravel, and clay were used to support the concrete on the left and right sides of the cavity, and at the same time, to reduce the change in the height of the cavity owing to a consolidation of the ground.

Figure 4b,c show the inversion using non-destructive electrodes, and the forward modeling results, respectively. In both results, the effect of the electrical resistivity was shown through the low resistivity effect at the location of the steel reinforcement and the high resistivity effect at the location of the cavity. Both effects also fitted well with the actual location. As a result of both models, an absolute difference in the resistivity value existed, although this occurred because the resistivity value of the actual concrete was different from the resistivity value used for the forward modeling. However, an ERT survey could be applied to the concrete face slab because the effects of the steel reinforcement and cavity of the concrete were quite visible.

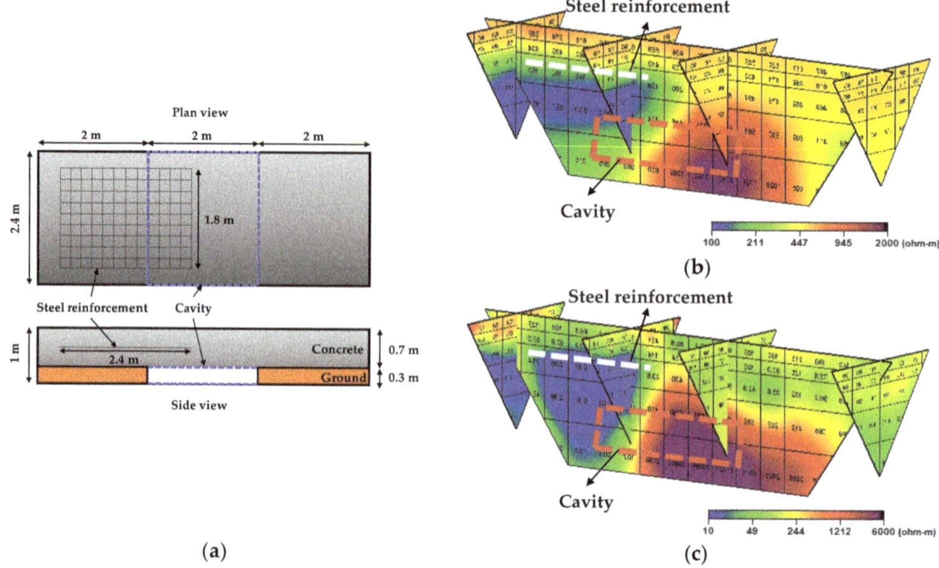

Figure 4. (**a**) A schematic of the concrete model used to apply the electrical resistivity tomography (ERT) survey using non-destructive electrodes. This model was designed to investigate the effects of the steel reinforcements and cavities in the interior and bottom of the concrete. (**b**) ERT survey results using non-destructive electrodes, and (**c**) ERT results using forward modeling. In both models, the steel reinforcements showed low resistivity effects, and the cavities showed high resistivity effects.

4. Results and Discussion

4.1. Impulse Response Results

The IR survey was applied to steps 1 and 2 of the upstream slope of the concrete face slab. The dynamic stiffness and average mobility could not be evaluated absolutely because each step had a different construction time and concrete thickness. Owing to the characteristics of the IR method, the parameters showed a difference in the average mobility as the thickness of the plate decreased. In addition, the dynamic stiffness decreased as the thickness of the reflected layer decreased. Therefore, the results of the IR survey were interpreted independently for each step.

Figure 5 shows dynamic stiffness and average mobility graphs for each step. Figure 5a,b show the dynamic stiffness and average mobility at CF-9, -13, -19, and -23 of the concrete face slab. This plot shows an inverse relationship between the general dynamic stiffness and average mobility. The overall dynamic stiffness decreased from top to bottom (IR points 1–19), and then tended to increase sharply at the bottom (IR points 20 and 21). In particular, at IR points 12–19, the entire concrete face slab showed a relatively low dynamic stiffness and high average mobility (red dotted box). In addition, throughout the concrete face slab shown in this box, CF-23 was expected to be the most vulnerable. The overall upstream slope of step 2 indicated that the state between the concrete face slab and the underlying support layer was more unstable at the bottom than at the top, and that a cavity was likely to exist at the bottom. IR points 20 and 21 exhibited a sharp high dynamic stiffness and low average mobility, which was presumed to maintain the high quality of the concrete because the supporting soil in the red dotted box was lost, accumulated at the bottom, or no space existed where the supporting soil would be lost near the bottom.

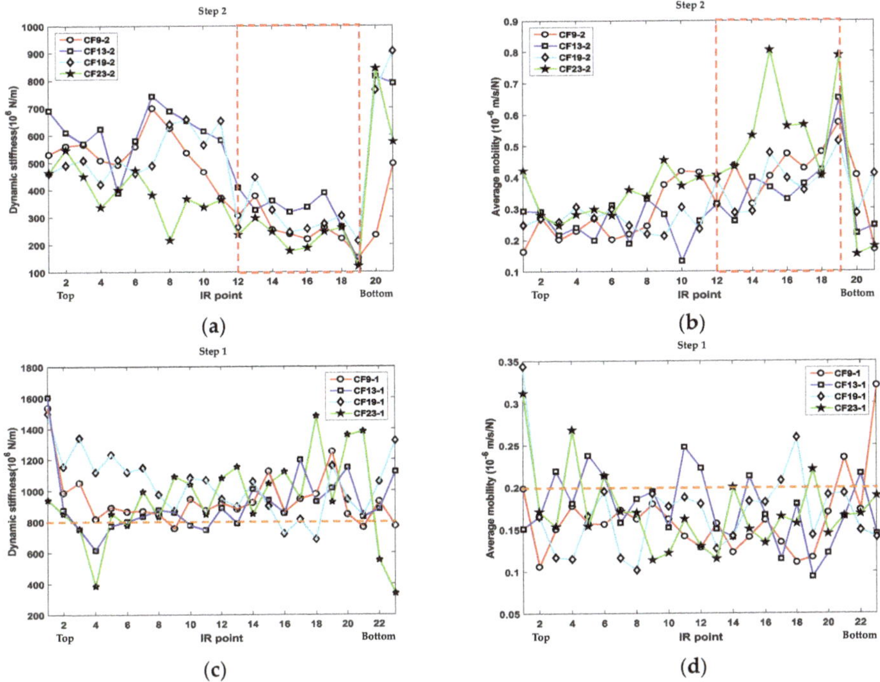

Figure 5. Dynamic stiffness and average mobility of the concrete face slab at CF-9, -13, -19, and -23 in steps 1 and 2. (**a**) The dynamic stiffness and (**b**) average mobility in step 2 showed a poor concrete quality owing to a low dynamic stiffness and high average mobility at the bottom (red dotted box). In step 1, (**c**) the dynamic stiffness was more than 800×10^6 N/m and (**d**) the average mobility was less than 0.2×10^{-6} m/s/N, demonstrating a constant trend but representing an unstable concrete condition at certain points.

Figure 5c,d showed the dynamic stiffness and average mobility at CF-9, -13, -19, and 23 of the concrete face slab. In steps 1 and 2, the concrete face slab showed a tendency to remain relatively constant in terms of the dynamic stiffness and average mobility, the former of which tended to remain at above 800×10^6 N/m. The average mobility tended to remain constant, usually at below 0.2×10^{-6} m/s/N, however, the average mobility of some of the IR points was relatively large. IR points 4, 22, and 23 showed a low dynamic stiffness and a poor concrete quality compared to the other points on the concrete face slab. In particular, IR point 4 of CF-23 tended to have a low dynamic stiffness and high average mobility. These points were relatively weak at supporting the concrete face slab and predicting potential cavities.

Figure 6 shows a mobility-frequency graph for the concrete face slab in step 2 according to the concrete conditions. IR point 15 of CF-23 was very low in dynamic stiffness and the highest in mobility. In contrast, IR point 7 of CF-13 was very low in average mobility and very high in dynamic stiffness. In general, the increase in mobility at below 80 Hz indicated that a cavity was likely to exist below the concrete face slab, and if a crack was present in the concrete, the mobility slope tended to increase [21]. IR point 15 of CF-23, which showed this tendency, demonstrated a relatively poor concrete quality, whereas IR point 7 of CF-13 was relatively stable before and after 80 Hz, indicating a high concrete quality. Therefore, the concrete condition can be distinguished according to the mobility-frequency characteristics of the raw data, and an IR survey is considered to be suitable for evaluating the condition of a concrete face slab.

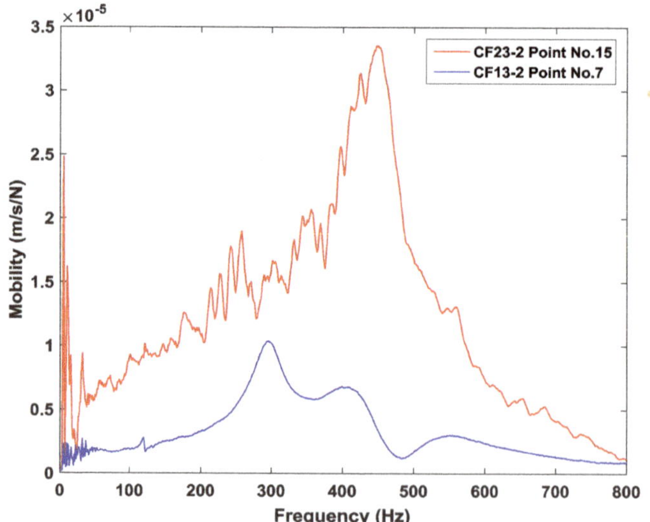

Figure 6. Mobility-frequency graph according to the conditions of the concrete. Impulse response (IR) point 15 of CF-23 had the highest average mobility and a low dynamic stiffness. IR point 7 of CF-13 was relatively high in dynamic stiffness and low in average mobility.

4.2. Electrical Resistivity Tomography Results

An ERT survey using non-destructive electrodes was applied to the concrete face slab at CF-13 and -19 in step 1. The cavity-estimated zone of CF-13 surveyed using GPR was located at 10 to 30 m from the dam berm, and no cavity was found at CF-19 [12]. Therefore, when an ERT is applied, the existence of a cavity can be identified based on an electrical resistivity distribution, and it can be confirmed whether an anomaly exists in the actual cavity estimation zone.

Figure 7 shows the ERT survey results of CF-13 and -19 and the cavity estimation zone using GPR. In Step 1, the thickness of the concrete face slab was 0.7–1 m, and the steel reinforcements were located at a depth 0.2–0.35 m. However, overall ERT results showed a lower resistivity range of up to 2 m in depth, the reason for which is that steel reinforcements have a greater impact. Therefore, the ERT results should be interpreted with caution in terms of depth of investigation when considering the steel reinforcements.

Considering the effects of the steel reinforcements, CF-13, which was in a cavity estimation zone, had a non-homogeneous high resistance effect within a depth of 2 m (Figure 7a). In addition, it was confirmed through GPR that the high resistivity zone coincided with the cavity estimation zone. However, CF-19 without a cavity estimation zone exhibited a homogeneously low resistivity effect horizontally up to about a 2 m depth (Figure 7b). Therefore, the effect of high resistivity within a 2 m depth can be interpreted as the possible existence of a cavity based on the ERT survey.

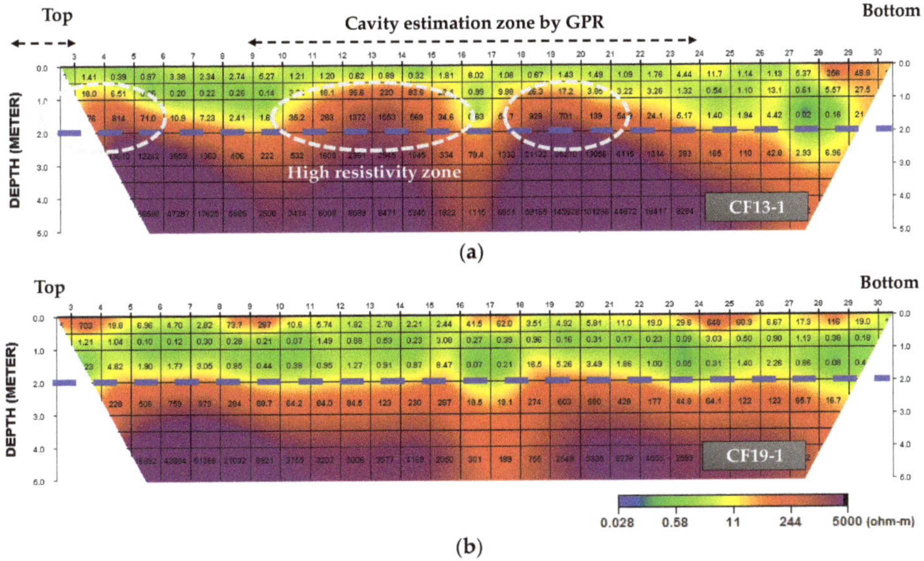

Figure 7. Results of the electrical resistivity tomography (ERT) survey of CF-13 and -19 in step 1. (**a**) CF-13 showed a heterogeneous high-resistivity effect in the cavity estimation zone based on ground penetrating radar (GPR). (**b**) CF-19 showed homogeneously low resistivity effects. Therefore, the high resistivity was expected to be due to the presence of a cavity.

4.3. Correlation of Impulse Response and Electrical Resistivity Tomography Results

The correlation between the IR and ERT results were analyzed based on an electrical resistivity distribution of a 2D section, the dynamic stiffness, and the average mobility of the concrete face slab. CF-13 and -19 of step 1 were considered for all surveys of the concrete face slab. The IR survey showed that the dynamic stiffness (greater than 800×10^6 N/m) and average mobility (less than 0.2×10^{-6} m/s/N) of step 1 tended to be relatively constant, but at some point an anomaly zone occurred (Figure 5c,d). The ERT survey showed an anomaly zone with a high resistivity effect at a depth of less than 2 m. Therefore, the correlation of each survey was analyzed based on the anomaly zone, which appeared to be due to a poor concrete quality or the presence of cavities.

Figure 8 shows the IR and ERT survey results for CF-13 and -19. In Figure 8a, the cavity estimation zone based on GPR was located at IR points 1–3 and 6–12. However, the IR points when including the IR and ERT results of the cavity estimation zone were 6–12. As a result of the ERT survey of IR points 6–12, it was possible to identify an anomaly zone with a high resistivity effect within a depth of 2 m. As a result of the IR survey in this section, it could be confirmed that the dynamic stiffness was low and the average mobility was higher than at the other points. Therefore, this section could predict the presence of cavities between the support layers and concrete with poor quality. In Figure 8b, no cavity estimation zones appeared, and relatively stable parameters were shown. To compare the IR and ERT results, IR points 4–14 were analyzed. As the results of the ERT survey show, the low resistivity effect within a depth of 2 m tended to be homogeneous. As a result of the IR survey, this section did not represent the anomaly zone shown in Figure 5. At certain points, the dynamic stiffness was low (IR point 9) and the average mobility was high (IR point 6), but the concrete was not of poor quality. Therefore, compared to CF-13, CF-19 could be confirmed from the results of the IR and ERT surveys to be a location of high-quality concrete.

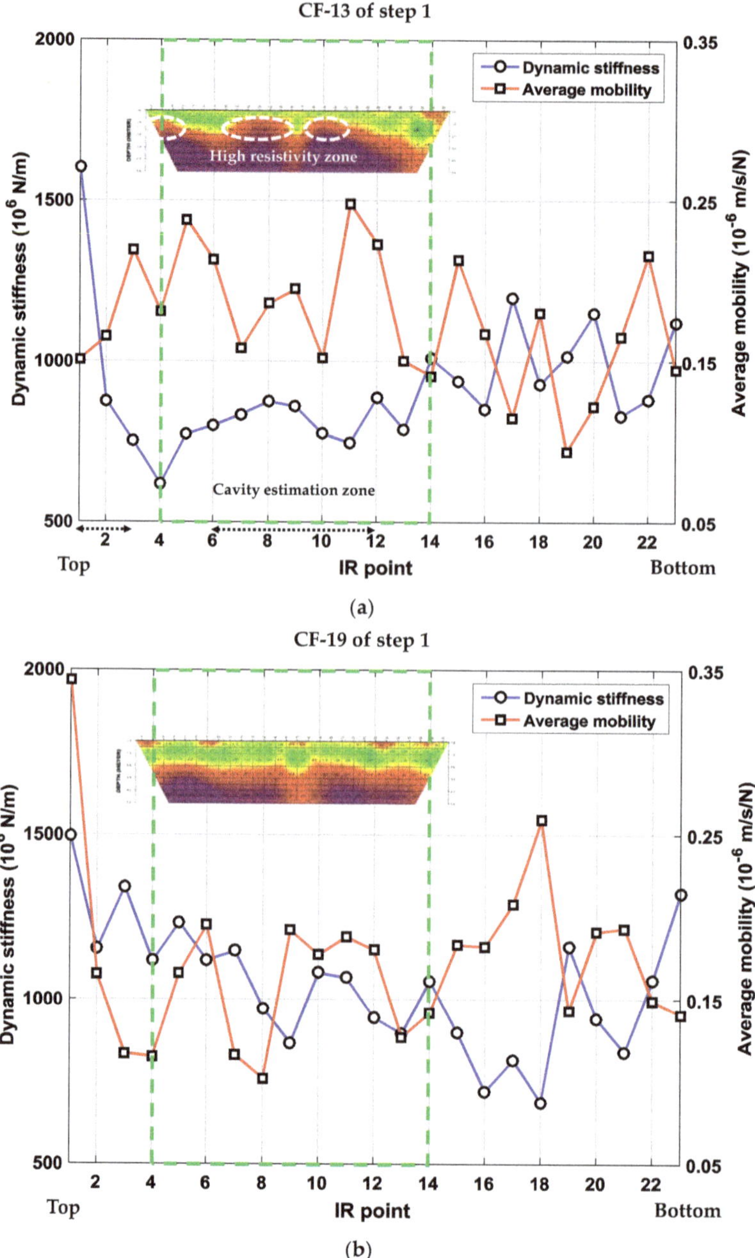

Figure 8. Impulse response (IR) and electrical resistivity tomography (ERT) survey results of CF-13 and -19 in step 1: (**a**) High resistivity appeared in the cavity estimation zone based on ground penetrating radar (GPR). The dynamic stiffness was relatively low, and the average mobility was high at certain points. (**b**) There was no cavity estimation zone and the low resistivity effect appeared to be homogeneous. In addition, the dynamic stiffness and average mobility tended to be stable.

Figure 9 shows a consistency analysis of IR point 4 with the lowest dynamic stiffness at CF-13 in step 1. A consistency analysis was conducted using two measurements at the same point for statistical consistency of the frequency sounding measurement. As a result of the analysis, the coherence factor showed a high reliability with a maximum value of 0.99 and an average value of 0.96 or more. However, when the frequency was higher than 300 Hz, the coherence factor was reduced to about 0.2, and a factor of between 0.8 and 1 indicated that the reliability of the average mobility was slightly lowered. As shown in Figure 9, the dynamic stiffness was very low, but the average mobility did not show such an anomaly. Therefore, it is expected that the average mobility will actually be higher, and poor concrete quality is expected.

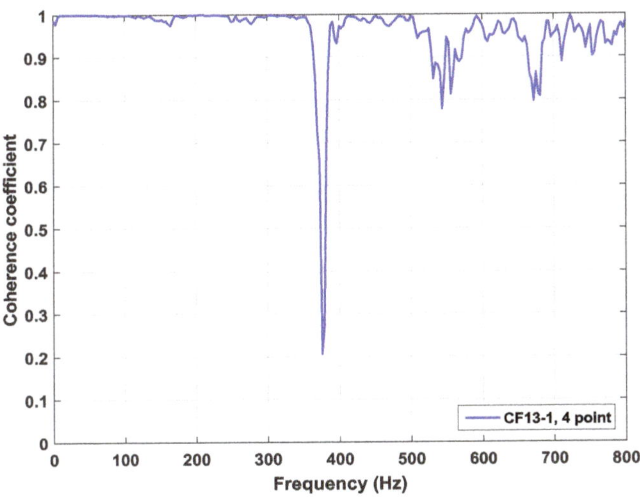

Figure 9. Consistency analysis results for impulse response (IR) point 4 of CF-13 in step 1.

5. Conclusions

IR and ERT surveys using non-destructive electrodes were applied to a concrete face slab to analyze the conditions of the concrete and investigate the presence of cavities between the underlying support layers. Unlike conventional NDT methods, an ERT survey can evaluate an area where a cavity is expected by visualizing the electrical resistivity distribution of the target object. It is also possible to analyze the vulnerability of concrete in combination with an IR survey. Each survey was analyzed by acquiring the dynamic stiffness, average mobility, and electrical resistivity distribution data using the NDT method, which could be easily measured without destroying the study object.

As the results of the IR survey indicated, step 2 showed a tendency toward a lower dynamic stiffness and higher average mobility at the bottom. This indicated the existence of a cavity or poor concrete quality between the concrete face slab and the underlying support layer. Step 1 showed a constant trend of dynamic stiffness and average mobility. However, at certain IR points, a zone judged to be an anomaly was present, and it indicated poor concrete quality. Therefore, it is possible to confirm the poor concrete quality at each step with a relatively low dynamic stiffness and high average mobility. These results indicated that a consistency analysis had high reliability and was suitable for evaluating the conditions of a concrete face slab.

The ERT survey was applied only to step 1, and the results indicated the anomaly of a high resistivity effect within a depth of 2 m. This effect was expected to be from a cavity between the concrete face slab and the underlying support layer, and was confirmed only in CF-13. In addition, it coincided with the cavity estimation zone as determined through GPR. Therefore, an ERT survey is considered suitable for investigating the presence of cavities.

As the results of the IR and ERT surveys showed, a high correlation existed between the two survey types for a cavity estimation zone because the dynamic stiffness was low and the average mobility was high in the high resistivity zone. Therefore, it is expected that cracking and a failure of the concrete face slab, as well as the loss of the underlying support layer, can be expected to be investigated in the IR and ERT method, and it is therefore appropriate to evaluate the stability of a concrete structure.

Consequentially, the application of the NDT method on a concrete face slab can provide useful information for investigating the state of the concrete and any cavities present, and it is expected that the NDT method can be used as a state and stability evaluation method for various types of concrete structures. Additionally, if it can be combined with the results of conventional NDT methods such as rebound hardness, ultrasonic pulse velocity, pull-out test, which can be applied to the assessment of the conditions of the concrete through integrated analysis of the study results, the analysis will be more confident.

Author Contributions: Investigation, J.H.; methodology, J.H.; supervision, S.O.; validation, E.I.; writing—original draft, J.H.; writing—review and editing, S.O.

Funding: This work was supported by the Nuclear Safety Research Program through the Korea Foundation Of Nuclear Safety (KoFONS) and granted financial resource from the Nuclear Safety and Security Commission (NSSC), Republic of Korea (No. 1705010).

Acknowledgments: This research was supported by the National Strategic Project, Carbon Upcycling of the National Research Foundation of Korea (NRF), funded by the Ministry of Science and ICT (MSIT), the Ministry of Environment (ME), and the Ministry of Trade, Industry, and Energy (MOTIE) (NRF-2017M3D8A2085342). This work was funded by the Korea Meteorological Administration Research and Development Program under Grant (KMI2018-09310).

References

1. Wieland, M.; Brenner, R.P. Earthquake aspects of roller compacted concrete and concrete-face rockfill dams. In Proceedings of the 13th World Conf. on Earthquake Engineering, Vancouver, BC, Canada, 1–6 August 2004.

2. Szelag, M. Influence of specimen's shape and size on the thermal cracks' geometry of cement paste. *Constr. Build. Mater.* **2018**, *189*, 1155–1172. [CrossRef]

3. Hutchinson, T.C.; Chen, Z. Improved image analysis for evaluating concrete damage. *J. Comput. Civ. Eng.* **2006**, *20*, 210–216. [CrossRef]

4. Fic, S.; Szelag, M. Analysis of the development of cluster cracks caused by elevated temperatures in cement paste. *Constr. Build. Mater.* **2015**, *83*, 223–229. [CrossRef]

5. Seo, M.W.; Im, E.S.; Kim, Y.S.; Ha, I.S. Centrifuge tests for simulating the behavior of CFRD with increasing water level. In Proceedings of the KGS Spring Conference, Seoul, Korea, 24-25 March 2006; pp. 784–793.

6. Cheng, S.; Zhang, J. Centrifuge modeling test of dynamic response and deformation law of concrete-faced rockfill dam under different input waves. In Proceedings of the Second International Conference on Mechanic Automation and Control Engineering (MACE), Inner Mongolia, China, 15–17 July 2011; pp. 2982–2985.

7. Arici, Y. Investigation of the cracking of CFRD face plates. *Comput. Geotech.* **2011**, *38*, 905–916. [CrossRef]

8. Zhang, Z.; Zhao, Y.; Ge, S.; Sun, P.; Liu, C. Analysis of hollow area beneath concrete slab of seawall by means of ground penetration radar. In Proceedings of the 8th International Workshop on Advanced Ground Penetrating Radar, Florence, Italy, 7–10 July 2015; pp. 1–4.

9. Arici, Y. Evaluation of the performance of the face slab of a CFRD during earthquake excitation. *Soil Dyn. Earthq. Eng.* **2013**, *55*, 71–82. [CrossRef]

10. Chen, S.S.; Han, H.Q. Impact of the '5.12' Wenchuan earthquake on Zipingpu concrete face rock-fill dam and its analysis. *Geomech. Geoeng.* **2009**, *4*, 299–306. [CrossRef]

11. Seo, K.; Kim, Y.S. Centrifuge test for simulating behavior of CFRD during initial impoundment. *J. Korean Geotech. Soc.* **2007**, *23*, 109–119.

12. Haselsteiner, R.; Ersoy, B. Seepage control of concrete faced dams with respect to surface slab cracking. In Proceedings of the 6th International Conference on Dam Engineering, Lisbon, Portugal, 15–17 February 2011; pp. 611–628.
13. Ha, I.S.; Seo, M.W.; Kim, H.S. Effects of stiffness of face supporting zone on face slab behaviors of CFRD. *J. Korean Soc. Civ. Eng.* **2006**, *26*, 351–358.
14. Korean Water Resources Association (KWRA). *Dam Design Criteria*; KWRA: Seoul, Korea, 2011.
15. Korea Water Resources Corporation (K-Water). *Report on the 3rd Precision Safety Diagnosis*; K-Water: Deajeon, Korea, 2014.
16. Davis, A.G. The non-destructive impulse response test in North America: 1985–2001. *NDT&E Int.* **2003**, *36*, 185–193.
17. Chung, H.J.; Lee, H.S.; Oh, S.H.; Song, S.H. Development of the impulse response measurement system for non-destructive test of slab structure. *Geophys. Geophys. Explor.* **2013**, *16*, 45–52. [CrossRef]
18. Clausen, J.S.; Zoidisand, N.; Knudsen, A. Onsite measurements of concrete structures using Impact-echo and Impulse response. In *Emerging Technologies in Non-Destructive Testing V*; CRC Press: Boca Raton, FL, USA, 2012; pp. 117–122.
19. Layssi, H.; Ghods, P.; Alizadeh, A.R.; Salehi, M. Electrical resistivity of concrete. *Concr. Int.* **2015**, *37*, 41–46.
20. Kim, J.H. *DIPRO for Windows, v. 4.0.*; Korea Institute of Geoscience and Mineral Resources: Daejeon, Korea, 2001.
21. Clausen, J.S.; Knudsen, A. Nondestructive testing of bridge decks and tunnel linings using impulse response. In Proceedings of the 10th ACI International Conference on Recent Advances in Concrete Technology, Seville, Spain, 13–17 October 2009; SP-261-19. pp. 263–275.

 *sustainability* 

Article

Optimizing the Design of a Vertical Ground Heat Exchanger: Measurement of the Thermal Properties of Bentonite-Based Grout and Numerical Analysis

Daehoon Kim and Seokhoon Oh *

Department of Energy and Resources Engineering, Kangwon National University, Chuncheon 24341, Korea; ibs0512@naver.com
* Correspondence: gimul@kangwon.ac.kr; Tel.: +82-33-250-6258

Received: 27 June 2018; Accepted: 27 July 2018; Published: 29 July 2018

Abstract: We prepared bentonite-based grouts for use in the construction of vertical ground heat exchangers (GHEs) using various proportions of silica sand as an additive, and measured the thermal conductivity (TC) and specific heat capacity (SHC) of the grouts under saturated conditions. Furthermore, we performed numerical simulations using the measured thermal properties to investigate the effects of grout-SHCs, the length of the high-density polyethylene (HDPE) pipe, the velocity of the working fluid, and the operation time and off-time during intermittent operation on performance. Experimentally, the grout TCs and SHCs were in the ranges 0.728–1.127 W/(mK) and 2519–3743 J/(kgK), respectively. As the proportions of bentonite and silica sand increased, the TC rose and the SHC fell. Simulation showed that, during intermittent operation, not only a high grout TC but also a high SHC improved GHE performance. Also, during both continuous and intermittent operation, GHE performance improved as the working fluid velocity increased, and there was a critical working fluid velocity that greatly affected the performance of the vertical GHE, regardless of operation mode, high-density polyethylene (HDPE) pipe length, or grout thermal properties; this value was 0.3 m/s. Finally, during intermittent operation, depending on the operation time and off-time, critical periods were evident when the ground temperature had been almost completely restored and any beneficial effect of intermittent operation had almost disappeared.

Keywords: vertical ground heat exchangers; thermal properties; numerical analysis

---

1. Introduction

Given the depletion of fossil energy and the need to reduce the carbon footprint, interest in new and renewable energy techniques has increased worldwide. Of the various renewable energy systems currently available, ground source heat pump (GSHP) systems have been widely used because of their efficiency, energy conservation, and low-level emissions [1–3]. Worldwide, GSHP capacity increased 1.52-fold from 2010 to 2015 at a compound annual rate of 8.69% [4].

GSHP systems are composed of a heat pump coupled with a GHE. The GHE, which is the most important component of the GSHP system, can be either vertical or horizontal in form, and exchanges heat with the ground by circulating a working fluid within a vertical or horizontal closed-loop HDPE pipe to transfer heat to/from the ground. Vertical GHEs, placed in boreholes of diameter 0.1–0.2 m and depth 20–200 m, usually afford better thermal performance than horizontal GHEs, and have been more widely used throughout the world [5–9].

When installing a vertical GHE, the empty space in the borehole is backfilled with grout. This prevents collapse of the borehole and contamination of the groundwater and aquifer, and enhances the thermal contact between the GHE and the ground [10]. GSHP performance is strongly affected by the thermal properties of the ground and grout, especially the TC of the grout [11]. However, neat bentonite–water or cement–water mixes, which are commonly used as grouts, exhibit relatively low TC. Various efforts have been made to solve this problem by mixing grout with additives to achieve superior TC.

Remound and Lund [12] reported that the use of quartzite sand as an additive improved the TC of bentonite-based grout. Lee et al. [13] measured the TC and viscosity of bentonite-based grouts using silica sand or graphite as an additive, and found that as the additive content increased, the TC rose, but the viscosity also increased. Allan and Philippacopoulos [14] evaluated the thermal, mechanical, and hydraulic properties of cement-based grouts, and developed a grout mix (MIX 111) containing silica sand. Lee et al. [15] developed a cement grout containing silica sand and graphite that afforded a TC of 2.6 W/(mK) when evaluated via in situ thermal response testing (TRT).

Numerical modeling, which is widely used to solve complex problems, has been employed in many studies to predict the performance of GSHP systems. Esen et al. [16] simulated the temperature distributions in boreholes of a vertical GHE using a two-dimensional (2D) finite element model. Lee et al. [17] simulated in situ TRT using a three-dimensional (3D) finite element model of a vertical GHE. The in situ TRT and numerical simulation results were in good agreement at a ground TC of 4 W/(mK). Jalaluddin and Miyara [18] numerically evaluated the thermal performances of three types of vertical GHEs operating in different modes. Intermittent operation improved GHE performance and may allow borehole depth to be reduced. Choi et al. [19] numerically investigated the effect of varying the thermal properties of partially saturated soil on the intermittent operation of vertical GHE. Li et al. [20] numerically evaluated the performance of vertical GHE with various HDPE pipe diameters and borehole parameters. As the borehole diameter increased or the borehole depth decreased, the influence of thermal interference was reduced. Bidarmaghz et al. [21] numerically investigated the effect of surface air temperature for the long-term operation, and reported that considering the surface air temperature fluctuation could reduce GHE length up to about 11%. Congedo et al. [22] analyzed the principal factors affecting heat transfer of horizontal GHEs via numerical simulation. The TC of the ground around the GHE and the velocity of the working fluid played important roles in energy saving.

Recently, Kim et al. [23] explored the TC and SHC of cement-based grouts with various proportions of silica sand. Numerical simulations were performed to investigate the effects of the thermal properties of such grouts on vertical GHE performance during either continuous or intermittent operation. During continuous operation, only TC positively influenced the GHE performance, but, during intermittent operation, both TC and SHC were significant. However, in earlier experimental studies [12–15] focused primarily on the TC of grout; SHC was usually ignored. Also, the SHC of grouts used in previous numerical analyses varied widely or was not measured [24–27]. During numerical modeling, invalid inputs may compromise the results. Therefore, to ensure accuracy, not only TC but also SHC (essential when performing numerical modeling) must be measured and entered.

Here, we prepared bentonite-based grouts using silica sand as an additive at various mixing ratios and measured TC and SHC under saturated conditions. Additionally, numerical simulations of vertical GHE performance during both continuous and intermittent operation were performed to explore the effect of grouts' SHC, working fluid velocity, and the length of the HDPE pipe, on measured thermal properties. During intermittent operation, the effects of operation time and off-time on vertical GHE performance were also investigated.

2. Materials and Methods

The Volclay bentonite used in this study is widely used in GSHP installation and for geophysical site investigations. The physical properties of the bentonite are listed in Table 1. The additive used to improve TC was silica sand, the grain size distribution of which is plotted in Figure 1. The specific gravity of the sand was 2.637.

Table 1. Physical properties of Volclay bentonite.

Permeability (cm/s)	Swelling Potential	Montmorillonite Content (%)	Thermal Conductivity (W/(mK))	Specific Gravity
$\leq 1 \times 10^{-7}$	25 mL/2.0 g	≤ 90	0.74	2.60

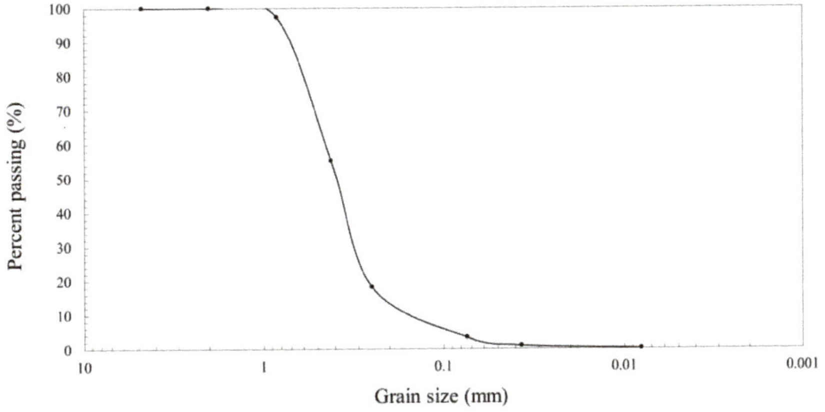

Figure 1. The grain size distribution curve of the silica sand.

The grout mix proportions were the same as in previous studies [13], as listed in Table 2. For each mix proportion, three specimens were formed by pouring the mixes into cylindrical molds (7 cm diameter × 7 cm height). After allowing for free swelling under sealed conditions for 48 h, the thermal properties were measured at about 21 °C using KD2 PRO and SH-1 sensors (Figure 2). The KD2 PRO, a handheld device, which uses a transient line heat-source method to measure thermal properties, consists of a handheld controller and sensors. Among the sensors, the dual needle SH-1 sensor used in the present study, which can measures TC, volumetric heat capacity (VHC), resistivity, and diffusivity, is equipped with two stainless steel needles mounted in parallel. One needle is a line-source heater and the other is a temperature sensor. After inserting the SH-1 sensor into to the specimen, a heat pulse is applied to the heater and the temperature at the temperature sensor is recorded as a function of time. The VHC and thermal diffusivity of the specimen are then determined from the measured temperature response with time at the temperature sensor [28–30]. TC is computed as the product of the VHC and thermal diffusivity. We used the following equation to calculate SHC:

$$c_m = \frac{c_v}{\rho},\tag{1}$$

where c_m (J/(kgK)) is the SHC, c_v (J/(m^{3}K)) the VHC and ρ (kg/m^3) the bulk density [31].

Table 2. Mix proportions for test specimens.

Specimen No.	Bentonite (wt %)	Silica Sand (wt %)	Water (wt %)
BS20-0		0	80
BS20-10	20	10	70
BS20-20		20	60
BS20-30		30	50
BS30-0		0	70
BS30-10	30	10	60
BS30-20		20	50
BS30-30		30	40

wt: weight percentage.

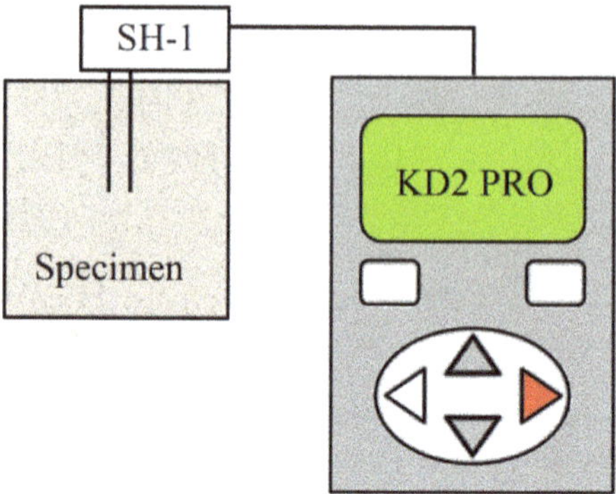

Figure 2. Schematic of the thermal property measurement setup.

3. Experimental Results and Discussion

Table 3 lists the mean saturated bulk density, porosity, TC, and SHC values. As the bentonite and silica sand levels increased, the saturated bulk density rose and the porosity fell.

Figure 3a plots the TCs of all specimens under saturated conditions. As the amounts of bentonite and silica sand increased, the TC also increased. Figure 4a shows the relationship between porosity and TC. Overall, as the porosity increased, the TC decreased. However, when the wt % values of soil (bentonite or bentonite + silica sand) were the same, the higher the ratio of silica sand to bentonite, the higher the TC at a similar porosity. Thus, BS20-30, BS20-20, and BS20-10 exhibited slightly higher TCs than BS30-20, BS30-10, and BS30-0, respectively, possibly attributable to differences in porosity and composition. Under saturated conditions, bentonite-based grouts consist of bentonite + water or bentonite + silica sand + water. The TC of water is 0.6 W/(mK) at 20 °C [32], thus lower than that of the specimens (Table 3). Therefore, the lower the porosity, the smaller the gaps between soil particles (bentonite and bentonite, or bentonite and silica sand), facilitating heat transfer and increasing the TC. Moreover, when the soil wt % values are the same, the higher the ratio of silica sand, the higher the TC at a similar porosity, because silica sand has a higher TC than bentonite.

Figure 3b plots the SHC of all specimens under saturated conditions. As the amounts of bentonite and silica sand increase, the SHC decreases. Figure 4b shows the relationship between porosity and SHC. As the porosity increases, the SHC increases, reflecting the high SHC of water, which is higher than that of any other common material (4182 J/(kgK), 20 °C) [33]. Therefore, under saturated conditions, the greater the amount of water retained within the void spaces (i.e., the higher the grout porosity), the greater the SHC. Thus, under saturated conditions, the SHC of bentonite-based grout is greatly affected by porosity.

Table 3. Physical and thermal properties of all specimens.

Specimen No.	Saturated Bulk Density (kg/m^3)	Porosity (%)	Thermal Conductivity (W/(mK))	Specific Heat Capacity (J/(kgK))
BS20-0	1096	91.57	0.728	3743
BS20-10	1170	86.56	0.791	3363
BS20-20	1298	80.17	0.855	2960
BS20-30	1354	74.19	0.988	2770
BS30-0	1158	86.64	0.775	3443
BS30-10	1256	80.78	0.846	3056
BS30-20	1359	74.06	0.976	2752
BS30-30	1439	67.11	1.127	2519

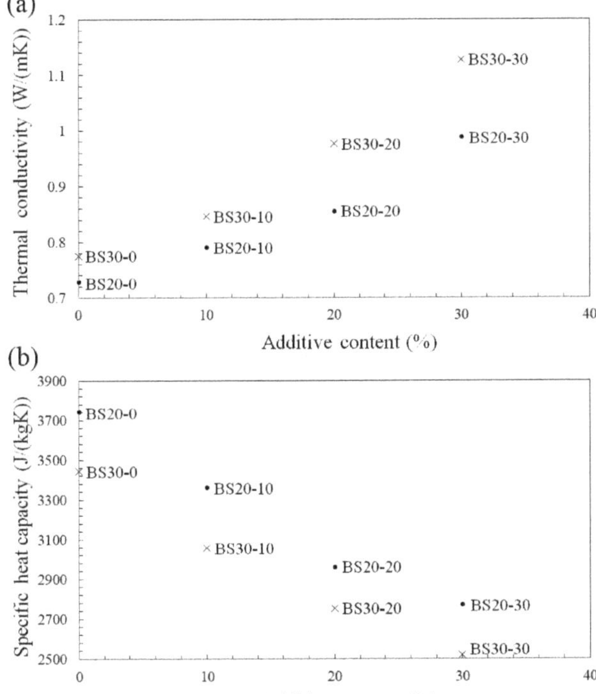

Figure 3. (a) Thermal conductivity and (b) specific heat capacity of specimens with various proportions of additives.

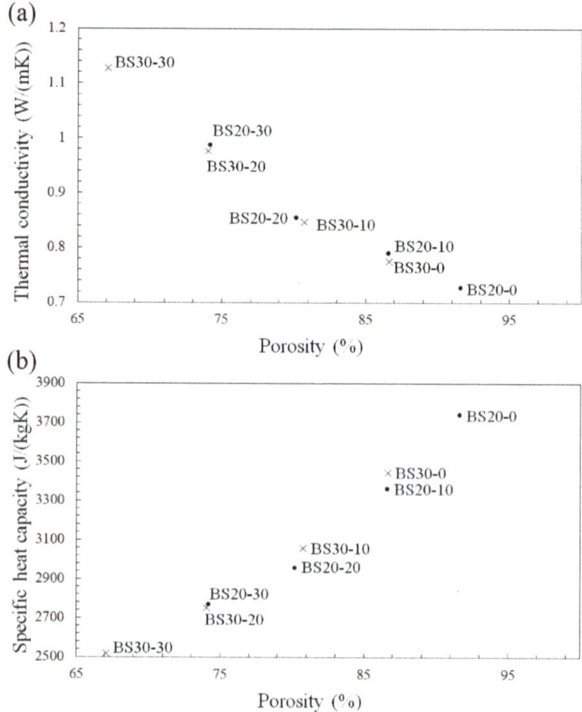

Figure 4. The relationships between porosity and thermal properties. (**a**) Thermal conductivity vs. porosity; (**b**) Specific heat capacity vs. porosity.

4. Numerical Analysis

4.1. Numerical Method

We used commercial FLUENT CFD software to simulate a 3D vertical GHE [34]. Figure 5 shows the configuration and mesh of the model, which is made up of an HDPE pipe, grout, and the ground. In the model, the length, diameter, and thickness of the HDPE pipe are 50 m, 0.025 m, and 0.005 m; the depth and diameter of the borehole are 50.3 m and 0.15 m; and the depth and diameter of the ground are 51 m and 3 m, respectively. The thermal and physical properties of the bentonite-based grouts derived above were used in modeling (Table 3). The properties of the working fluid, the HDPE pipe, and the ground; and the boundary conditions; are listed in Table 4. Additional boundary conditions used in each section of the numerical analysis were as follows:

- To investigate the effects of the SHC of bentonite-based grouts, HDPE pipe length, and working fluid velocity, the model simulated the cooling mode during intermittent operation (3 h operation time and 3 h off-time) and during continuous 24-h operation.
- To investigate the effects of the operation time and off-time on vertical GHE, the model simulated the cooling mode during intermittent operation (3 h operation time and 3 h off-time) over three days.
- To simulate the intermittent operation mode, user-defined operation times and off-times were modeled.
- When exploring the effects of HDPE pipe length and working fluid velocity, and operation time and off-time, numerical simulations were performed using the thermal properties of the most

diverse grout specimens (i.e., BS20-0, which has the lowest TC and the highest SHC, and BS30-30, which has the highest TC and the lowest SHC).

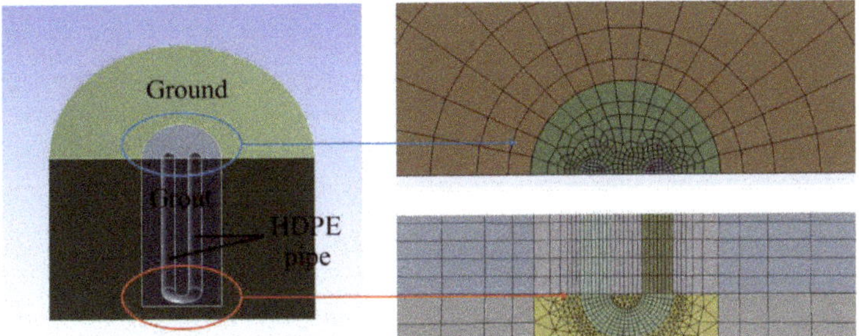

Figure 5. The configuration and mesh of the 3D vertical ground heat exchanger (GHE).

Table 4. Parameters for simulation analysis [17,23].

Item				Description
Inlet water velocity				0.3 m/s
Inlet water temperature				35 °C (308.15 K)
Turbulence intensity				5%
Initial ground temperature				18.2 °C (291.35 K)
Physical Parameters	**Density (kg/m^3)**	**Thermal Conductivity (W/(mK))**	**Specific Heat Capacity (J/(kgK))**	**Viscosity (kg/(ms))**
Working fluid	998.2	0.6	4182	0.001003
HDPE pipe	955	0.4	525	
Ground	2600	4.0	790	

HDPE, high-density polyethylene.

We considered the mass, momentum, and turbulence of the working fluid; and energy conservation by the grout and the ground. The SIMPLEC method, second-order upwind scheme, and realizable k-ε model were used [35]. The relevant equations are listed in Table 5.

Table 5. Summary of CFD model equations [34].

Continuity equation		$\frac{\partial \rho}{\partial t} + \nabla\left(\rho \vec{v}\right) = 0$
Momentum equation		$\frac{\partial \rho \vec{v}}{\partial t} + \nabla\left(\rho \vec{v} \times \vec{v}\right) - \nabla\left(\mu_{eff}\nabla \vec{v}\right) = -\nabla p + \nabla\left(\mu_{eff}\nabla \vec{v}\right)^T + S$
Energy equation		$\frac{\partial(\rho h)}{\partial t} - \frac{\partial(p)}{\partial t} + \nabla\left(\rho \vec{v} h\right) = \nabla\left[\left(\mu + \frac{\mu_t}{\sigma_t}\right)\nabla h\right] - S_h$
Turbulence equations	Turbulent energy equation	$\frac{\partial(\rho k)}{\partial t} + \nabla\left(\rho \vec{v} k\right) = \nabla\left[\left(\mu + \frac{\mu_t}{\sigma_k}\right)\nabla k\right] + G_k + G_b - \rho\varepsilon$
	Turbulence dissipation rate equation	$\frac{\partial(\rho\varepsilon)}{\partial t} + \nabla\left(\rho \vec{v}\varepsilon\right) = \nabla\left[\left(\mu + \frac{\mu_t}{\sigma_\varepsilon}\right)\nabla\varepsilon\right] + \frac{\varepsilon}{\lambda}\left(c_{\varepsilon 1}G_k - c_{\varepsilon 2}\rho\varepsilon\right)$

$c_{\varepsilon 1}$: 1.44, $c_{\varepsilon 2}$: 1.92
G_b, G_k: turbulence kinetic energies (m^2/s^2)
h: enthalpy of the fluid (J)
k: turbulence kinetic energy (m^2/s^2)
p: pressure (Pa)
S: source term
S_h: volumetric heat source (kJ/(m^{3}s))
t: time (h)
T: Temperature (°C)

ε: turbulent dissipation rate (m^2/s^2)
μ: viscosity ((Ns)/m^2)
μ_{eff}: effective viscosity ((Ns)/m^2)
μ_t: turbulence viscosity ((Ns)/m^2)
$\vec{v}$: velocity (m/s)
ρ: density (kg/m^3)
σ_ε: 1.2, σ_k : 1, σ_t: constant

4.2. Validation

We used the experimental data of Lee et al. [17] for verifying the accuracy of the 3D GHE model under the same conditions of the cited work. Lee et al. performed in situ TRT in Wonju, South Korea, and simulated the results. To simulate in situ TRT using bentonite-based grouts, we applied two user-defined functions accounting for variation in the ground temperature with depth (i.e., the geothermal gradient) and the difference in temperature between the working fluid inflow and outflow. Figure 6 compares the numerical simulations of the present study to the in situ TRT data and the numerical simulations of Lee et al. Overall, the results are in good agreement. Therefore, it can be concluded that our numerical model well predicts the heat transfer characteristics.

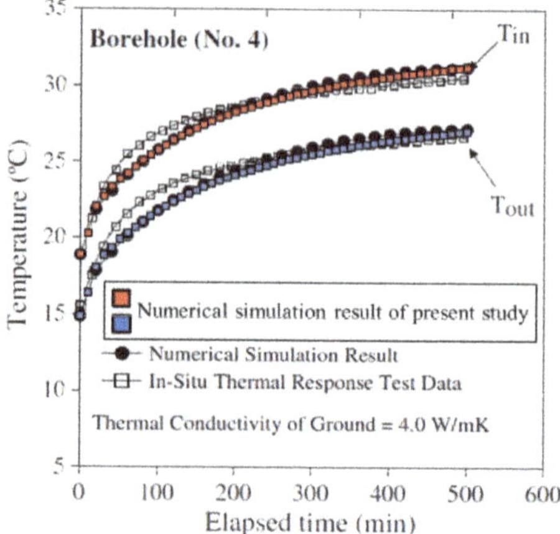

Figure 6. Numerical simulation of the present study, the in situ thermal response testing (TRT) data, and the numerical simulation of Lee et al. [17].

4.3. Numerical Results and Discussion

4.3.1. Effects of the Specific Heat Capacity of Bentonite-Based Grouts

When investigating the performance of a vertical GHE, we used the heat exchange rate (HER, W/m), calculated as:

$$\text{HER} = \frac{c_m \times \dot{m} \, (T_{in} - T_{out})}{l}, \tag{2}$$

where c_m (J/(kgK)) is the SHC of the fluid, $\dot{m}$ (kg/s) the mass flow rate of the fluid, T_{in} (°C) the inlet temperature of the fluid, T_{out} (°C) the outlet temperature of the fluid, and l (m) the length of the HDPE pipe.

The HERs for each specimen operating in the different modes are plotted in Figure 7a. In the continuous and intermittent modes, the higher the TC, the greater the HER. Moreover, the HERs during intermittent operation were higher than during continuous operation, reflecting ground temperature recovery during the off period. Therefore, use of the intermittent operation mode improved vertical GHE performance.

To investigate the effect of the SHC of grout, the percentage change in the HERs of the continuous and intermittent operation modes was calculated as:

$$\text{Percent change} = \frac{\text{HER}_{int} - \text{HER}_{con}}{\text{HER}_{con}} \times 100(\%) \tag{3}$$

where HER_{int} (%) and HER_{con} (%) are the HERs during intermittent and continuous operation, respectively. The percentage changes in the HER under different operation modes are shown in Figure 7b. Overall, the higher the grout-SHC, the greater the change; the mean value for bentonite-based grout was 27%. These results are similar to those of a previous study on the effects of the SHC of cement-based grouts with silica sand as an additive [23]. However, as the numerical simulations were performed under the same conditions as applied in a previous study [23], the mean percentage changes associated with bentonite-based grouts were about 3% higher than those associated with cement-based grouts. To further investigate these effects during intermittent operation, we performed a numerical simulation using the highest TC and SHC of bentonite-based grouts measured in the present study. Then, the HER was 62.72 W/m, 2.39 W/m higher than that of BS30-30, which exhibited the same TC but the lowest SHC. Thus, during intermittent operation, use of a grout with a high SHC improves the vertical GHE performance; a bentonite-based grout would be preferred to a cement-based grout. However, when silica sand is used as an additive, the SHC of bentonite-based grout decreases as the amount of silica sand increases. Therefore, to develop a grout with not only high TC but also high SHC, improving vertical GHE performance, various other additives should be investigated and experimental studies conducted. Especially as an increase in the proportion of additive decreases the SHC, an additive affording high-level TC is required, even at small proportions.

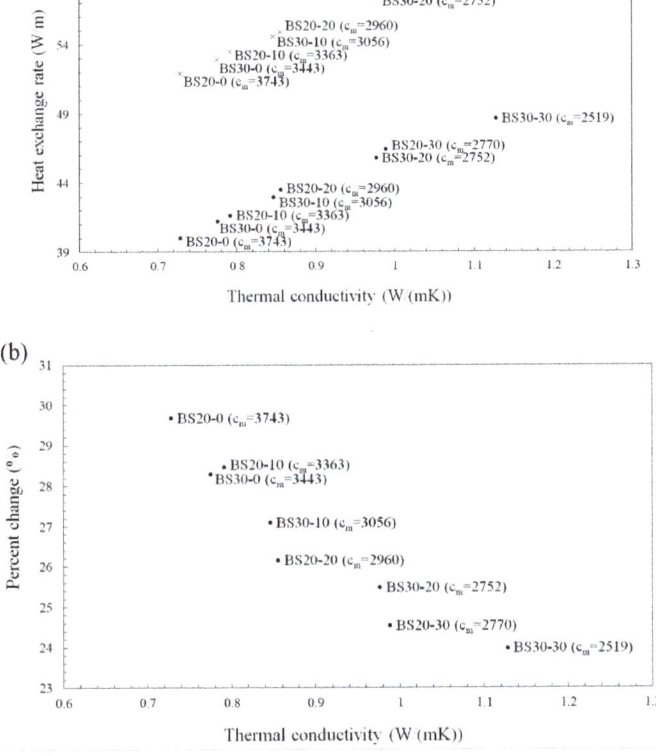

Figure 7. (a) Mean heat exchange rates in different operation modes; (b) Percentage changes in heat exchange rates between different operation modes (c_m (J/(kgK)): the specific heat capacity of each specimen).

4.3.2. Effects of Working Fluid Velocity and HDPE Pipe Length

Figure 8 shows the changes in the HERs of BS20-0 and BS30-30 during continuous and intermittent operation at various fluid working velocities and HDPE pipe lengths. Overall, regardless of the pipe length, the HERs were higher during intermittent than continuous operation, and the HERs of BS30-30 (with a higher TC) were greater than those of BS20-0. In addition, the HERs increased as the working fluid velocity increased, similar to what was noted in a previous study [25]. However, when the working fluid velocity was <0.3 m/s, the HERs increased markedly, but when the velocity exceeded 0.3 m/s, the changes were small. This result indicates that there is a critical velocity for the working fluid that has a large effect on the performance of a vertical GHE, regardless of the operation mode, HDPE pipe's length, and thermal properties of the grout; that flow rate is 0.3 m/s. Nevertheless, when the velocity was >0.3 m/s, the HERs increased slightly as the length of the HDPE pipe increased, but little difference in the HERs was evident at a velocity of 1.2 m/s. For example, during intermittent operation, the HER of BS30-30 was about 70 W/m at a fluid velocity of 1.2 m/s, regardless of HDPE pipe length, whereas the HERs at a fluid velocity of 0.3 m/s were 64.9 W/m, 60.5 W/m, and 56.5 W/m when the pipe length was 30 m, 50 m, and 70 m, respectively. This result indicates that there are optimum velocities for the working fluid depending on the length of the HDPE pipe (i.e., velocities in the range 0.3 to 1.2 m/s). Moreover, as shown in Figure 8, given the increase in HERs with increased HDPE pipe length and working fluid velocity, the optimum velocity for the working fluid is expected to increase as the length of the HDPE pipe increases. However, since increasing the velocity of the working fluid can increase energy consumption, additional field studies should be carried out to determine the optimum velocity for a vertical GHE for a given length of HDPE pipe.

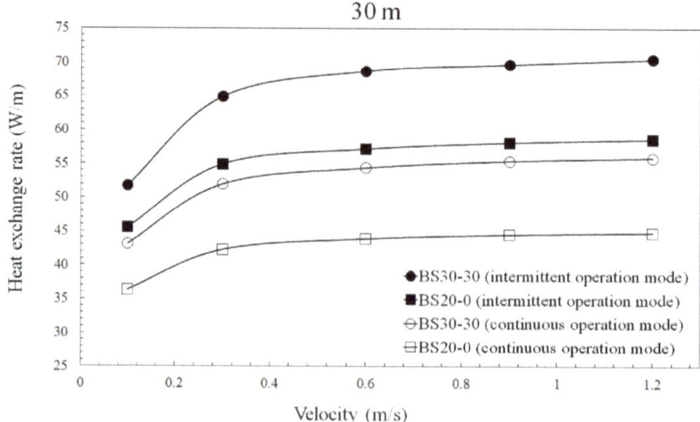

Figure 8. *Cont.*

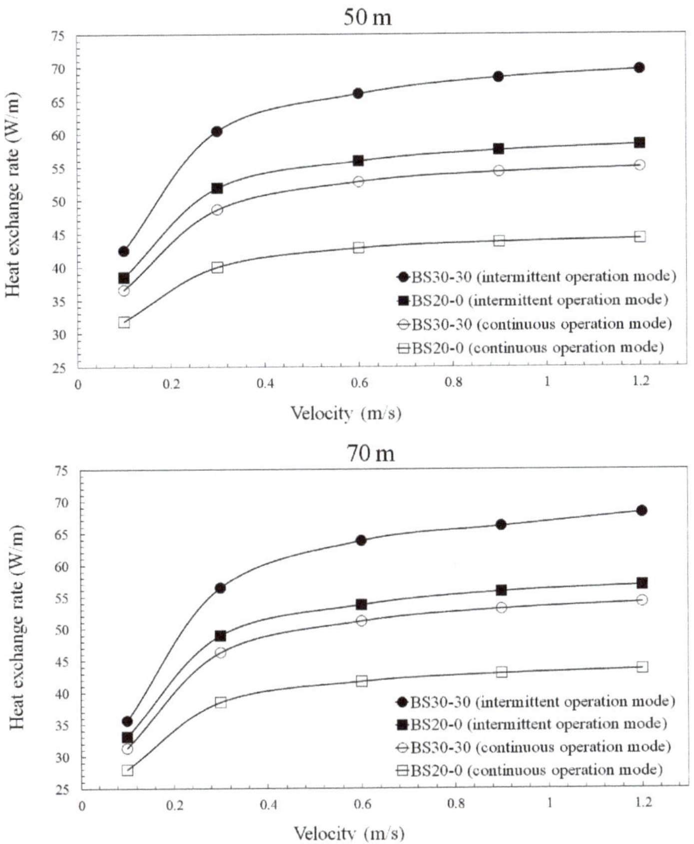

Figure 8. The effects of working fluid velocity and high-density polyethylene (HDPE) pipe length in different operation modes.

4.3.3. Effects of Various Operation Times and Off-Times during Intermittent Operation

Figure 9a shows the changes in HERs of BS20-0 and BS30-30 grouts during intermittent operation over three days, when the operation time was 3 h and the off-time varied. Figure 9b shows the changes in HERs of BS20-0 and BS30-30 grouts during intermittent operation over three days, when the off-time was fixed at 3 h and the operation time varied. Overall, the higher the grout TC, the higher the HER; as the operation time or off-time increased, the HERs were similar regardless of the thermal properties of the grout. However, as shown in Figure 9a, when the off-time exceeded 9 h, the HERs differed only marginally, attributable to recovery of the ground temperature during the long off-time. In Figure 10, which is the simulation for BS20-0, panel (a) shows the ground temperature distribution when the operation time is 3 h and panels (b), (c), and (d) show the ground temperature distributions after 3 h of operation at off-times of 3, 6, and 9 h, respectively. The ground temperature recovered gradually as off-time increased; after 9 h, the ground temperature had almost completely recovered. This identifies a critical time when the temperature of the ground has become almost completely restored; this time depends on the operation time and off-time, not on the thermal properties of the grout. Also, as shown in Figure 9b, when the operation time exceeded 9 h, little change in the HERs was evident, probably because the ground temperature had not recovered sufficiently during the off-time given the long operation time. The HERs of BS20-0 and BS30-30 during continuous operation over three days were 37.86 W/m and 46.49 W/m, respectively, similar to those associated with operation times ≥9 h (Figure 9b). This indicates that there is a critical interval at which any beneficial effect of intermittent operation almost disappears if the operation time is too long, regardless of the thermal properties of the grout. Therefore, after installation of a vertical GHE, it may be possible to use our method to estimate the critical times for optimizing intermittent operation. Using our present results as an example, if the operation time is 3 h, it is not necessary to set the off-time to ≥9 h, because the ground temperature recovers almost completely. Also, if the off-time is 3 h, the operation time should be set to <9 h; otherwise, the beneficial effect of intermittent operation almost disappears. Therefore, if the optimal operation time and off-time are thus determined, the vertical GHE will operate more efficiently. However, the critical times (when the ground temperature is almost completely restored and the benefit of intermittent operation almost disappears) may depend on the working fluid velocity, borehole depth, ground thermal properties, and operation time and off-time. Therefore, to efficiently apply our method, additional field investigations are required.

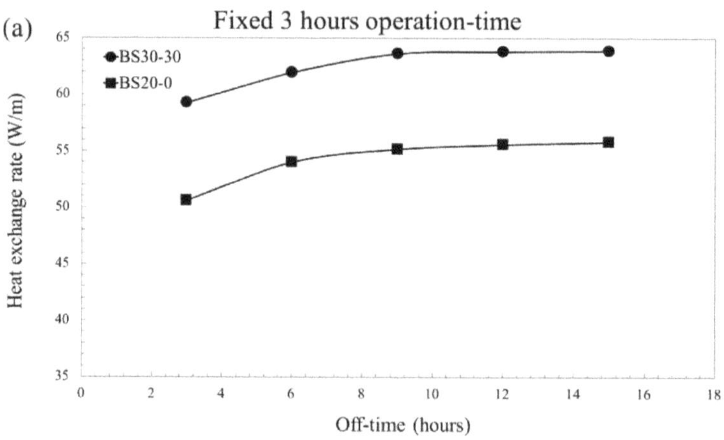

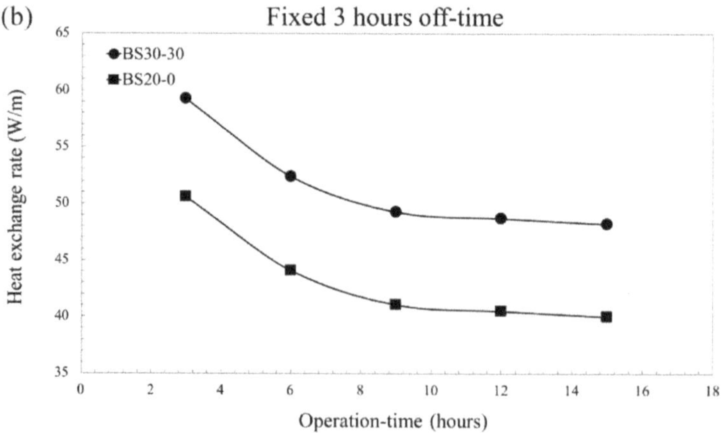

Figure 9. (**a**) Heat exchange rates by off-time; (**b**) Heat exchange rates by operation time.

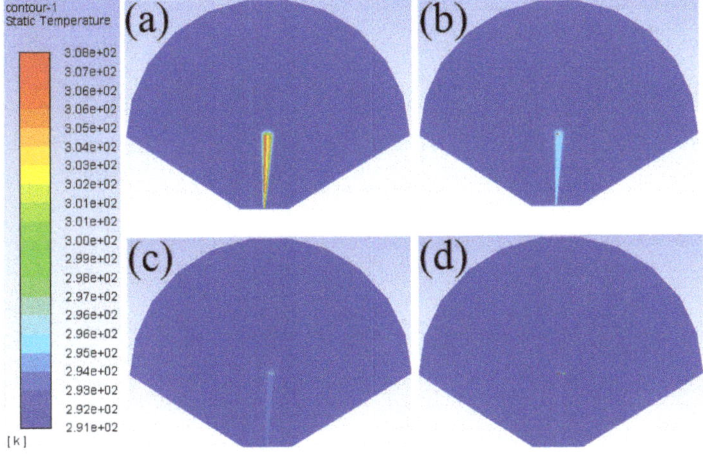

Figure 10. (**a**) Ground temperature distribution 3 h after operation; (**b**) Ground temperature distribution after 3 h of off-time; (**c**) Ground temperature distribution after 6 h of off-time; (**d**) Ground temperature distribution after 9 h of off-time.

5. Conclusions

We studied the thermal properties of bentonite-based grouts used to construct vertical GHEs and explored the effects on GHE performance of grout-SHC, HDPE pipe length, working fluid velocity, and operation time and off-time during intermittent operation. We also performed numerical simulations using the measured thermal properties.

(1)　Under saturated conditions, the TC and SHC of bentonite-based grouts ranged from 0.728–1.127 W/(mK) and 2519–3743 J/(kgK), respectively. As the proportion of silica sand increased, the TC of the bentonite-based grouts increased, but the SHC decreased. The thermal properties of bentonite-based grouts were affected principally by composition and porosity.

(2)　For bentonite-based grouts, the mean HER was 27% higher during intermittent operation than during continuous operation. Also, during intermittent operation, grout with high TC and high SHC improved GHE performance. Because the SHC of bentonite-based grout is higher than that of cement-based grout, the effect of bentonite-based grout was more significant.

(3)　During both continuous and intermittent operation, GHE performance improved as the working fluid velocity increased. However, there was a critical working fluid velocity that greatly affected the performance of the vertical GHE, regardless of the operation mode, HDPE pipe length, or grout thermal properties; this value was 0.3 m/s. Therefore, it is recommended to set the velocity for operation of the vertical GHE to 0.3 m/s or higher.

(4)　During intermittent operation, depending on the operation time and off-time, we found critical time intervals at which the ground temperature was almost completely restored, or any benefit of intermittent operation almost disappeared. Moreover, the method to estimate the critical time intervals for optimizing intermittent operation was proposed. Our results can be used as input data for analyses of the thermal behavior of vertical GHEs to improve GHE performance and operation.

Author Contributions: D.K. and S.O. conceived the study plan and contributed to the analysis and to the experiment. All authors read and approved the final manuscript.

Funding: This research was supported by the National Strategic Project-Carbon Upcycling of the National Research Foundation of Korea (NRF) funded by the Ministry of Science and ICT (MSIT), the Ministry of Environment (ME) and the Ministry of Trade, Industry and Energy (MOTIE) (NRF-2017M3D8A2085342).

Conflicts of Interest: The authors declare no conflict of interest.

References

1. Blazquez, C.S.; Martin, A.F.; Nieto, I.M.; Garcia, P.C.; Sanchez Perez, L.S.; Gonzalez-Aguilera, D. Analysis and study of different grouting materials in vertical geothermal closed-loop systems. *Renew. Energy* **2017**, *114*, 1189–1200. [CrossRef]

2. Wang, H.; Lu, J.; Qi, C. Thermal conductivity of sand bentonite mixtures as a backfill material of geothermal boreholes. *GRC Trans.* **2011**, *35*, 1135–1138.

3. Indacoechea-Vega, I.; Pascual-Muñoz, P.; Castro-Fresno, D.; Calzada-Pérez, M.A. Experimental characterization and performance evaluation of geothermal grouting materials subjected to heating-cooling cycles. *Constr. Build. Mater.* **2015**, *98*, 583–592. [CrossRef]

4. Lund, J.W.; Boyd, T.L. Direct utilization of geothermal energy 2015 worldwide review. *Geothermics* **2016**, *60*, 66–93. [CrossRef]

5. Yang, H.; Cui, P.; Fang, Z. Vertical-borehole ground-coupled heat pumps: A review of models and systems. *Appl. Energy* **2010**, *87*, 16–27. [CrossRef]

6. Cui, P.; Yang, H.; Fang, Z. Numerical analysis and experimental validation of heat transfer in ground heat exchangers in alternative operation modes. *Energy Build.* **2008**, *40*, 1060–1066. [CrossRef]

7. Lee, J.Y. Current status of ground source heat pumps in Korea. *Renew. Sustain. Energy Rev.* **2009**, *13*, 1560–1568. [CrossRef]

8. Hu, J. An improved analytical model for vertical borehole ground heat exchanger with multiple-layer substrates and groundwater flow. *Appl. Energy* **2017**, *202*, 537–549.

9. Chong, C.S.A.; Gan, G.; Verhoef, A.; Garcia, R.G. Comparing the thermal performance of horizontal slinky-loop and vertical slinky-loop heat exchangers. *Int. J. Low-Carbon Technol.* **2014**, *9*, 250–255. [CrossRef]

10. Choi, W.; Ooka, R. Effect of natural convection on thermal response test conducted in saturated porous formation: Comparison of gravel-backfilled and cement-grouted borehole heat exchangers. *Renew. Energy* **2016**, *96*, 891–903. [CrossRef]

11. Desmedt, J.; Van Bael, J.; Hoes, H.; Robeyn, N. Experimental performance of borehole heat exchangers and grouting materials for ground source heat pumps. *Int. J. Energy Res.* **2012**, *36*, 1238–1246. [CrossRef]

12. Remund, C.P.; Lund, J.T. *Thermal Enhancement of Bentonite Grouts for Vertical GSHP Systems in Heat Pump and Refrigeration Systems*; Den Braven, K.R., Mei, V., Eds.; American Society of Mechanical Engineers: New York, NY, USA, 1993; Volume 29, pp. 95–106.

13. Lee, C.H.; Lee, K.J.; Choi, H.S.; Choi, H.P. Characteristics of thermally-enhanced bentonite grouts for geothermal heat exchanger in South Korea. *China Technol. Sci.* **2011**, *53*, 123–128. [CrossRef]

14. Allan, M.L.; Philippacopoulos, A.J. *Properties and Performance of Thermally Conductive Cement-Based Grouts for Geothermal Heat Pumps*; Department of Applied Science, Brookhaven National Laboratory: New York, NY, USA, 1999.

15. Lee, C.H.; Park, M.S.; Nguyen, T.B.; Sohn, B.H.; Choi, H.S. Performance evaluation of closed-loop vertical ground heat exchangers by conducting in-situ thermal response tests. *Renew. Energy* **2012**, *42*, 77–83. [CrossRef]

16. Esen, H.; Inalli, M.; Esen, Y. Temperature distributions in boreholes of a vertical ground-coupled heat pump system. *Renew. Energy* **2009**, *34*, 2672–2679. [CrossRef]

17. Lee, C.; Park, M.; Park, S.; Won, J.; Choi, H. Back-analyses of in-situ thermal response test (TRT) for evaluating ground thermal conductivity. *Int. J. Energy Res.* **2013**, *37*, 1397–1404. [CrossRef]

18. Jalaluddin; Miyara, A. Thermal performance investigation of several types of vertical ground heat exchangers with different operation mode. *Appl. Therm. Eng.* **2012**, *33*, 167–174. [CrossRef]

19. Choi, J.C.; Lee, S.R.; Lee, D.S. Numerical simulation of vertical ground heat exchangers: Intermittent operation in unsaturated soil conditions. *Comput. Geotech.* **2011**, *38*, 949–958. [CrossRef]

20. Li, Y.; An, Q.S.; Liu, L.X.; Zhao, J. Thermal performance investigation of borehole heat exchanger with different U-tube diameter and borehole parameters. *Energy Procedia* **2014**, *61*, 2690–2694. [CrossRef]

21. Bidarmaghz, A.; Narsilio, G.A.; Johnston, I.W.; Colls, S. The importance of surface air temperature fluctuations on long-term performance of vertical ground heat exchangers. *Geomech. Energy Environ.* **2016**, *6*, 35–44. [CrossRef]

22. Congedo, P.M.; Colangelo, G.; Starace, G. CFD simulations of horizontal ground heat exchangers: A comparison among different configurations. *Appl. Therm. Eng.* **2012**, *33–34*, 24–32. [CrossRef]

23. Kim, D.; Kim, G.; Kim, D.; Baek, H. Experimental and numerical investigation of thermal properties of cement-based grouts used for vertical ground heat exchanger. *Renew. Energy* **2017**, *112*, 260–267. [CrossRef]

24. Pu, L.; Qi, D.; Li, K.; Tan, H.; Li, Y. Simulation study on the thermal performance of vertical U-tube heat exchangers for ground source heat pump system. *Appl. Therm. Eng.* **2015**, *79*, 202–213. [CrossRef]

25. Chen, S.; Mao, J.; Han, X.; Li, C.; Liu, L. Numerical analysis of the factors influencing a vertical u-tube ground heat exchanger. *Sustainability* **2016**, *8*, 882. [CrossRef]

26. Chen, J.; Xia, L.; Li, B.; Mmereki, D. Simulation and experimental analysis of optimal buried depth of the vertical U-tube ground heat exchanger for a ground-coupled heat pump system. *Renew. Energy* **2015**, *73*, 46–54. [CrossRef]

27. Jin, G.; Zhang, X.; Guo, S.; Wu, X.; Bi, W. Evaluation and analysis of thermal short-circuiting in borehole heat exchangers. In Proceedings of the 8th International Conference on Applied Energy (ICAE2016), Beijing, China, 8–11 October 2016; pp. 1677–1682.

28. Welch, S.M.; Kluitenberg, G.J.; Bristow, K.L. Rapid numerical estimation of soil thermal properties for a broad class of heat-pulse emitter geometries. *Meas. Sci. Technol.* **1996**, *7*, 932–938.

29. Bristow, K.L. Measurement of thermal properties and water content of unsaturated sandy soil using dual-probe heat-pulse probes. *Agric. Forest Meteorol.* **1998**, *89*, 75–84. [CrossRef]

30. KD2 Pro Manual. Available online: http://manuals.decagon.com/Manuals/13351_KD2%20Pro_Web.pdf (accessed on 20 July 2018).

31. Cengel, Y.A. *Heat and Mass Transfer: A Practical Approach*; McGraw-Hill: New York, NY, USA, 2006.

32. Hens, H. *Applied Building Physics: Boundary Conditions, Building Performance and Material Properties*; Wiley: Berlin, Germany, 2010.

33. Engineeringtoolbox. Available online: http://www.engineeringtoolbox.com/specific-heat-capacity-d_391.html (accessed on 20 July 2018).

34. Fluent 6.3 User's Guide. Available online: https://www.sharcnet.ca/Software/Fluent6/html/ug/main_pre.htm (accessed on 20 July 2018).

35. Gao, J.; Zhang, X.; Liu, J.; Li, K.S.; Yang, J. Numerical and experimental assessment of thermal performance of vertical energy piles: An application. *Appl. Energy* **2008**, *85*, 901–910. [CrossRef]

 *sustainability* 

Article

Pressurization Ventilation Technique for Controlling Gas Leakage and Dispersion at Backfilled Working Faces in Large-Opening Underground Mines: CFD Analysis and Experimental Tests

Van-Duc Nguyen [1,*], Won-Ho Heo [2], Rocky Kubuya [1] and Chang-Woo Lee [1,*]

[1] Department of Energy and Mineral Resources, College of Engineering, Dong-A University, Saha-gu, Busan 49315, Korea; pakrockykiro@gmail.com
[2] Mining Tech Co., Ltd, Saha-gu, Busan 49315, Korea; hoya4016@naver.com
* Correspondence: nguyenduc.imsat@gmail.com (V.-D.N.); cwlee@dau.ac.kr (C.-W.L.)

Received: 30 April 2019; Accepted: 13 June 2019; Published: 15 June 2019

Abstract: Pressurization ventilation techniques, originally designed to control building fires, have never been applied to the mines. The working face, backfilled with fly-ash-based materials, is likely to be contaminated during, and even after, the curing period of the backfill materials. Gases such as NH_3 and CO_2 may leak out prolongedly from the backfilled sites. Proper ventilation schemes should be implemented to control toxic gas leakage and thus minimize the workers' exposure. This study aims at evaluating the applicability of a pressurization ventilation scheme at backfilled working faces in large-opening limestone mines. To pressurize the working face, two different fans (15 kW and 37 kW) were developed and two ventilation scenarios were tested. Computational Fluid Dynamics (CFD) analysis was also carried out for comparison purposes. There is no established standard for differential pressure between the inside and outside of working faces to prevent gas leakage at mines. However, taking the differential pressure of 50 Pa in British standards for controlling building fires (where a relatively stronger dissipation force than the gas leakage of a mining face occurs), the pressure differential created by two blowing fans seems to be sufficient to control the gas leakage and dispersion within the work space.

Keywords: pressurization system; ventilation fan; differential pressure; gas leakage; backfilled area

1. Introduction

Backfilling of voids generated by mining is a fundamental component of most underground stopping operations. Without backfill, these operations would be unsafe, less productive, and shorter in life [1]. It is expected that large quantities of NH_3 and CO_2 will leak into the atmosphere, as well as the working space, in the mining area backfilled with the composite carbonate-based material. At most thermal power plants, a large amount of NH4 is injected into the denitrification unit of the flue gas cleaning system, thereby reducing NOx to N_2 and water through catalytic reduction. As a result, NH_4 is retained in the fly ash bulk and is likely to leak out for a long time. This is why some countries, such as Poland, regulate NH4 levels in the air at backfilled mine sites. Therefore, it is necessary to control the thermal environment and the air quality within the underground working space during, and even after, the curing period for backfilled materials. In large-opening underground limestone mines in Korea with typical airway dimensions of 10 m (W) and 8 m (H), the conventional ventilation system of large-capacity axial-flow fans with ducts may be insufficient to control the toxic gas which escapes during the curing period, since those mines are developed with multiple interconnected levels and the intake and return airways are not always well-maintained.

Thus, the ventilation system proposed in this study has the twofold purpose of minimizing the ventilation cost while isolating the contaminated zone near the vicinity of the face in order to reduce the workers' exposure. To achieve this goal, we suggest the implementation of the pressurization ventilation system which was originally designed to control building fire dispersion and secure safe escape routes by generating a pressure difference of at least 50 Pa between the escapees and the fire-affected space. The same concept is applied to the backfilled site to generate a pressurized zone just in front of the face to minimize the leakage and confine the polluted air within the pressurized zone. Therefore, the ultimate goal of this study is to secure a safe working environment at backfilled sites by applying a low-cost pressurization ventilation technique.

2. Pressurization Ventilation Technique

2.1. Background of the Pressurization Ventilation Technique

Pressurization ventilation techniques were developed for the purpose of creating safe zones within burning buildings for a variety of purposes, like maintaining escape, and firefighting access routes in lobbies and stairwells, and creating refuge areas for trapped victims. The aim of the pressure differential system is to establish airflow paths from protected spaces at high pressure to spaces at lower or ambient pressure, preventing the spread of toxic gas released during a fire [2]. The successful operation of the pressurization system remains a standard feature of high rise building codes in USA, UK, Australia, China, India, the UAE, and many other locations. A pressurization system applied in buildings is intended to prevent smoke leaking past closed doors into stairwells by injecting clean air into the stair enclosure at sufficient pressure that the air pressure in the stairwell is greater than that in the adjacent fire compartment [3]. Studies of pressurization systems by Tamura 1989, 1992 [4,5], Budnick 1987 [6], Wang et al. 2004 [7], Jo et al. 2007 [8], Gai et al. 2017 [2], and Li et al. 2018 [9] form the theoretical foundation for the development of ventilation schemes for gas leakage and dispersion controlling at backfilled underground mine sites. Figure 1a illustrates the basic concept of a pressure differential system for generating and maintaining safe zones within buildings. Generally, the positively pressurized zone can be generated by the operation of a fan in an enclosed space. The purpose of a positively pressurized compartment is to prevent the spread of smoke from the compartment of first ignition to other areas. Thus, during a fire, a refugee can move to a high-pressure compartment for safety.

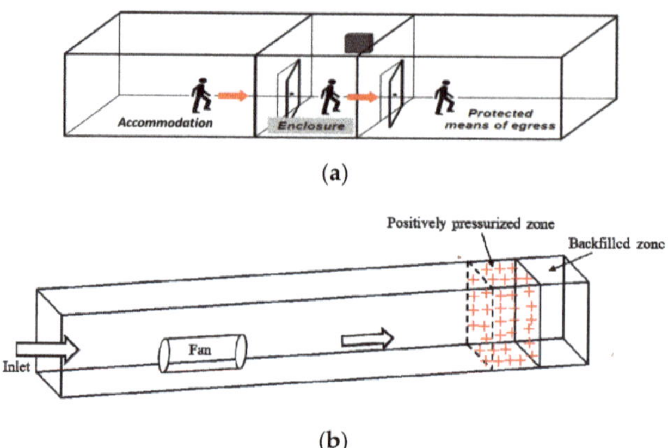

Figure 1. Background of pressurization ventilation technique. (**a**) Basic concept of the pressure differential system [2]; (**b**) application of the pressurization ventilation concept at a backfilled area.

Figure 1b shows the application of the pressurization ventilation technique at a mine working space. It can be seen that a positively pressurized zone near the backfilled area is be generated by blowing fans. The pressurized zone is expected to prevent hazardous gas leakage from the backfilled zone. According to the British standards BS EN 12101-6 "Specification for pressure differential systems" [10] and BS 5588-4 "Code of practice for smoke control using pressure differentials" [11], there are two requirements to maintain a pressurization system. Firstly, the minimum pressure differential between inside and outside of the pressurized zone is 50 Pa. Secondly, for an open door condition, the airflow discharged from the pressurized zone through the doorway should be not less than 2.0 m/s.

2.2. Description of Fans for the Pressurization Ventilation System

Several studies at the National Institute for Occupational Safety and Health (NIOSH) by Grau et al. 2002 [12,13], Krog et al. 2006 [14,15], and Chekan et al. 2006 [16] indicated that large diameter, low-pressure fans provide better regional air coverage in large-opening mines than high-pressure fans. However, while low-pressure fans installed in a large-opening airway can generate a relatively large amount of slow-moving airflow at a low cost, the pressure differentials created by those fans are not sufficiently high. This limitation is due to the fan locations; fans must be installed at a distance far enough away from the working face to allow space for the many equipment employed there. Thus, for the pressurization purpose at backfilled areas in large-opening mines in this study, two types of fans were developed—a 37 kW high-pressure fan and a 15 kW low-pressure fan. Figure 2 shows the two axial-flow high-pressure and low-pressure fans used for the pressurization ventilation experiment in this study. Their specifications are summarized in Table 1.

 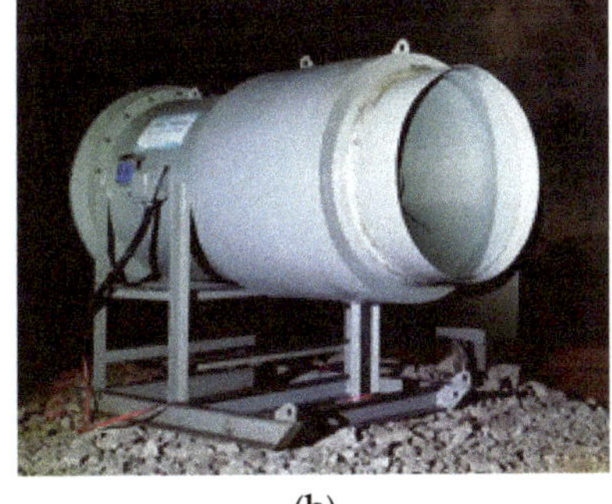

(a)   (b)

Figure 2. Two fans developed for the pressurization ventilation experiments: (a) 37 kW high-pressure fan; (b) 15 kW low-pressure fan.

Table 1. Fan specifications.

Categories	37 kW High-Pressure Fan	15 kW Low-Pressure Fan
Flow quantity (m^3/s)	47.1	16.67
Diameter (m)	1.4	0.95
Pressure (Pa)	555.1	235
Discharge velocity(m/s)	30.6	23.5
Noise level (dB(A))	105	103
Power (kW)	37	15
Length (m)	3.0	2.23
Weight (kg)	998	792

3. Mine Site Study of the Pressurization Ventilation Technique

3.1. Description of Mine Site Study and Experiment Method

In Korea, most underground limestone mines have been developed through the multi-level room-and-pillar mining method with large openings in steeply-dipping and faulted ore bodies. In general, the entries are 6–9 m high and 10–15 m wide and the rampways connecting levels with a vertical difference of approximately 20 m are inclined at 10%–13%. The number of entries in each level depends on the vein width; two to five entries being typical. Figure 3a shows a 3D map of a large-opening underground mine, the D limestone mine in Chungbuk province. Pressurization ventilation experiments were carried out at a blind entry development site in Level 5; the development entry was 8 m (W) × 7 m (H) and the experiment segment was 188 m long as shown in Figure 3b.

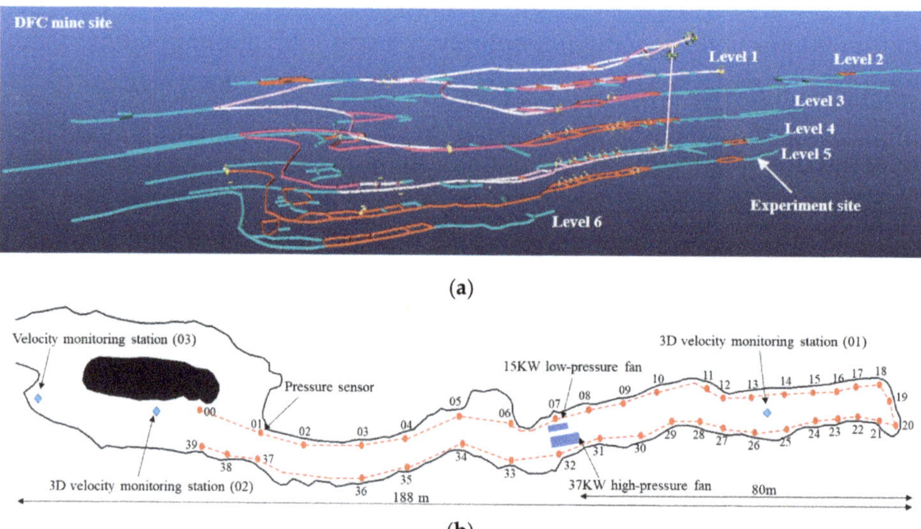

Figure 3. D large-opening mine and layout of pressurization ventilation experiment. (**a**) D limestone mine in Chungbuk province; (**b**) layout of the experiment site at the fifth level.

Figure 3b shows the locations of the test equipment installation and monitoring stations. Two pressurizing ventilation fans were installed 80 m from the working face, a hypothetical backfilled working face. The 37 kW high-pressure fan was installed on the right side of the entry facing toward the workface, while the 15 kW low-pressure fan was installed on the opposite side. Both fans were installed in the blowing mode. At the same time, the 40 pressure sensor modules, as shown in Figure 4, developed for this study were installed along the side of the entry, as shown in Figure 3b, to monitor pressure changes during the tests. A controller area network (CAN) was employed as the pressure sensors' communication method. CAN is the international standardization organization-defined serial communications bus originally developed for the automotive industry to replace complex wiring harnesses with a two-wire bus [17]. Pressure distributions during the fan operation were monitored and the results were analyzed to identify the existence of the pressurized zone near the face. In addition, three velocity monitoring stations were installed to evaluate the exhaust efficiency of contaminated air to the areas outside of the working face, as shown in Figure 3b and Table 2.

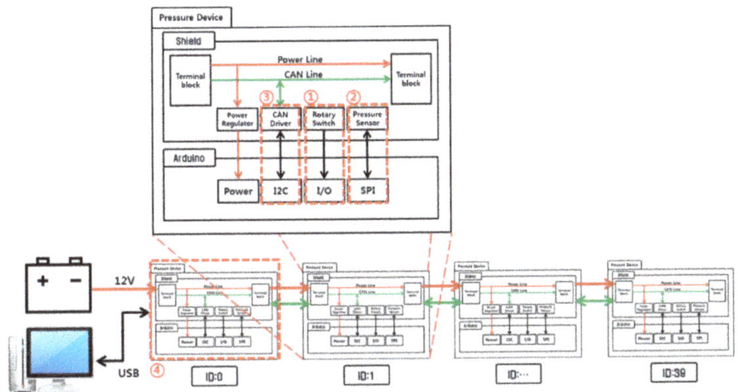

Figure 4. Pressure monitoring and communication module. CAN: controller area network.

Table 2. Experimental and numerical scenario description.

Experiment Scenarios	Airway Layout	Description
Scenario I	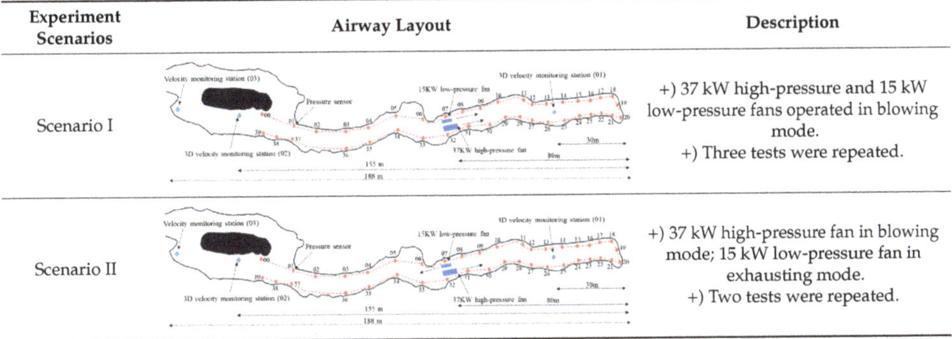	+) 37 kW high-pressure and 15 kW low-pressure fans operated in blowing mode. +) Three tests were repeated.
Scenario II	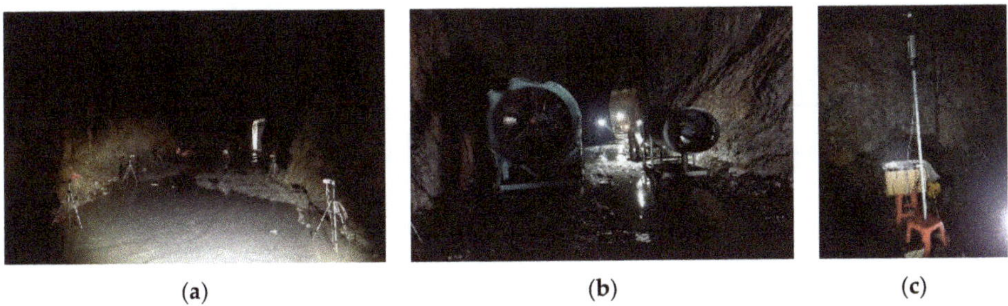	+) 37 kW high-pressure fan in blowing mode; 15 kW low-pressure fan in exhausting mode. +) Two tests were repeated.

Figure 5 shows the pressure sensors (a) and fans (b) installed at the test site. In addition, three velocity monitoring stations were set up (c). Two experimental scenarios of the pressurization ventilation technique application described in Table 2 were evaluated. In Scenario I, both fans were operated in blowing mode, while in Scenario II the 37 kW fan was operated in blowing mode and the 15 kW fan was operated in exhaust mode.

(a) (b) (c)

Figure 5. Fans, pressure sensors, and 3D velocity monitoring installation in experiment site. (**a**) Pressure sensors; (**b**) fans; (**c**) velocity station.

3.2. Experimental Results

3.2.1. Scenario I: 15 kW and 37 kW Fans in Blowing Mode

Figure 6 shows the pressure data collected by 40 pressure sensors during the Scenario I tests. The first test was conducted for 49 min, while the second and third tests were carried out for 60 min. As shown in Figure 6a,c,e, the pressure data show that the initial pressure right at the beginning of fan operation increases dramatically and then the pressures at all sensor locations fluctuate. However, the pressure differentials among all the sensor locations remain almost constant. As clearly observed in Figure 6, the pressure data were divided into two zones—the outlet side of the fans showing higher pressures and the inlet side of the fans showing lower pressures. At all sensors in the outlet side (nos. 8–31), relatively higher pressures were monitored, while the sensors showing lower pressures (nos. 0–6 and 32–39) were located on the inlet side. The same trends were observed in all three tests as described in Figure 6a,c,e; the pressure differentials between those two groups of pressure data sensors remained constant even with serious fluctuations of the pressure within the working space due to the fan operations. This result indicates that a positively pressurized zone can be created and maintained continuously near the face during the fan operation to prevent gas leakage from the backfilled zone.

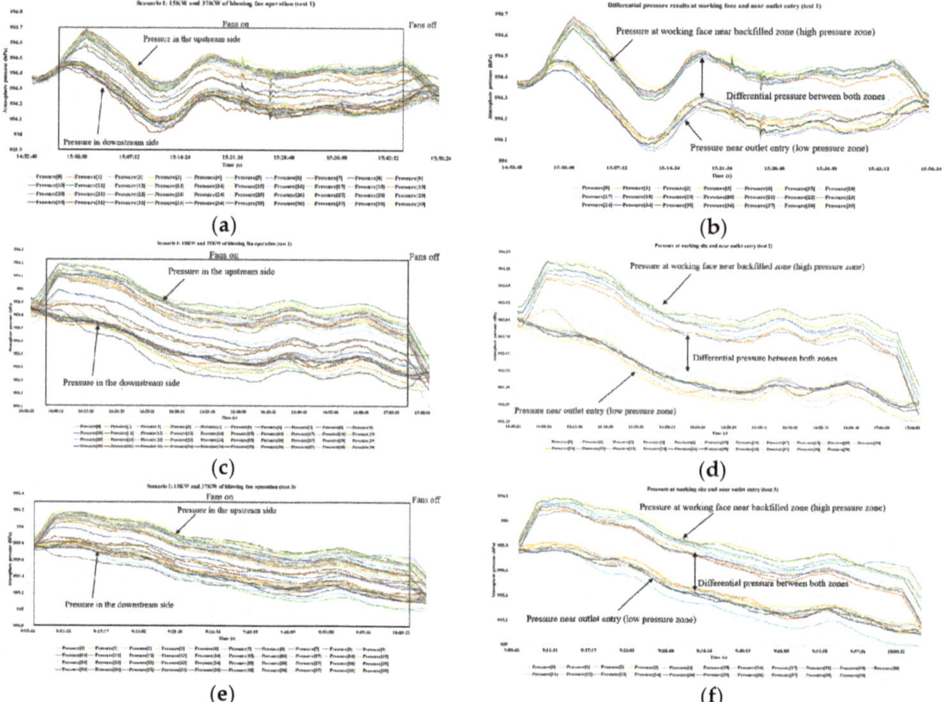

Figure 6. Temporal pressure distributions for Scenario I. (**a**) Test 1—49 min of fan operation; (**b**) differential pressure in high- and low-pressure zone (test 1); (**c**) test 2—60 min of fan operation; (**d**) differential pressure in high- and low-pressure zone (test 2); (**e**) test 3—60 min of fan operation; (**f**) differential pressure in high- and low-pressure zone (test 3).

Figure 7a,c,e show the average pressures of these two groups and the pressure differentials between them. The differential pressure between the outlet and inlet sides of the fan was 22.30 Pa in the first test, while the second and third tests showed differences of 32.78 Pa and 30.50 Pa, respectively. However, considering the differential pressure of 50 Pa specified in the British standards of BS EN

12101-6 and BS 558-4 to control building fires, which has relatively larger dissipation force than the gas leakage in the mining face, the pressure difference of approximately 30 Pa created by two blowing fans in this study may be sufficient to control the gas leakage and dispersion within the working space. Further research is planned for determining the minimum level of pressure differential required to control the gas leakage at the backfilled sites.

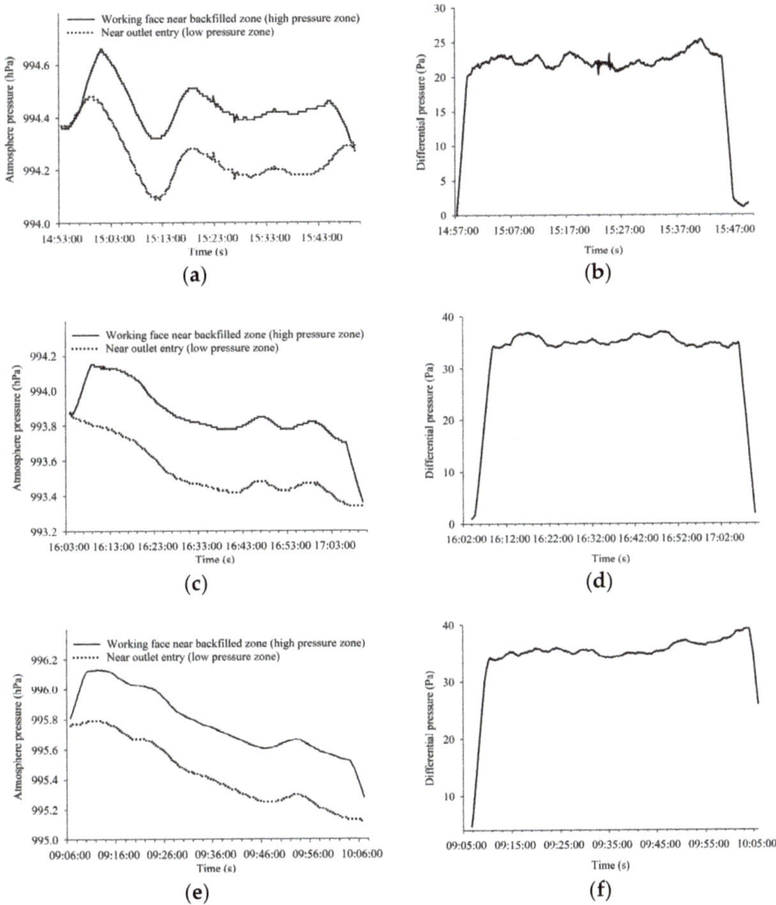

Figure 7. Differential pressure at high- and low-pressure zone in Scenario I. (**a**) Average differential pressure (test 1); (**b**) differential pressure in high- and low-pressure zone (test 1); (**c**) average differential pressure (test 2); (**d**) differential pressure in high- and low-pressure zone (test 2); (**e**) average differential pressure (test 3); (**f**) differential pressure in high- and low-pressure zone (test 3).

The monitored pressure data were rearranged in Figure 8 to show the spatial distribution from the face to the end of the experiment space. During three tests of scenario I with two blowing fans operational, positive high-pressure zones can be observed within a distance of 40 m from the working face, while in the inlet side between 60 and 140 m, the low-pressure zone can be seen clearly. This result indicates that with the fan operation in blowing mode, a positive high-pressure zone can be generated continuously near the face.

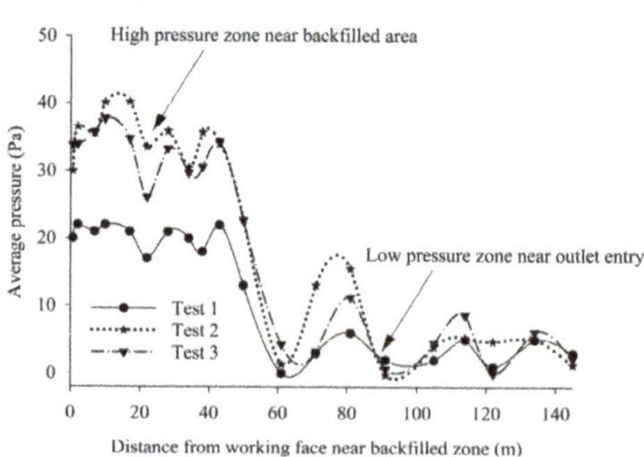

Figure 8. Differential pressure by distance in Scenario I.

Figure 9 shows the velocity distribution measured at the velocity monitoring Station 2. During the fan operation, the velocity of the airflow moving out of the working space was kept under 0.01 m/s; this implies that the air was circulating only within the pressurized working space and that contaminants generated were also expected to remain within the pressurized zone.

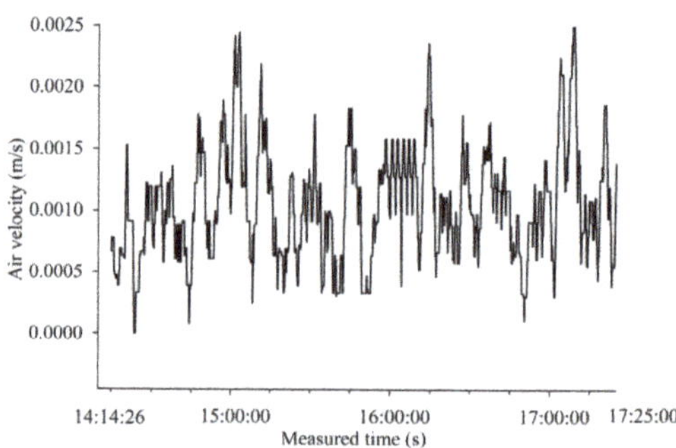

Figure 9. Velocity profiles at velocity monitoring Station 2.

3.2.2. Scenario II: 15 kW Fan in Exhausting Mode and 37 kW Fan in Blowing Mode

The test results for Scenario I with 15 kW and 37 kW fans in blowing mode show that the whole area down to the working face can be well pressurized to approximately 30 Pa higher than the opposite side of the fans and this pressurized zone can keep the toxic gases that have leaked from the backfilled face within the zone. In Scenario II, the 15 kW low-pressure fan was operated in exhausting mode and the 37 kW high-pressure fan in blowing mode. This scenario was designed to see if a certain portion of the contaminated air within the pressurized zone can be exhausted through the safe airflow path to reduce workers' exposure. Two tests were carried out for Scenario II. Figure 10 shows the pressure distributions monitored by 40 sensors. Figure 10a,c illustrate the recorded pressure data during 60 min of fan operation for the first test and 30 min for the second test. Compared to Scenario I, it can be observed that the high- and low-pressure zones are not well differentiated. However, as in Figure 6b,d,f, the pressure data from the sensors near the working face and near the left boundary of the experiment space were compared in Figure 10b,d. A similar pattern indicating the pressurization can be observed but the differential pressure between both zones is much smaller than in Scenario I. This result indicates that positive pressurization can still be created in this experiment, but the pressurization effect is not significant.

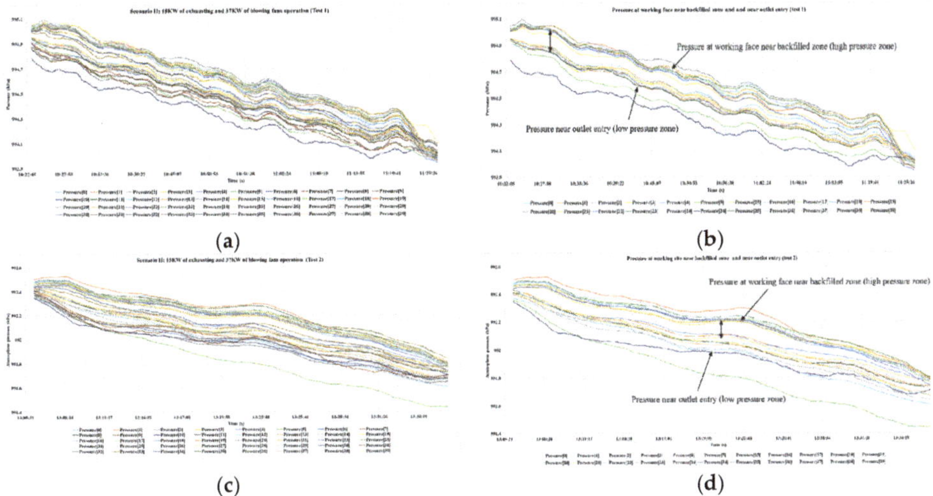

Figure 10. Temporal pressure distributions for Scenario II. (**a**) Test 1—60 min of fan operation; (**b**) differential pressure in high- and low-pressure zone (test 1); (**c**) test 2—30 min of fan operation; (**d**) differential pressure in low- and high-pressure zone (test 2).

The average pressures and the pressure differentials on both sides are plotted in Figure 11. In this regard, the high positive pressure zone located at the working face can be clearly observed in Figure 11c,d. The average pressure differentials of the two tests were 15.38 and 17.56 Pa, respectively. Thus, even with the 15 kW fan in exhausting mode, the working zone near the backfilled area can still be pressurized by only the 37 kW high-pressure fan in blowing mode. However, whether the pressure differentials created in Scenario II are enough to control the leakage and dissipation will be analyzed further in the following section.

Figure 12 shows the pressure distribution by distance. During two tests of scenario II, the high-pressure zone exists within 40 m from the working face, while within 60 to 140 m, the low-pressure zone can be located.

Based on Figure 13, which shows the velocity profiles at Station 2 during the fan operation, the airflow discharged by the 15 kW exhaust fan was almost stagnant with the average velocity less than 0.01 m/s. This implies that as in Scenario I, the contaminated air in the working zone can be hardly exhausted even with a fan in exhaust mode.

The test results for the two scenarios are summarized in Table 3. The results show that the pressure differentials on each side of the fans were well maintained, above 22 Pa in Scenario I and 15 Pa in Scenario II. This pressurized zone near the working face and the extremely low exhaust efficiency shown by two-fan operation implies the applicability of the pressurization ventilation technique to control gas leakage from the backfilled working face and dissipation of the polluted air.

Table 3. Summary of the experimental results.

Categories		Differential Pressure (Pa)	Average Velocity at Station 2 (m/s)
	Test 1	22.30	0.001
Scenario I	Test 2	32.78	0.001
	Test 3	30.50	0.001
Scenario II	Test 1	15.38	0.012
	Test 2	17.56	0.011

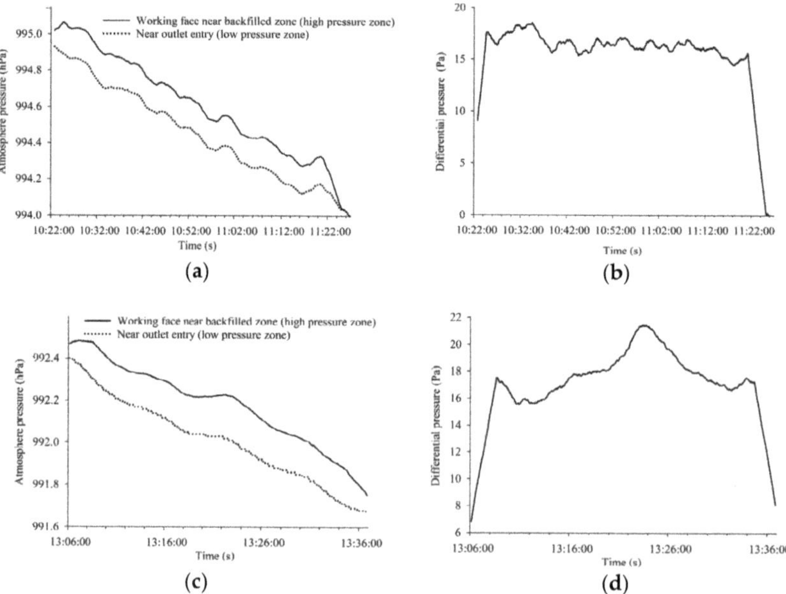

Figure 11. Differential pressure in high- and low-pressure zone in Scenario II. (**a**) Average differential pressure (test 1); (**b**) differential pressure in high- and low-pressure zone (test 1); (**c**) average differential pressure (test 2); (**d**) differential pressure in high- and low-pressure zone (test 2).

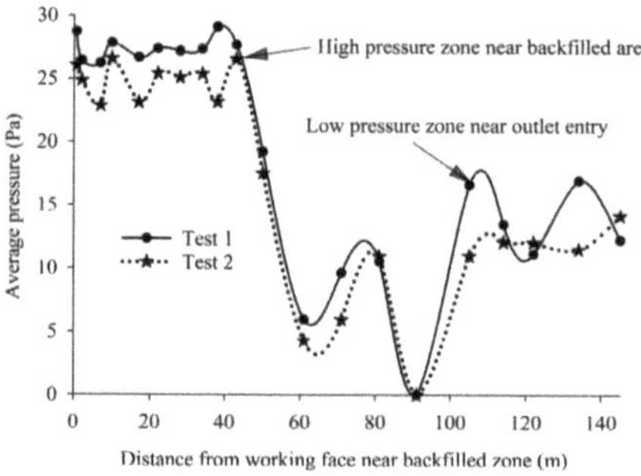

Figure 12. Differential pressure by distance in Scenario II.

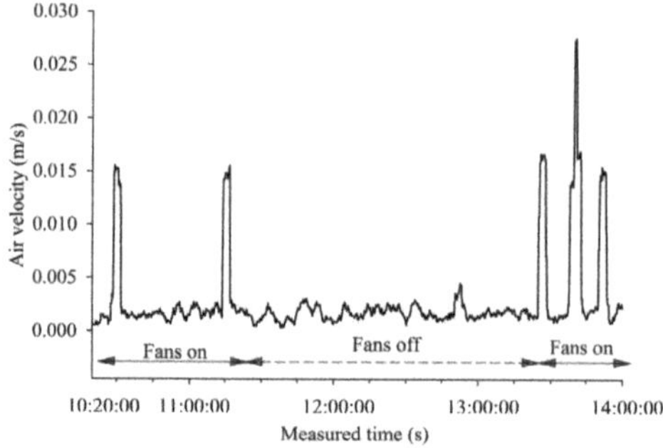

Figure 13. Velocity profiles at velocity monitoring Station 2.

4. Comparison through CFD Analysis

4.1. CFD Analysis

Applications of the CFD technique in the underground mine environment can be found in numerous studies. In this study, the CFD analysis was carried out using ANSYS FLUENT, which has commonly been used by researchers to study various fluid-flow and heat transfer problems in the underground mine environment [18]. Based on the real scale of the experiment site in Figure 3b, the 3D geometry layout for the CFD analysis is illustrated in Figure 14. The initial condition of the CFD simulation can be shown in Table 4. The $k - \varepsilon$ model was employed as the turbulence model in this study and all computational iterations were solved implicitly. The defaulted convergence absolute criteria were 10^{-4} for the continuity and momentum equation, and 10^{-6} for the energy equation. Moreover, to evaluate the possibility of contaminant dispersion under the effect of fan operations, a CO source of 0.05 g/s was included at the working face for CFD analysis.

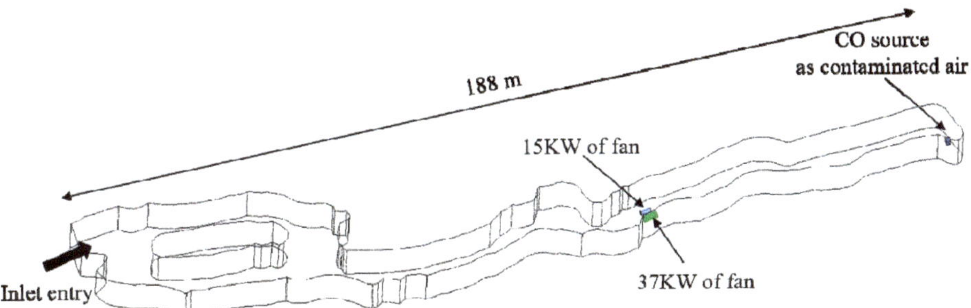

Figure 14. 3D geometry layout for CFD analysis.

Table 4. Initial conditions of the CFD simulation.

Parameters	CFD Model
Inlet boundary	Pressure inlet
Wall boundary	Friction wall
Ventilation resistance (k)	0.014 kg/m^3
Wall temperature	20 °C
Mesh type	Tetrahedron elements
Solution model	Turbulence model ($k - \varepsilon$)
Mesh size function	Proximity and curvature
Number of mesh elements	500,000
Simulation condition	Transient-state conditions

For the CFD simulation, a 15 kW low-pressure fan and a 37 kW high-pressure fan were installed at the same location as in the site tests. All input information for CFD simulation was collected based on the mine site study. The fan specifications summarized in Table 1 were applied in CFD analysis and Table 5 shows the fan input data. For the purpose of comparison, both Scenario I and II in the site study were simulated in the CFD analysis. In Scenario I, 15 kW and 37 kW fans were operated in blowing mode, while in Scenario II the 15 kW low-pressure fan was operated in exhaust mode. The pressure profiles and differential pressures were compared.

Table 5. Fan characteristics in the CFD analysis.

	Fan Type	High-Pressure Fan	Low-Pressure Fan
Fan dimension	Discharge diameter (m)	1.4	0.95
	Length (m)	3.0	2.2
	Power (kW)	37	15
	Fan efficiency (h)	0.7	
	Fan Pressure (Pa)	551	235
	Outlet velocity (m/s)	30.6	23.5

4.2. Discussion and Comparison of the Site Study with the CFD Analysis

Figure 15 shows the pressure profiles of the CFD analysis. It can be clearly observed that the high-pressure zone was distributed between the fan location and the working face in Scenario I and II. In Scenario I, due to the effects of two blowing fans, the pressure differentials were relatively higher than in Scenario II. The differential pressure in Scenario I estimated by CFD analysis was found to be approximately 30 Pa as shown in Figure 15a. A similar result can be obtained in Scenario II, as plotted in Figure 15b, which shows a pressure differential of approximately 15 Pa. Since the operation of the 15 kW fan was reversed, the differential pressure between the fan front and rear was decreased.

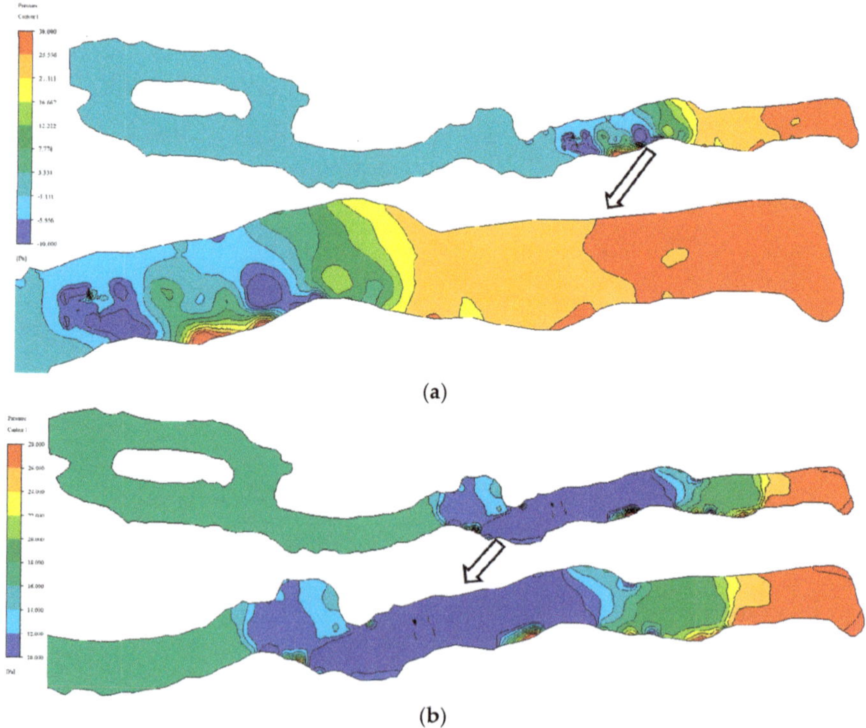

Figure 15. Pressure profiles by CFD analysis. (**a**) Scenario I; (**b**) Scenario II.

Figure 15 shows that the whole working space located downstream of the fan is pressurized. Additionally, the site test results indicated the airflow within this pressurized zone was vigorously recirculated. To prove this, the air velocity contours demonstrated in the CFD analysis are plotted in Figure 16. It is clearly seen in Figure 16a that the jet streams discharged from the two blowing fans lose most of the momentum when they collide with the entry wall. Especially, the high discharged air velocity of the 37 kW fan of high-pressure has collided directly with the curved wall, 30 m from the fan location. In Figure 16b of Scenario II, the same jet stream collision occurs even for the 15 kW fan in exhausting mode. In brief, after the collision, due to this collision, the air was moving at extremely low velocity and eventually sucked into the fan again. This led to the recirculation illustrated clearly in Figure 16. This recirculation is the main reason for the observation made in the site study that most of the airflow remains trapped within the pressurized zone during fan operation.

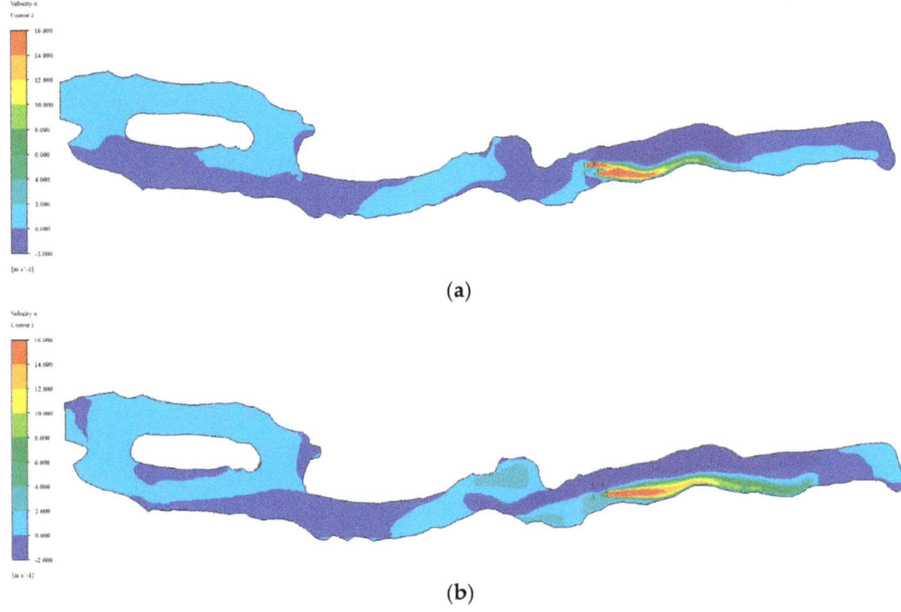

(a)

(b)

Figure 16. Air velocity contours by CFD analysis. (**a**) Scenario I; (**b**) Scenario II.

Figure 17 illustrates the differential pressure between high- and low-pressurization zones shown in the CFD results after 60 min of simulation time. After the fans were turned on, a high-pressurization zone was created in both scenarios. The average differential pressure was 29.23 and 16.15 Pa for Scenario I and II, respectively. These results are very similar to those obtained in the site experiments as shown in Figure 7b,d,f of Scenario I and Figure 11b,d of Scenario II. Consequently, the site experiments show, and the CFD analysis confirms, that by the operation of two fans, a positively pressurized zone can be created continuously near the face to prevent gas leakage from the backfilled site.

To compare the experimental and numerical results in more detail, the average pressures by distance from the face were plotted in Figure 18. The two sets of data from the site experiments and the CFD studies are extremely similar; in Scenario I, a highly pressurized zone can be observed as far as 40 m from the working face as in Figure 18a, while a low-pressure zone is located from 60 to 140 m away from the working face. A similar pattern can be found in the results of Scenario II as shown in Figure 18b. The comparisons are summarized in Table 6.

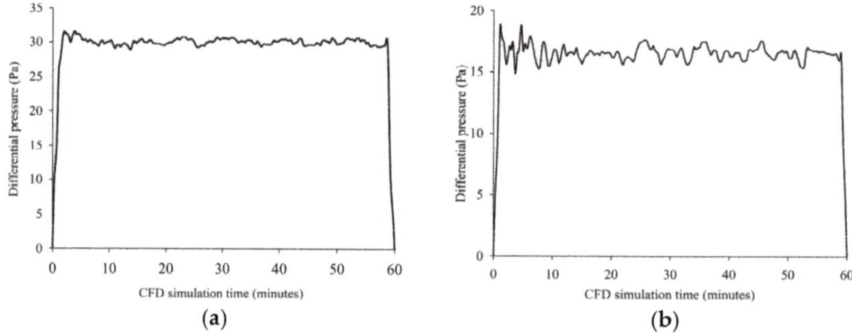

Figure 17. Differential pressures by CFD results. (a) Scenario I; (b) Scenario II.

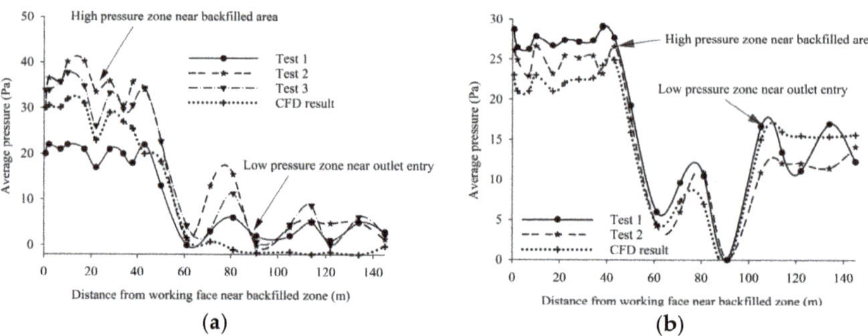

Figure 18. Average pressures by distance between CFD and experimental results. (a) Scenario I; (b) Scenario II.

Table 6. Comparisons between the site experiments and the CFD analysis.

Scenarios of Experiment		Differential Pressure by Experimental Result (Pa)	Differential Pressure by CFD Results (Pa)
Scenario I	Test 1	22.3	
(15 kW and 37 kW of	Test 2	32.78	29.23
blowing fans operation)	Test 3	30.5	
Scenario II	Test 1	15.38	
(15 kW of exhausting and			16.15
37 kW of blowing fan)	Test 2	17.56	

4.3. Possibility of Contaminant Dispersion

The pressurizing ventilation can be designed to have twofold effects. Once a strong airstream discharged from the fans collides with the working face, then the positively pressurized boundary created on the face will deter the leakage of gases entrapped and adsorbed in the backfilled zone pore space. In addition, even though some gas is leaked, it can be confined within the pressurized zone between the face and fan installation. This will minimize workers' exposure. The first effect can hardly be evaluated since the pore pressure is not known. However, in this paper, the second effect was studied by assuming a gas source at the face having an emission rate of 0.05 g/s. This emission rate is similar to that of the methane emission rate at an abandoned coalbed methane wells reported by Johnson and Heltzel [19]. A series of CFD analyses were carried out to account for the gas dispersion behavior during the fan operation.

Figure 19 shows the CO gas concentration profiles by CFD analysis after 20, 40, and 60 min of simulation time. It can be observed that the leaked-out CO was dispersed mainly between the fan location and the working face in Scenario I and II. Especially in Scenario I, due to the effects of two blowing fans, the pressure differentials between the pressurized zone and the outside are relatively higher than in Scenario II. As a result, less CO was dissipated outside the pressurized zone, and this resulted in higher CO concentrations at the space within 30 m from the face after 20 min of fan operation as shown in Figure 19a. In Scenario I, it is well demonstrated that CO leaked out of the backfilled zone can be efficiently confined to the small space within approximately 30 m of the face. Even when the smaller 15 kW fan was reversed in Scenario II, a single 37 kW fan seems to be effective to pressurize the working space and control the dispersion of CO outside the working area as shown in Figure 19b.

Figure 20 illustrates the CO gas concentration variations at three monitoring stations, Point 1, 2, and 3, during the 60-min simulation. The three stations are 2, 40, and 150 m away from the face, respectively. Due to the two blowing fans in Scenario I, the concentrations at Point 1 fluctuate more significantly than in Scenario II. This fluctuation close to the face does not influence the workers' exposure. Even though the CO level at Point 2 continued to increase linearly during the CFD analysis, it was 8.3 and 7.3 ppm after 60 min in Scenario I and II, respectively. If the linear trend was assumed to continue even after 60 min, the CO levels at Point 2 would have been 16.6 and 14.6 ppm after 2 h, still far less than 30 ppm which is the 8 h TWA permissible level of CO in Korea. These results imply that due to the pressurized zone created by two blowing fans installed as in Scenario I and II, the gases leaked from the backfilled zone can be well-confined near the working face and thus minimize the workers' exposure.

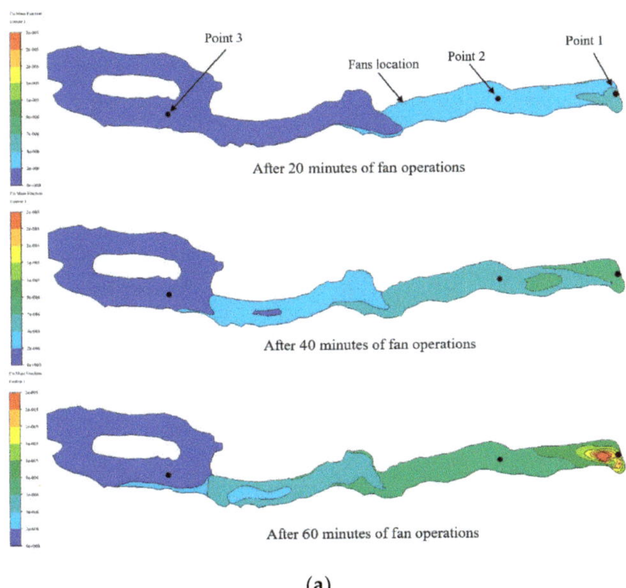

(a)

Figure 19. *Cont.*

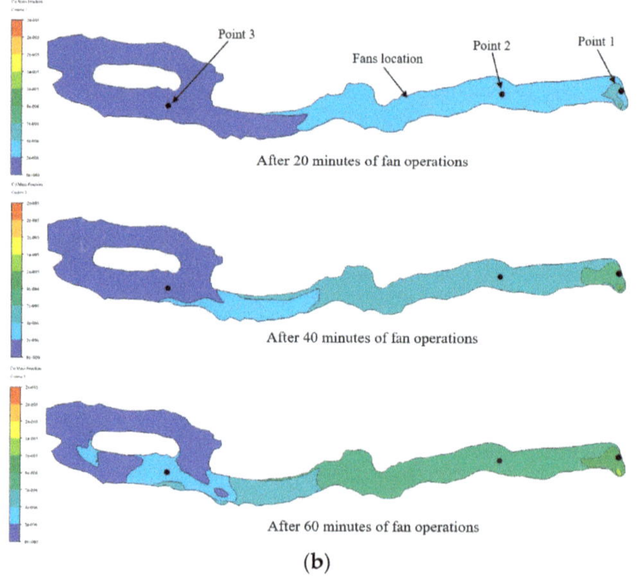

Figure 19. CO gas concentration profile by CFD analysis. (**a**) Scenario I; (**b**) Scenario II.

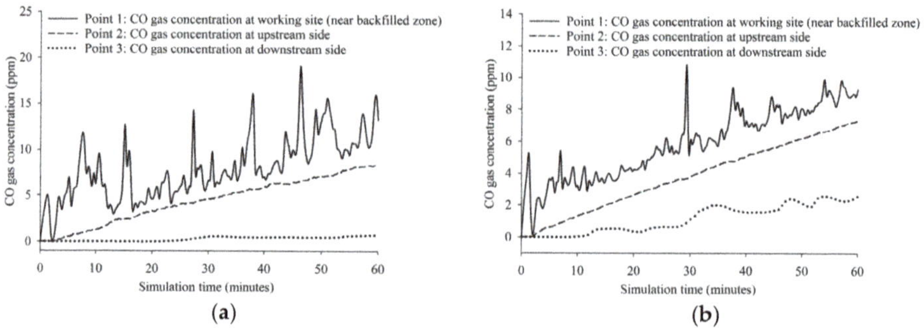

Figure 20. CO gas concentration at working face, upstream side, and downstream side of airway. (**a**) Scenario I; (**b**) Scenario II.

5. Conclusions

The so-called pressurization ventilation technique was evaluated as a ventilation scheme to control the leakage and dissipation of hazardous gases from the working site backfilled by fly-ash-based materials. The method originally developed for securing a safe escape route in a building fire was designed to create a certain level of pressure differentials between the fire zone and the escape paths. The test site for this site was a limestone mine development heading with a large cross-sectional area (8 m (W) × 7 m (H)). To pressurize the area within the vicinity of the working face and thus confine the contaminated zone to minimize the workers' exposure, two different types of fans—15 kW and 37 kW jet-fan-type ventilation fans—were developed and several ventilation scenarios were tested at the test site and also by CFD analysis. Several important results are summarized as follows:

1. With two fans in blowing mode, a positively pressurized zone can be generated continuously near the face, and the pressure differentials between the downstream and upstream measured

by 40 pressure sensors communicated by CAN ranged between 22.3 and 32.78 Pa. The pressure differential simulated by CFD analysis was 29.23 Pa.

2. With the 37 kW fan in blowing mode and the 15 kW fan in exhausting mode, relatively smaller pressure differentials of 15.38–17.56 Pa were observed and are comparable to the CFD analysis result of 16.15 Pa.

3. Since the differential pressure of 50 Pa specified in British standards to control building fires which have a relatively larger dissipation force than the gas leakage in the mining face, the pressure difference of approximately 30 Pa created by two blowing fans and 16 Pa by one-blowing and one-exhausting fan in this study seems to be sufficient to control the gas leakage and dispersion within the working space.

4. The above conclusion was supported by the velocity distributions measured at both within and outside of the pressurized zones. The air in the pressurized zone was vigorously circulated, while the outside airflow was almost stagnant. This implies that contaminated air can be well confined within the pressurized zone near the face.

5. Since most of the limestone mines in Korea are developed within the steeply dipping veins, the developed entries are not straight but curved irregularly. This makes the jet stream discharged from the fan collide with the nearby sidewalls and reduce the fan efficiency considerably. Therefore, to install the fans for this ventilation scheme, the fan location must guarantee the minimum loss of jet stream momentum in the downstream side.

6. Even though the time required for curing depends on the type of backfilling materials, this ventilation system can be turned on only during the curing period to have a high possibility of containing gas leakage. In addition, it was shown that the pressurization system with low-pressure fans can be operated at low cost. These are the economic advantages of the system discussed in this study.

7. With this pressurization system, the positively pressurized zone can be generated continuously near the face to restrain the gas leakage and also confine the leaked-out gases near the backfilling face. However, the efficiency of preventing gas leakage from the backfilled zone was not analyzed in this study, due to the lack of basic knowledge about the permeability characteristics of backfilled zones. The total efficiency of the system is a topic requiring further study in the future.

Author Contributions: Data collection and experimental works: V.-D.N., W.-H.H., R.K.; Writing, discussion, analysis: V.-D.N., C.-W.L.

Funding: This research was funded by the Ministry of Science and ICT (MSIT), the Ministry of Environment (ME) and the Ministry of Trade, Industry and Energy (MOTIE), grant number 2017M3D8A2090024 under the National Strategic Project-Carbon Upcycling Program of the National Research Foundation of Korea (NRF).

Acknowledgments: The authors would like to thank to Daesung Mining Development INC, S. Korea; the engineers and leaders of DFC large-opening underground limestone mine, Chungbuk province, S. Korea for their help and cooperation.

References

1. Bloss, M.L. An operational perspective of mine backfill. In *Proceedings of the Eleventh International Symposium on Mining with Backfill*; Australian Centre for Geomechanics: Perth, Australia, 2014; pp. 15–30. Available online: https://papers.acg.uwa.edu.au/p/1404_0.2_Bloss/ (accessed on 13 June 2019).

2. Gai, G.; Cancelliere, P. Design of a pressurized smokeproof enclosure: CFD analysis and experimental tests. *Safety* **2017**, *3*, 13. [CrossRef]

3. Lay, S. Pressurization systems do not work & present a risk to life safety. *Case Stud. Fire Saf.* **2014**, *1*, 13–17. [CrossRef]

4. Tamura, G.T. Assessment stair pressurization systems for smoke contorol. *ASHRAE Trans.* **1992**, *98*, 66–72.

5. Tamura, G.T. Stair pressurization for smoke contorol: Design considerations. *ASHRAE Trans.* **1989**, *95*, 184–192.

6.   Budnick, E.K.; Klote, J.H. The Capabilities of SMOKE Control: Part II-System Performance and Stairwell Pressurization. In NFPA Meeting 1987. Available online: https://www.iccsafe.org/wp-content/uploads/ctc/Capabilities_of_smoke_control_Bukowski.pdf (accessed on 13 June 2019).

7.   Wang, Y.; Gao, F. Tests of stairwell pressurization systems for smoke control in a high-rise building. *ASHRAE Trans.* **2004**, *110*, 185.

8.   Jo, J.H.; Lim, J.H.; Song, S.Y.; Yeo, M.S.; Kim, K.W. Characteristics of pressure distribution and solution to the problems caused by stack effect in high-rise residential buildings. *Build. Environ.* **2007**, *42*, 263–277. [CrossRef]

9.   Li, M.; Gao, Z.; Ji, J.; Li, K. Modeling of positive pressure ventilation to prevent smoke spreading in sprinklered high-rise buildings. *Fire Saf. J.* **2018**, *95*, 87–100. [CrossRef]

10.   British Standards Institute. *Smoke and Heat Control Systems. Specification for Pressure Differential Systems*; BS EN 12101-6; BSI: London, UK, 2005.

11.   British Standards Institute. *Fire Precautions in the Design, Construction and Use of Buildings Part 4: Code of Practice for Smoke Control Using Pressure Differentials*; BS EN 5588; BSI: London, UK, 2016; Part 4.

12.   Grau, R.H., III; Mucho, T.P.; Robertson, S.B.; Smith, A.C.; Garcia, F. Practical techniques to improve the air quality in underground stone mines. In Proceedings of the 9th US/North American Mine Ventilation Symposium, Kingston, ON, Canada, 8–12 June 2002; pp. 123–129.

13.   Grau, R.H., III; Krog, R.B.; Robertson, S.B. Maximizing the ventilation of large-opening mines. In Proceedings of the 11th US/North American Mine Ventilation Symposium, University Park, PA, USA, 5–7 June 2006; pp. 53–60.

14.   Krog, R.B.; Grau, R.H., III. Fan selection for large-opening mines: Vane-axial or propeller fans-which to choose? In Proceedings of the 11th US/North American Mine Ventilation Symposium, University Park, PA, USA, 5–7 June 2006; pp. 527–534.

15.   NIOSH. Using propeller fans to improve ventilation in large-entry stone mines. *NIOSH Tech News* **2002**, *2*, 499.

16.   Chekan, G.J.; Colinet, J.F.; Grau, R.H., III. Impact of fan type for reducing respirable dust at an underground limestone crushing facility. In Proceedings of the 11th US/North American Mine Ventilation Symposium, University Park, PA, USA, 5–7 June 2006; pp. 203–210.

17.   HPL, S.C. Introduction to the Controller Area Network (CAN). Available online: https://barrgroup.com/Embedded-Systems/How-To/Controller-Area-Network-CAN-Robustness (accessed on 13 June 2019).

18.   ANSYS, Inc. *FLUENT User's Guide, Version 17.0*; ANSYS, Inc.: Canonsburg, PA, USA, 2019.

19.   Derek, J.; Robert, H. Methane emissions measurements of natural gas components using a utility terrain vehicle and portable methane quantification system. *Atmos. Environ.* **2016**, *144*, 1–7. [CrossRef]

Communication

Water Environment Policy and Climate Change: A Comparative Study of India and South Korea

Mohd Danish Khan [1,2], **Sonam Shakya** [3], **Hong Ha Thi Vu** [2], **Ji Whan Ahn** [2,*] and **Gnu Nam** [4]

[1] Resources Recycling Department, University of Science and Technology (UST), 217, Gajeong-ro, Yuseong-gu, Daejeon 34113, Korea; danish0417@ust.ac.kr
[2] Center for Carbon Mineralization, Mineral Resources Research Division, Korea Institute of Geosciences and Mineral Resources (KIGAM), 124 Gwahak-ro, Yuseong-gu, Daejeon 34132, Korea; hongha@kigam.re.kr
[3] Department of Chemistry, Aligarh Muslim University, Aligarh, Uttar Pradesh 202002, India; sonamshakya08@gmail.com
[4] National Institute of Chemical Safety, Ministry of Environment, 90 Gajeongbuk-ro, Yuseong-gu, Daejeon 34111, Korea; gnunam@korea.kr
* Correspondence: ahnjw@kigam.re.kr; Tel.: +82-42-868-3578

Received: 30 April 2019; Accepted: 11 June 2019; Published: 14 June 2019

Abstract: Climate change is considered to be a potential cause of global warming, which leads to a continuous rise in the global atmospheric temperature. This rising temperature also alters precipitation conditions and patterns, thereby causing frequent occurrences of extreme calamity, particularly droughts and floods. Much evidence has been documented by the Intergovernmental Panel on Climate Change, illustrating fluctuations in precipitation patterns caused by global climate change. Recent studies have also highlighted the adverse impact of climate change on river flow, groundwater recovery, and flora and fauna. The theoretical political approach and scientific progress have generated ample opportunities to employ previously allusive methods against impacts caused by varying climatic parameters. In this study, the current state of India's water environment policy is compared with that of South Korea. The "3Is"—ideas, institutions, and interests—which are considered pillars in the international field of political science, are used as variables. The concept of "ideas" highlights the degree of awareness regarding climate change while formulating water environment policy. Here, the awareness of India's management regarding emerging water issues related to climate change are discussed and compared with that of South Korea. The concept of "institutions" illustrates the key differences in water environment policy under the umbrella of climate change between both countries within the associated national administrations. India's administrations, such as the Ministry of Environment, Forests, and Climate Change; the Ministry of Water Resources, River Development, and Ganga Rejuvenation; the Ministry of Rural Development; and the Ministry of Housing and Urban Affairs, are used as a case study in this work. Finally, the concept of "interest" elaborates the prioritization of key issues in the respective water environment policies. Common interests and voids in the policies of both countries are also briefly discussed. A comparison of India's water environment policies with that of South Korea is made to expose the gaps in India's policies with respect to climate change, thereby seeking to identify a solution and the optimal direction for the future of the water environment policy of India.

Keywords: climate change; sustainability; environment; water policy; India; South Korea; precipitation

---

1. Introduction

The Sustainable Development Goals (SDGs) were consensually adopted in September 2015 by the United Nation member states during the 2030 Agenda for Sustainable Development. These SDGs are considered to be a blueprint to achieve a sustainable future for our planet after the expiration of the Millennium Development Goals in 2015. There is a total of 17 SDGs, which must be accomplished by the international community within 15 years, i.e., between 2016 and 2030. The SDGs are treated as "integrated and unified", thereby endorsing distinctive priorities and potentials for countries regarding the fulfillment of these goals. These distinctive characteristics of the SDGs are considered to be axiomatic and corroborate assessments of 'policy congruity', establishing a foundation for consideration of where incongruity would be most valuable [1]. Among all the SDGs, four goals deal specifically with water-related issues: Goal 6 ensures the availability of clean water and improved sanitation; Goal 13 deals with climate change; Goal 14 preserves and sustains the use of water resources; and Goal 15 protects and restores biodiversity on land [2,3]. This signifies the importance of water management and highlights the urgent need for the rectification of all possible water-related issues to achieve a sustainable future.

Climate change refers to a statistical shift of climatic variables at a regional or global level over considerable periods. Over millennia, climatic variables, including global mean temperature were relatively stable, but in the last few decades, catastrophic changes have been observed [4]. According to the European Environmental Agency, an approximately 2.7% reduction per decade with respect to ice in the Arctic Sea, an expansion of the global sea level by 1.8 mm per year since 1961, and a rise in the global mean temperature by 0.74 °C have been witnessed in the 20th century [5]. Moreover, the Intergovernmental Panel on Climate Change (IPCC) has also predicted severe floods and droughts in the 21st century due to a calamitous rise in the global mean temperature from 1.5 to 5.8 °C [6]. The IPCC has also provided evidence illustrating a significant shift in global precipitation patterns within the period from 1900 to 2005. Many regions of Central Asia and Europe, the eastern and southern parts of South America, and areas of North America have witnessed a considerable rise in precipitation levels, whereas South Asia, Southern Africa, and the Mediterranean face reduced precipitation [7,8]. Recent studies have also pointed out that climate change-derived changes in precipitation patterns are having adverse impacts on the water environment, such as river flow and the recovery rate of groundwater [9,10], and inimical alterations in the ecosystem [11].

India is the seventh largest country in the world, encompassing a population of around 1.2 billion in 29 states and seven union territories. Due to its large area, India demonstrates variable precipitation patterns depending on the geographical region. India's monsoon season precipitation is mainly concentrated in July, but since 2000, a significant shift towards August rainfall has been observed. For example, in the states of Jharkhand and Karnataka, the average August rainfall has increased by approximately 50.1% and 20.5%, respectively from 2000 to 2014, while in the same period, the states of Punjab and Uttar Pradesh have witnessed a reduction in average August rainfall by 27.3% and 13.3%, respectively, as shown in Figure 1a,b [12]. Moreover, the states of Punjab and Uttar Pradesh are facing a continuous reduction in average rainfall in both July and August by a considerable margin, as shown in Figure 1c,d [12]. The monthly rainfall (mm) of the major states of India for July and August from 2000 to 2014 are illustrated in Figure S1. Evidence shows that the consequences of climate change will continue in the future [9,13]. Therefore, frequent modifications and possible changes in water environment policies with regular monitoring for any change in the water environment due to climatic variables are necessary for the accomplishment of the water-related SDGs.

India is a fortunate country, having 20 river basins; however, most of them are water stressed owing to the high water demand for agriculture, domestic, and industrial purposes. The uneven distribution of water demand across the country, the growing population, and climate change further intensify the already stressed resources and thereby complicate water environment management. With all these water environment-related issues, India's water management appears to be fragile and requires some urgent modifications and revisions to its water policies [14]. To establish and improve the comprehensive management of the water regulation system and water policies, rectification of urban and rural water environment issues holds utmost importance. As a developing country, India is in a state of transition; thus, its potential market economy, organized economic process, and statutory system are comparatively fragile. Therefore, it is paramount that the Indian government focuses significant attention and funding on water environmental policy. It is well accepted that bolstering the interface between public participation, policy, and science and technology is critical for effective and reliable water resource management. However, in India, there seems to be a big communication gap between policymakers and scientists. To the best of our knowledge, no researcher has pointed out this critical blind spot. In last few decades, South Korea has been able to rectify most of its water environment-related issues owing to its skillfully planned strategies and policies and nearly perfect implementations of such approaches [14]. Hence, a comparative study between India's and South Korea's environment policies is useful to identify previously unidentified solutions for the development of India's policies. The results of this study can also indirectly serve other underdeveloped and developing countries around the world by fostering more awareness and identifying executable strategies. Therefore, the present study aims to highlight the gaps in India's water policies and the potential causes for the differences in the water environment policies between India and South Korea through qualitative analysis of scientific facts particularly using the 3Is—which represent ideas, institutions, and interests—as variables. The findings of this study can also set an example for other countries regarding the significance of water environment policies related to climate change.

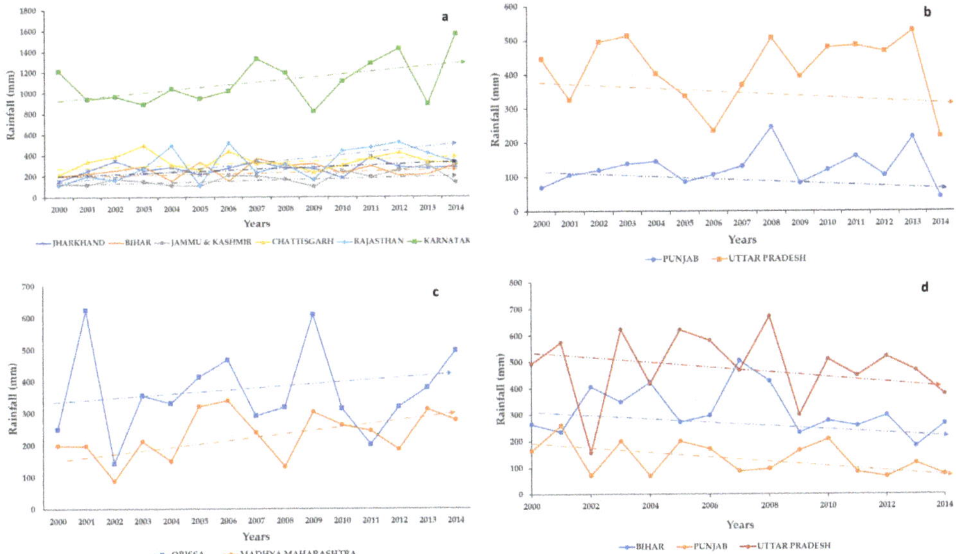

Figure 1. Rainfall patterns (mm) illustrating shifts in the amount of rainfall occurring between 2000 and 2014 in (**a,b**) August, (**c,d**) July [12].

2. Methodology

Theoretical political approaches and advancing scientific technology have generated ample opportunities for the employment of distinctive methods against climatic concerns. One of the most significant contributions of distinctive political strategies is to provide a broader understanding of analytical aspects and to highlight the validity of findings wherever required.

In the present study, the recent state of India's water environment policy was compared with that of South Korea [15]. Literature searches were conducted using PubMed, PLOS, Nature.com, ScienceDirect, MDPI, and published government reports, such as those by the IPCC and United Nations and individual ministry reports. The primary focus was to highlight the gaps in the water environment policies of India with scientific facts from peer-reviewed papers and government reports from 2000 to the present. Keywords, such as climate change, precipitation, water, policy, environment, India, and South Korea, were used separately and in combination to obtain relevant information. Recent government reports (since year 2000) were searched to analyze recent progress and the present situation. The obtained data were categorized according to the three variables—the 3Is, idea, institutions, and interest. The first variable, idea, highlights the degree of awareness regarding climate change, while formulating water environment policy. The awareness of India's management regarding emerging water issues relating to climate change were analyzed and compared with that of South Korea. The institutions variable illustrates the major differences in water environment policies under the umbrella of climate change between India and South Korea within the associated national administrations. India's administrations, such as the Ministry of Environment, Forests, and Climate Change; the Ministry of Water Resources, River Development, and Ganga Rejuvenation; the Ministry of Rural Development; and the Ministry of Housing and Urban Affairs, were considered in a case study. The third and last variable, interests, elaborates the prioritization of key issues associated with water environment policy. Common gaps in the policy of both countries were also analyzed.

3. Results and Discussion

3.1. India's Water Environment Status under the Umbrella of Climate Change

3.1.1. Ideas

By 1980s, it was well understood that water is a crucial national resource, and a shortfall in national water policy can be a severe impediment for the formulation of rational water policies. This led to the emergence of India's first National Water Policy in 1987, which was further reformed in 2002 [16]. Since then, the core motive of policymakers has been to improve institutional frameworks through improved institutional performance, encouragement of rehabilitation schemes for the displaced, ensuring effective monitoring systems, the embodiment of private parties in water management, and ensuring the sustainable use of joint rivers by states. However, state water policies are often supplementing to the national policy. The introduction of water rights was another major step towards the effective management of water issues [17]. Those rights are considered to be a necessary premise for the encouragement of the 'management' of water resources, organizing water user associations, and induction of trading. Moreover, the idea of 'incentives' for the sustainable use of water was also encouraged. This can increase the involvement of private sectors in water control and utilization, thereby improving the planning, management, and development of the administering authority of water resources projects [17].

Recently, the government has initiated a mammoth project called the 'National River-Linking Project', which aims to link rivers throughout the country. The rationale behind this project is the fact that some states in India are suffering from water shortages while others have excess [18]. Through this project, excess water in the form of floods caused by heavy rains can be transferred to water-deficient basins, which can indirectly serve as stores and means of capturing those excess rain waters. The desirable distribution of water, the promotion of economic activities in water-deficient areas, and drought and flood protection are some of the arguments in favor of the River-Linking Project [19]. However, arguments against this project include the huge expenditure, soil contamination through salination and seepage, and the absence of surplus water for most of the year [20,21]. Furthermore, apart from the criticisms against big dams, activists within India claim that there are various ecological, economic, and social shortcomings inherent to the project [21]. Hence, clear and properly defined criterion should be met to justify such schemes.

3.1.2. Institutions

Currently, a total of 40 ministries are functioning under the government of India. Among them, four ministries manage all distinctive parts of the water environment: the Ministry of Environment, Forest, and Climate Change (MoEFCC); the Ministry of Water Resources, River Development, and Ganga

Rejuvenation; the Ministry of Rural Development; and the Ministry of Housing and Urban Affairs. More specifically, the water quality, quantity, and other environment-related anthropogenic aftermath (e.g., coastal areas and river cleaning, and municipal sewage) and natural calamities (e.g., flood, drought, tsunami, heavy precipitation, and earthquake) are managed by the government of India. The MoEFCC is the nodal administrative organization responsible for the designing, promotion, coordination, and monitoring of the implementation and execution of environment-related policies and programs in India. Its total budget in 2018–2019 was US $370 million, of which US $110 million was allocated to the water environment, including protection and sustainable development projects, while US $5.7 million was distributed for managing the climate change action plan [22]. The Ministry of Water Resources, River Development, and Ganga Rejuvenation is the central administrative agency responsible for the sustainable development of water resources through the regulation of the country's water resources-related policy guidelines and programs. The total budget of this ministry in 2018–2019 was US $1.28 billion, which was used mainly in the irrigation sector, flood control management, and river conservation projects [23]. The distribution of the total budgets allocated to the Ministry of Environment, Forest, and Climate Change; the Ministry of Water Resources, River Development, and Ganga Rejuvenation; the Ministry of Rural Development; and the Ministry of Housing and Urban Affairs in 2018–2019 are depicted in Table S1. The Ministry of Rural Development and the Ministry of Housing and Urban Affairs hold a small portion of the budget regarding the water environment in the form of water supply management under a scheme called 'Pradhan Mantri Awas Yojna' (PMAY) in rural and urban regions, respectively. This scheme under the Ministry of Rural Development provides US $3.03 billion, which accounts for 18.60% of the total budget of this ministry in 2018–2019, while the Ministry of Housing and Urban Affairs is allocated US $44.67 million, a mere 0.75% of the total budget on water environment issues [24,25]. The budget for the Ministry of Water Resources, River Development, and Ganga Rejuvenation was almost thrice that of the Ministry of Environment, Forest, and Climate Change for 2018–2019, thereby managing all major water environment projects, such as irrigation, dam projects, and river conservation projects. Focusing particularly on the budget of water-related projects, the Ministry of Environment, Forest, and Climate Change is attempting to maximize their management, manpower, and budget through water resources-related projects, which currently account for 29.73% of the total budget, thereby contending with the jurisdiction budget [22]. The Ministry of Water Resources, River Development, and Ganga Rejuvenation allocates almost all of its budget to water resources-related projects, thus securing hegemony in water environment business, although other environmental issues are mostly managed by the Ministry of Environment, Forest, and Climate Change.

Such conflicts are generally common in large, territorial, democratic countries, which leads to duplicate, similar, or overlapping project and policy objectives, thereby causing wastage of the budget [15]. This further encourages the emergence of competition for the quantitative results that these ministries are expected to display. Moreover, enforcement of different policies by ministries for similar issues exposes the inconsistency in the policy objectives and applications. For example, 'Pradhan Mantri Awas Yojna' (PMAY) and 'National Livelihood Mission–Ajeevika' are the schemes underlying both the Ministry of Rural Development and the Ministry of Housing and Urban Affairs. Separate funds are allocated to these schemes under both ministries (refer to Table S1). Therefore, there is an urgent need for the integration of policies wherever required. These should be designed with extreme clarity along with accurate future predictions and should also include possible solutions for all similar water environment issues. Proper division and simplification of water management projects among ministries are highly desirable. Although individual projects by ministries may prosper, a high likelihood of failure always beleaguers the overall management.

3.1.3. Interests

India is known to be composed of agricultural land, and hence, irrigation and river projects are sanctioned and implemented on a priority basis. In 2015, the government of India created an umbrella program called "Namami Gange". The allocated budget of this program was US $2.86 billion for 2015–2020 with the aim of cleaning India's biggest river 'Ganga' and its tributaries. In mid-2017, the Ministry of Water Resources, River Development, and Ganga Rejuvenation reported that from all 163 projects sanctioned under Namami Gange, which deals with the rejuvenation of the >1500-miles-long Ganga river, 41 projects have been completed. A sewage capacity of 223.13 MLD and 1339 km of sewer network was reported as an accomplishment of those completed projects [26]. However, a project such as the 'National River-Linking Project' often encounters more complications and is critical to successfully implement, as research shows many possible catastrophic post-effects. Studies provide evidence for a significant reduction in inflow and sediment deposits in the northern rivers [27,28]. For example, the Ganga can witness a 24% flow reduction, while its tributaries such as Gandak and Ghaghara can face a reduction in flow by 68% and 55%, respectively. A minor loss of 6% can be observed in the Brahmaputra, another major river in northern India, while its tributaries, such as Manas, Sankosh, and Raidhak, can face a massive flow reduction of 72%, 73%, and 53%, respectively [27]. This not only reveals a critical knowledge gap, but also highlights significant lacking in comprehensive pre-planning and surveys by the Ministry of Water Resources, River Development, and Ganga Rejuvenation.

3.2. South Korea's Water Environment Status under the Umbrella of Climate Change

The influence of climate change in South Korea is similar to the rest of the world. Apart from the rising temperature, South Korea's monsoon season is also concentrated in the month of July. It has been reported that since 1980, South Korea's precipitation pattern, particularly in August, has increased by over 25%, resulting in frequent localized downpours [15]. With the recognition of the adverse effects of climate change and to mitigate climate change-related water environmental problems, the government of South Korea has modified their following five ministries: the Ministry of Environment (MOE); the Ministry of Land, Infrastructure, and Transport (MOLIT); the Ministry of the Interior and Safety (MOIS); the Ministry of Trade, Industry, and Energy (MOTIE); and the Ministry of Agriculture, Food, and Rural Affairs (MAFRA). They together manage all the water quality, quantity, and natural environment-related calamities, such as flood, storm, drought, heavy rain, and earthquakes. Being a central administrative agency, the Ministry of Environment, with an allocated budget of US $23 billion (in 2015), managed all projects related to water quality improvement in response to climate change [29]. With regard to the water environment, the ministry of Land mostly manages water resources projects, such as the management of rivers and groundwater with an allocated budget (in 2015) of US $2 billion [30]. Of the overall budget of the South Korean government, 1.2% was allocated for all water environment projects in 2015.

Unlike India's management policies, South Korea's management strategy is more precise, accurate, and well implemented. Being a small country compared to India, the allocated budget to the respective ministries is much higher, and there is almost no overlapping of similar projects among ministries. The distribution of respective allocated budgets and the implementation strategies are superior to those of India. South Korea is ahead with respect to their policies, designs, and progress regarding climate change and is almost on the verge of meeting the SDGs. For example, in 2013, the Ministry of Environment asked the Korea Environment Institute to organize a sustainable layout for waterfront regions, which was accomplished by early 2014. As per the plan, the 1-km area surrounding the banks of the rivers, which accounts for a total length of >1300 miles for just five major rivers of South Korea, was considered to be the highest priority transition area and thus required conservation and restoration management. As another example, the Ministry of Environment is putting strenuous effort into enhancing tap water drinking rates. It is trying to achieve this by changing people's vague anxiety towards tap water through various promotional activities and regular water quality tests. Since 2013, the "Tap Water Loving Villages" project has been implemented and is playing a key role in improving the reputation of tap water. A remarkable rise and much progress have been reported in only one year as the number of Tap Water Loving Villages have increased to 50 [29].

In 2017, the Ministry of Environment reported on South Korea's water management progress, which mainly highlights improvements in the quality and sustainable utilization of groundwater and surface water [31]. The government is also focusing on setting stringent guidelines to conserve groundwater ecosystems. The inclusion of groundwater, via the extension of groundwater ecosystem conservation guidelines, in the water environment policy, which is centered on surface water, illustrates that South Korea's management is progressing towards an integrated water environment policy. The integration of water environment policies can be a driving step in achieving water-related Sustainable Development Goals.

4. Common Voids

Slight changes in climate in the form of temperature rise or alteration in evaporation and precipitation can dramatically influence climate-associated hydrological behavior globally [32]. The hyporheic zone, a region where surface water and groundwater interact, is characterized by unique biogeochemical reactions. Recent studies have revealed that these hyporheic zones are very sensitive to climatic parameters, thereby working as a valuable gauge in climate and water environment research [32,33]. Traditionally, surface water and groundwater have been researched independently under hydrology; due to this, a definitive understanding of hyporheic zones is missing. However, the European Commission (EC) and other European studies have been focusing on hyporheic zones in response to climate change in recent years [15,34]. Contrastingly, South Korea and India both manage surface water and groundwater separately; hence, a coherent concept and distinctive management of hyporheic zones are lacking. There is no policy, scheme, or plan for managing these zones in both countries. Even if both countries wish to incorporate these zones into their management strategies, there will be a debate regarding in which sector, i.e., groundwater or surface water, they should be included. For example, if the concept of hyporheic zones is perceived as being more associated with groundwater, then such zones should be managed through the laws concerning groundwater by the Ministry of Water Resources, River Development, and Ganga Rejuvenation in India and by the Ministry of Land in South Korea, respectively. However, if these hyporheic zones are perceived as a hydro-ecological gauge, then such zones should be managed through the laws concerning hydro-ecological conservation and water quality by the Ministry of Environment, Forest, and Climate Change in India and by the Ministry of Environment in South Korea respectively. This ambiguity around the perception of hyporheic zones exposes the inadequacy of the management policies and laws of the managing institutions, placing the management of climate change-related water environment in a blind spot.

5. Concerns in the Formulation of Water Environment Policy

It is evident that each country possesses a unique set of social, environmental, institutional, physical, economic, and legal conditions that influences its water environment policies and strategies. Global monitoring and experience can be beneficial while searching for options, and the solutions to any country's problems must be tailored to its specific needs. The designing and implementation of any national water environment policy or strategy can be an intricate task as it involves numerous factors, such as its hydrological conditions, size, diversity of stakeholders, and political organization of the country. For example, floods may be an important issue in the countries of North America, South America, and Asia, whereas water scarcity is more catastrophic in the Middle East and Africa. While formulating any new policies or modifying existing water environment policies, consideration of the following four points can be critical for respective organizations and their governments. First and foremost is the formulation or shifting towards integrated water environment strategies. This will not only lead to sustainable solutions for present issues, but also will cover a wide variety of unpredictable circumstances related to the water environment. Second, comprehensive pre- and post-surveys are very much required in figuring out the consequences of any water environment policy. Third, increased attention and focus are required in order to avoid any overlapping of schemes, policies, or projects, which not only help economically, but also minimize various disputes among ministries. Lastly, the implementation of any strategy or policy can be a defining step for any organization. A great policy without proper implementation is worthless.

6. Conclusions

India's water environment policy may have several drawbacks when compared with South Korea's, but it surely consists of some unique idea, which may suit its specific needs. The cluster of shortcomings found in India's water environment policy in the present study can be compiled into three points: i) The concept of integrated water environment policy, which can deal with the whole water environment as an interlinking unit, has not been introduced in Indian water policy. Therefore, the conservative attitude towards traditional bifurcated practices concerning the water environment is anticipated to persist. ii) This ambiguity led to the overlapping of policies and schemes, resulting in hindered implementation and disputes among administrators, which was further deepened by a lack of comprehensive surveys on the water environment regarding climate change. iii) The concept of sustainability is very limited in Indian water environment projects. As opposed to South Korea, India still is not focusing on hyporheic zones and other ecological maintenance, which are critical strategies in response to climate change. Eventually, the Indian government should expand its water management system by including cultural and recreational objectives, prioritizing the need for clean water while perpetuating the ecosystem.

Supplementary Materials: The following are available online at http://www.mdpi.com/2071-1050/11/12/3284/s1, Figure S1: Rainfall patterns (mm) of major states of India. (**a**) illustrates shifts in the amount of rainfall occurred between the years 2000–2014 in the month of July; (**b**) in the month of August [12], Table S1: The total budgets allocated to various ministries and their distribution in water management sectors [22–25].

Author Contributions: M.D.K., S.S., and H.H.T.V. planned and designed the presented work and analyzed the data; M.D.K. wrote the manuscript; G.N. and J.W.A. reviewed and revised the manuscript.

Acknowledgments: This research was supported by the National Strategic Project—Carbon Mineralization Flagship Center of the National Research Foundation of Korea (NRF) funded by the Ministry of Science and ICT (MSIT), the Ministry of Environment (ME), and the Ministry of Trade, Industry, and Energy (MOTIE) (2017M3D8A2084752).

Conflicts of Interest: The authors declare no conflicts of interest.

References

1. McGowan, P.J.K.; Stewart, G.B.; Long, G.; Grainger, M.J. An imperfect vision of indivisibility in the sustainable development goals. *Nat. Sustain.* **2019**, *2*, 43–45. [CrossRef]
2. *The Sustainable Development Goals Report*; United Nations: New York, NY, USA, 2018; Available online: https://unstats.un.org/sdgs/files/report/2018/TheSustainableDevelopmentGoalsReport2018-EN.pdf (accessed on 1 April 2019).
3. Schmidt-Traub, G.; Kroll, C.; Teksoz, K.; Durand-Delacre, D.; Sachs, J.D. National baselines for the sustainable development goals assessed in the SDG index and dashboards. *Nat. Geosci.* **2017**, *10*, 547–555. [CrossRef]
4. Solomon, S.; Qin, D.; Manning, M.; Chen, Z.; Marquis, M.; Averyt, K.B.; Tignor, M. Summary for policymakers. Contribution of Working Group 1 to the Fourth Assessment Report of the Intergovernmental Panel on Climate Change. In *Intergovernmental Panel on Climate Change (IPCC). Climate Change 2007: The Physical Science Basis*; Miller, H.L., Ed.; Cambridge University Press: Cambridge, UK; New York, NY, USA, 2007; Available online: https://www.ipcc.ch/publications_and_data/publications_ipcc_fourth_assessment_report_wg1_report_the_physical_science_basis.htm (accessed on 2 March 2019).
5. Christiansen, T.; Voigt, T. *Impact of Europe's Changing Climate—2008 Indicator-Based Assessment*; Joint EEA-JRC-WHO Report; European Environment Agency: Copenhagen, Denmark, 2008; Available online: https://ec.europa.eu/jrc/sites/jrcsh/files/jrc_reference_report_2008_09_climate_change.pdf (accessed on 2 March 2019).

6. Watson, R.T. A Contribution of Working Groups I, II, and III to the Third Assessment Report of the Intergovernmental Panel on Climate Change. In *Intergovernmental Panel on Climate Change (IPCC). Climate Change 2001: Synthesis Report*; Team, C.W., Ed.; Cambridge University Press: Cambridge, UK; New York, NY, USA, 2001; Available online: https://www.ipcc.ch/pdf/climate-changes-2001/synthesis-syr/english/front.pdf (accessed on 12 March 2019).

7. Pachauri, R.K.; Reisinger, A. Intergovernmental Panel on Climate Change. In *IPCC 4AR (2007) Fourth Assessment Report (4AR) of the Intergovernmental Panel on Climate Change (IPCC)*; IPCC: Geneva, Switzerland, 2007; p. 104. Available online: http://www.ipcc-wg2.gov/publications/AR4/index.html (accessed on 31 January 2019).

8. Intergovernmental Panel on Climate Change. Summary for Policymakers. In *Climate Change 2013: The Physical Science Basis. Contribution of Working Group I to the Fifth Assessment Report of the Intergovernmental Panel on Climate Change*; Stocker, T.F., Qin, D., Plattner, G.-K., Tignor, M., Allen, S.K., Boschung, J., Nauels, A., Xia, Y., Bex, V., Midgley, P.M., Eds.; Cambridge University Press: Cambridge, UK; New York, NY, USA, 2013; Available online: https://www.ipcc.ch/site/assets/uploads/2018/02/WG1AR5_SPM_FINAL.pdf (accessed on 31 January 2019).

9. Ficklin, D.L.; Luedeling, E.; Zhang, M. Sensitivity of groundwater recharge under irrigated agriculture to changes in climate, CO_2 concentrations and canopy structure. *Agric. Water Manag.* **2010**, *97*, 1039–1050. [CrossRef]

10. Goderniaux, P.; Brouyère, S.; Fowler, H.J.; Blenkinsop, S.; Therrien, R.; Orban, P.; Dassargues, A. Large scale surface-subsurface hydrological model to assess climate change impacts on groundwater reserves. *J. Hydrol.* **2009**, *373*, 122–138. [CrossRef]

11. Lafforgue, M. Climate Change Impacts on Forest-Water Interactions and on Forest Management. In *Forest and the Water Cycle: Quantity, Quality, Management*; Lachassagne, P., Lafforgue, M., Eds.; Cambridge Scholars Publishing: Newcastle upon Tyne, UK, 2016; pp. 612–649.

12. Open Government Data (OGD) Platform India. *Area Weighted Monthly, Seasonal and Annual Rainfall (in mm) for 36 Meteorological Subdivisions*; Ministry of Earth Science: New Delhi, India, 2014. Available online: https://data.gov.in/catalog/area-weighted-monthly-seasonal-and-annual-rainfall-mm-36-meteorological-subdivisions (accessed on 1 April 2019).

13. Park, S.K.; Ha, K. A Study on the Characteristics of the Sediments of the Rainy Season and the Changes in the Forms of the Ranger. *Climate* **2002**, *12*, 348–351.

14. Choi, I.; Shin, H.; Nguyen, T.T.; Tenhunen, J. Water Policy Reforms in South Korea: A Historical Review and Ongoing Challenges for Sustainable Water Governance and Management. *Water* **2017**, *9*, 717. [CrossRef]

15. Kim, H.; Lee, K. A comparison of the water environment policy of Europe and South Korea in response to climate change. *Sustainability* **2018**, *10*, 384. [CrossRef]

16. Iyer, R.R. The New National Water Policy. *Econ. Polit. Wkly.* **2002**, *37*, 1701–1705.

17. Cullet, P.; Gupta, J. India: Evolution of Water Law and Policy. In *The Evolution of the Law and Politics of Water*; Dellapenna, J.W., Gupta, J., Eds.; Springer: Dordrecht, The Netherlands, 2009; pp. 157–173, ISBN 978-1-4020-9866-6.

18. Briscoe, J.; Malik, R.P.S. *India's Water Economy: Bracing for a Turbulent Future*; The World Bank: New Delhi, India; Oxford University Press: Oxford, UK; Available online: https://openknowledge.worldbank.org/handle/10986/7238 (accessed on 22 March 2019).

19. Iyer, R.R. River-linking project: Many questions. In *River Linking: A Millennium Folly?* Patkar, M., Ed.; National Alliance of People's Movements: Mumbai, India, 2004; pp. 9–19.

20. Bandyopadhyay, J.; Perveen, S. Interlinking of Rivers in India: Assessing the Justifications. *Econ. Polit. Wkly.* **2004**, *39*, 5307–5316.

21. Gupta, J.; Van Der Zaag, P. Interbasin water transfers and integrated water resources management: Where engineering, science and politics interlock. *Phys. Chem. Earth Part A/B/C* **2008**, *33*, 28–40. [CrossRef]

22. Ministry of Environment, Forest and Climate Change (MoEFCC). Union Budget, Demand No. 27. 2019–2020. Available online: https://www.indiabudget.gov.in/ub2019-20/eb/sbe27.pdf (accessed on 22 March 2019).

23. Ministry of Water Resources. River Development and Ganga Rejuvenation. Union Budget 2018–2019. Available online: https://openbudgetsindia.org/dataset/ministry-of-water-resources-river-development-and-ganga-rejuvenation-2018-19 (accessed on 22 March 2019).

24. Ministry of Rural Development. Union Budget 2018–2019. Available online: https://openbudgetsindia.org/dataset/department-of-rural-development-2018-19 (accessed on 22 March 2019).
25. Ministry of Housing and Urban Affairs. Demand No. 56. 2018–2019. Available online: https://www.indiabudget.gov.in/ub2018-19/eb/sbe56.pdf (accessed on 22 March 2019).
26. National Mission for Clean Ganga. Ministry of Water Resources, River Development & Ganga Rejuvenation. Available online: https://nmcg.nic.in/NamamiGanga.aspx (accessed on 17 March 2019).
27. The Impact of the River Linking Project. Available online: https://www.thehindu.com/opinion/op-ed/the-impact-of-the-river-linking-project/article24857371.ece (accessed on 17 March 2019).
28. Sahoo, N.K.; Rout, C.; Khuman, Y.S.C.; Prasad, J. Sustainability issues of river linking. In Proceedings of the National Specialty Conference on River Hydraulics, Haryana, India, 29–30 October 2009.
29. Ministry of Environment. Available online: http://www.me.go.kr/home/web/main.do (accessed on 17 March 2019).
30. Ministry of Land. Infrastructure and Transport. Available online: http://www.molit.go.kr/portal.do (accessed on 17 March 2019).
31. Ministry of Environment: Experiences and Achievements. Available online: https://www.ib-net.org/docs/Water_Management_in_Korea(2017,_MOE).pdf (accessed on 17 March 2019).
32. Zhou, S.; Yuan, X.; Peng, S.; Yue, J.; Wang, X.; Liu, H.; Williams, D.D. Groundwater-surface water interactions in the hyporheic zone under climate change scenarios. *Environ. Sci. Pollut. Res.* **2014**, *21*, 13943–13955. [CrossRef] [PubMed]
33. Hyun, Y.; Kim, Y.; Lee, H.; Ahn, J.; Lee, H. *Sustainable Hyporheic Zone Management*; Korea Environmental Policy Evaluation Institute: Korea, 2014; pp. 1–130. Available online: http://repository.kei.re.kr/handle/2017.oak/20201 (accessed on 16 March 2019).
34. European Commission (EC). Directive 2000/60/EC of the European Parliament and of the council of 23 October 2000 establishing a framework for Community action in the field of water policy. *Off. J. Eur. Commun.* **2000**, *327*, 1–22.

 *sustainability*

Article

Process Design Characteristics of Syngas (CO/H$_2$) Separation Using Composite Membrane

Jeeban Poudel Ja Hyung Choi and Sea Cheon Oh *

Department of Environmental Engineering, Kongju National University, 1223-24 Cheonan-Daero, Seobuk, Chungnam 31080, Korea; jeeban1985@gmail.com (J.P.); jah9206@naver.com (J.H.C.)
* Correspondence: ohsec@kongju.ac.kr; Tel.: +82-41-521-9423

Received: 24 December 2018; Accepted: 28 January 2019; Published: 29 January 2019

Abstract: The effectiveness of gas separation membranes and their application is continually growing owing to its simpler separation methods. In addition, their application is increasing for the separation of syngas (CO and H$_2$) which utilizes cryogenic temperature during separation. Polymers are widely used as membrane material for performing the separation of various gaseous mixtures due to their attractive perm-selective properties and high processability. This study, therefore, aims to investigate the process design characteristics of syngas separation utilizing polyamide composite membrane with polyimide support. Moreover, characteristics of CO/H$_2$ separation were investigated by varying inlet gas flow rates, stage cut, inlet gas pressures, and membrane module temperature. Beneficial impact in CO and H$_2$ purity were obtained on increasing the flow rate with no significant effect of increasing membrane module temperature and approximately 97% pure CO was obtained from the third stage of the multi-stage membrane system.

Keywords: composite membrane; polyamide; membrane module temperature; hydrogen; carbon monoxide; stage cut

---

1. Introduction

In recent years, there have been tremendous advancements in the fields of industry, service, and commerce. Although this growth has brought substantial economic and social benefits, such as improving people's lives, it has also provoked numerous side effects. Municipal solid waste (MSW) is one of the crucial problems that is of major concern globally. The quantity of MSW is rising significantly in industrialized and developing countries bringing into question its sustainable disposal management [1–3]. The desire to protect the environment has resulted in the development of various ideas for the legitimate management of MSW. For instance, technologies, such as incineration, gasification, generation of biogas, and utilization in a combined heat and power (CHP) plants, have been investigated for the production of energy from MSW [4,5]. In spite of the advantages derived from incineration of MSW, such as heat recovery, there are numerous disadvantages associated with the production of large flue gas volumes and hazardous waste streams accompanying the fly ash. However, gasification is an attractive way to treat MSW with fewer pollution emissions in contrast to other treatment methods and offers better competency in terms of energy production [6].

The major components obtained from gasification of waste are CO, H$_2$, CO$_2$, N$_2$, and CH$_4$ although the higher percentage is constituted by CO, CO$_2$, and H$_2$. CO$_2$ from natural gas and post-combustion can be removed through effective and low-cost absorption processes comprising chemical solvents [7]. Among various processes, adsorption technology holds the potential to effectively separate CO$_2$ due to its increased flexibility to adapt to varying feed stream specifications and operating conditions along with its simplicity, lower energy requirement, and low cost [8]. CH$_4$ and N$_2$ are present in an insignificant amount and can be removed through appropriate control of the operating conditions.

Moreover, CH_4 and N_2 can be removed through layered pressure swing adsorption (PSA) as reviewed by Simone et al. [9]. Synthesis gas (syngas) is a mixture of H_2 and CO which is used in diverse petrochemical processes and can be obtained after separating CO_2. It can be used as a mixture or can be separated into its constituents before use [10]. State-of-the-art shows that the most efficient method for production of syngas is the reforming processes [11]. Thus, one of the aims of this study is to highlight the syngas produced from MSW gasification. In addition, the separation and purification of carbon monoxide are of great significance with respect to its importance as the starting material for the synthesis of a variety of basic chemicals in C_1 chemistry processes [12] and capitalize on H_2 as a source of fuel. However, the separation of gases by the membrane is one of the most flexible separation processes. Amongst many available techniques, such as distillation, crystallization, absorption, or solvent extraction for chemical separations, membrane technology is economically competitive for administering alternative separation technologies [11,13]. Opportunities thrive to extend membrane markets for gas separations although the existing membrane materials, membrane structures, and formation processes are inadequate for thorough utilization of these opportunities, and the requirements for the viability of membranes vary considerably with each application [14]. PSA and Cryogenic distillation processes are commercially available separation techniques among which cryogenic distillation is energy intensive [9,15,16]. The high operating costs associated with low productivity are the main disadvantages of adsorption separation method, such as PSA [10]. The interest on membrane technology continues to grow as the membrane process equipment, and methods of operation are comparatively easy compared to conventional separation methods. This is especially true for the separation of gaseous mixtures, such as CO and H_2, which require cryogenic temperatures when separated by condensation, distillation, or adsorption [17]. Lu et al. [18] utilized inorganic membranes for hydrogen production and purification. References [19,20] portray the benefits of palladium and Pd-alloy membranes as an excellent means of hydrogen purification, especially while merging a catalytic reactor and separation in a single unit. However, the application of dense polymeric membrane dominates the gas separation field where the separation of gases with similar particle size can be obtained through diffusion if a variation in solubility exists [21]. Nevertheless, the polymeric membrane still needs further research due to "upper bound" where the membranes result in lower selectivity due to its open structure [22,23]. Even though selective permeation exhibits great potential for the separation of gases, it was not economically feasible due to the large membrane surface areas required to carry out industrial-scale separation. Considering the trade-off relation between cost and efficiency in selecting the membranes, this study utilizes the rubbery composite membrane with braid to provide mechanical strength to the membrane.

Along with the development of new and efficient membrane materials with enhanced gas separation efficiencies, process design is also indispensable as it directly affects the economics of membrane-based separation processes. The application of a rubbery composite membrane was carried out to separate two main components of syngas (CO and H_2) obtained from MSW gasification as the working pressure was low and the cost associated with the membrane is comparatively lower than specific metallic membrane for hydrogen gas permeation. The manufacturing and commercialization of other inorganic materials are still expensive compared to polymeric membranes although inorganic membranes demonstrate higher performance [24]. The composite polyamide membrane with polyimide support was used to enhance the mechanical strength of the membrane. Characteristic analysis of syngas separation using gas membrane through CO/H_2 permeability and selectivity was conducted along with a detailed study of multistage membrane design. Characteristic analysis of syngas separation using gas membrane through CO/H_2 permeability and selectivity was conducted along with a detailed study of multistage membrane design. Multistage membrane design is of great importance while obtaining syngas of varying composition, and a high purity level is one of the novel achievements of this study. The comprehensive characteristic analysis was determined using various gas flow rate, stage cut, membrane-module temperature, and input pressure. The polyamide composite membrane with polyimide support has not been studied which further highlights the possibility of incorporating polyamide (rubbery polymer) with polyimide (glassy polymer) to obtain better separation efficiency.

2. Experimental

A polyamide composite membrane with polyimide support was used in this experiment. The membrane module was manufactured by SepraTek Co. (Republic of Korea). The schematic diagram of the overall membrane system used in this work and the specific features of the membrane are illustrated in Figure 1. It also demonstrates the SEM image of the outer surface and the active membrane layer where the inner polyamide membrane (active layer) is supported by the polyimide layer (middle layer). The outermost layer is the braid that provides support and strength to the membrane which comprises metallic wires and polymer fibers. The outermost layer is braided providing support and strength to the membrane and is made of a high strength polyester fiber, the porous intermediate substrate is of polyetherimide, and the active inner layer is of rubbery polyamide. The rubbery active layer has a dense structure. The integration of an individual braid-reinforced hollow fiber membrane aided to establish the complete membrane as demonstrated in Figure 2. The outer diameter of the final membrane is 7 cm with an inner surface area of 0.55 m^2.

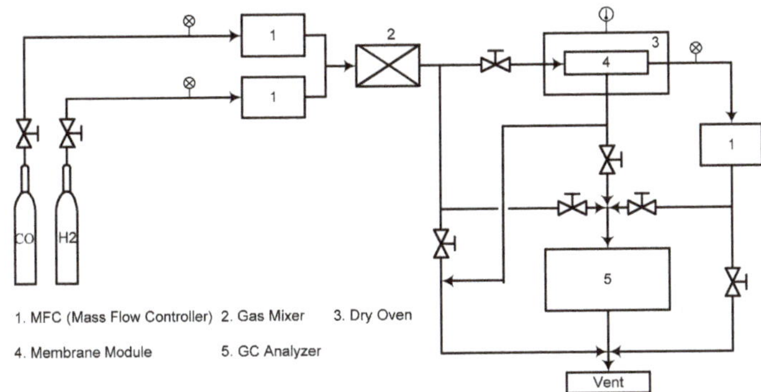

Figure 1. Schematic diagram of the membrane system.

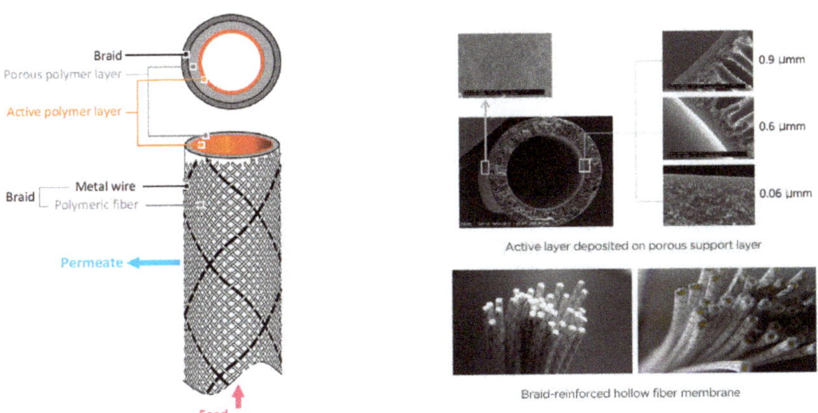

Figure 2. Schematic Representation of Braid-reinforced hollow fiber pervaporation membrane.

The pseudo syngas mixed from CO with purity of 99.95 vol% and H$_2$ with purity of 99.999 vol% was used in this study. The composition of pseudo syngas was controlled by adjusting inlet flow rates of CO and H$_2$ using mass flow controller (MFC). The CO and H$_2$ gases from cylinders are uniformly mixed in a gas mixer and injected into membrane module. The gas inlet pressure was also controlled by the gas regulator connected to the gas cylinders. The stage cut was controlled by adjusting the flow rate of retentate using MFC, and the permeate pressure was atmospheric. To analyze the composition of gases, each gas flow line was connected to gas chromatography (GC) analyzer (*i* GC 7200, DS SCIENCE, Republic of Korea). Six foot by an eighth of an inch stainless steel packed column with 80/100 mesh molecular sieve 5A as column material was used in the GC analyzer. Argon (25 mL/min) was used as the carrier gas with a column temperature of 130 °C at a heating rate of 10 °C/min. The detector temperature was set at 130 °C/min. The kinetic diameter of CO and H_2 was 3.6 and 2.9 Å, respectively [25] due to which H_2 was obtained as permeate and CO as retentate. Furthermore, cumulative data of CO separation characteristics at various CO:H_2 concentrations were studied to estimate the multistage membrane number required to obtain varying CO purity.

The permeability is related to the gas permeation rate through the membrane, the surface area of the membrane, and the pressure difference across the membrane and is expressed as:

$$P_{H_2} = \frac{Q_{H_2}}{A \Delta P} \tag{1}$$

$$P_{CO} = \frac{Q_{CO}}{A \Delta P} \tag{2}$$

where, P_{H2} and P_{CO} are the permeability (cm^3 cm^{-2} s^{-1} cm Hg^{-1}) of H_2 and CO, respectively, and Q_{H2} and Q_{CO} are the flow rate (cm^3 s^{-1}) of H_2 and CO, respectively. In addition, in Equation (1), A denotes the surface area (cm^2) of the membrane, and ΔP is pressure difference (cm Hg) between retentate and permeate. The selectivity of H_2 in permeate is defined as

$$\alpha_{CO} = \frac{P_{H2}}{P_{CO}} \tag{3}$$

In this work, the permeability of each gas was calculated by Equations (1) and (2). And the selectivity of H_2 in permeate was calculated by Equation (3). Furthermore, the stage cut (%) used in this work was calculated as follows:

$$Stage\ cut = \frac{Q_R}{Q_F} \times 100 \tag{4}$$

where, Q_R and Q_F are the flow rate (cm^3 s^{-1}) of retentate and feed gas, respectively.

3. Results and Discussions

3.1. Effect of Varying Gas Flow Rate for Various Stage Cut

The characteristics of CO/H_2 separation for various stage cut (15, 30, and 50%) with an increase in flow rate of the input gas at 6 bars inlet pressure is exhibited in Figure 3a–c. The actual working pressure (the gauge pressure of the membrane) was much lower in comparison to the inlet gas pressure and was maintained by controlling the stage cut. The stage cut is directly related to the recovery percentage and the economic feasibility of the process, thereby, making it one of the most important parameters while selecting membrane technology. From the figure, it can be observed that the working pressure of the membrane, which is the pressure difference, increased with a decrease in the stage cut, thereby, increasing the actual driving pressure of the membrane. A decrease in the stage cut reduced the free flow of gas through the retentate which increases the pressure inside the membrane volume. In addition, the CO purity was elevated while the H_2 purity deteriorated with the lowering of the stage cut and vice versa. The decrease in the stage cut relates to the increase in driving pressure which tends to push the gas mixture more towards permeate forcing more CO to pass through the permeate. This increases the purity of CO in the retentate while the passage of CO in permeate decreases the purity of H_2. A gradual increase in the input flow rate increases the CO and H_2 purity which further is affected by the stage cut.

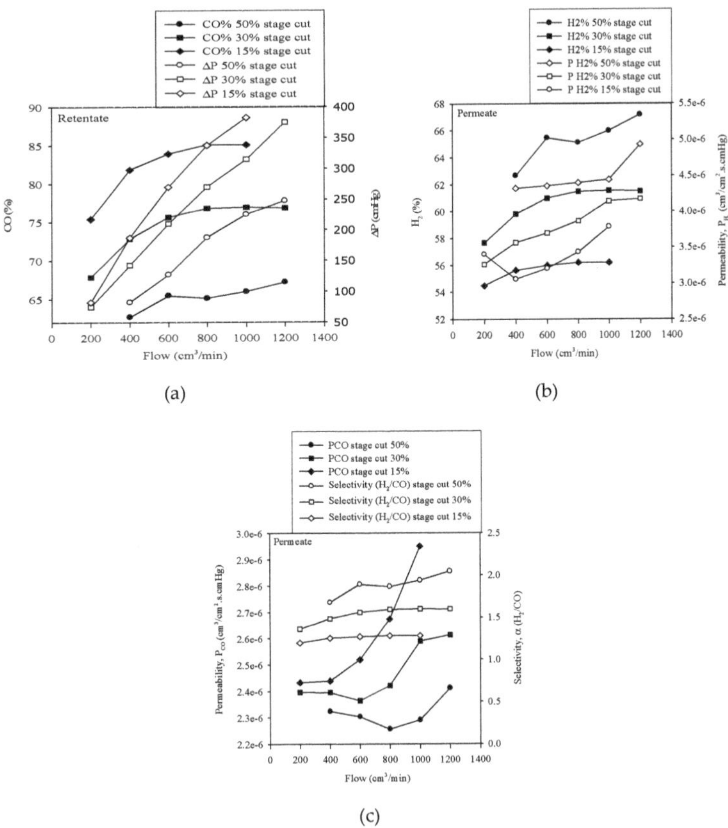

Figure 3. Characteristics of CO/H_2 separation with increasing flow rates at 6 bars inlet pressure for various stage cut (**a**) CO% & ΔP, (**b**) H_2% & P_{H2}, and (**c**) P_{CO} & $\alpha(H_2/CO)$.

In addition, the permeability of H_2 and CO increased with an increase in flow rate which is directly correlated with the amount of gas. The permeability of H_2 was higher while permeability of CO was lower for increased stage cut. An increase in the stage cut permitted the natural flow of CO towards retentate and H_2 towards permeate. However, a reduction in the stage cut increased the working pressures, thus, forcing CO towards permeate increasing CO permeability. Initially, on increasing the gas flow rate, the separation efficiency of the membrane increased with an increase in working pressure. On further increasing the flow rate, CO was forced to pass through permeate thereby increasing the CO permeability. However, the increase in CO permeability reduced the CO/H_2 separation efficiency. Nevertheless, the purity of CO can be increased although the separation efficiency is reduced. However, reduction in the stage cut reduces the recovery while reducing the economic feasibility. The selectivity of the membrane for H_2 as it is the target product of permeate slightly increases with the increase in flow rate and decreases with a decrease in the stage cut.

3.2. Effect of Varying Gas Flow Rate for Various Gas Inlet Pressures

The characteristics of CO/H_2 separation at various inlet pressures with an increase in inlet gas flow rate are shown in Figure 4a–c. Although the trend obtained for an increase in flow rate is similar to the discussion in Section 3.1 for gas flow rate, the characteristics were not significantly affected by the variation in gas inlet pressure. CO and H_2 purity along with H_2 selectivity somewhat reduced with an increase in inlet pressure thereby resulting in the forced passage of CO in permeate and H_2 in the retentate. However, the pressure difference, which is the primary driving force of the membrane, was not affected at all by the change in the inlet pressure. Therefore, the characteristics are indifferent to the variation in inlet gas pressure. Furthermore, CO and H_2 permeability failed to depict a particular trend with variation in gas inlet pressure. The presence of a mixer for homogeneous mixing of the inlet gases before entering the membrane as shown in Figure 1 significantly reduced the effect of inlet gas pressure.

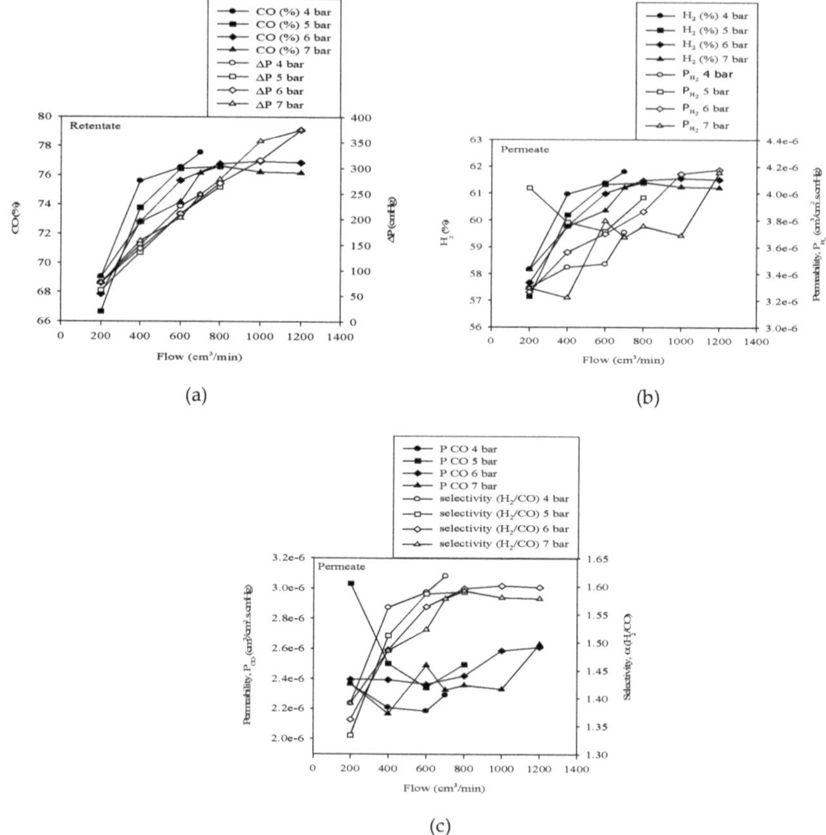

Figure 4. Characteristics of CO/H_2 separation with increasing flow rates at 70% stage cut for various inlet pressures (**a**) CO% & ΔP, (**b**) H_2% & P_{H2}, and (**c**) P_{CO} & $\alpha(H_2/CO)$.

3.3. Effect of Varying the Stage Cut for Various Gas Flow Rate

The effect of varying the stage cut for various gas flow rates is shown in Figure 5a–c. The increase in the stage cut refers to the increase in the gas flow amount from the retentate as described by Equation (4). CO purity and the pressure difference gradually dropped with an increase in the stage cut. On varying the gas flow rate, a notable difference was obtained for 200 and 400 cm^3/min while the difference was insignificant for 400 and 600 cm^3/min. An increase in the stage cut reduced the working pressure thereby forcing H_2 to pass through the retentate and reducing CO purity. However, the reduction of working pressure significantly increased the H_2 purity in the permeate. The permeability of H_2 did not vary for 400 and 600 cm^3/min which however increased after 25% stage cut. Oana et al. [26] also demonstrated a gradual increase in the permeability of H_2 on increasing the gas flow rate to 6 cm^3(STP)/min at 30 °C and 4 bar feed pressure. However, on further increasing the flow rate, the H_2 permeability remained constant. Very low working pressure at this point resulted in a very low flow of CO in permeate thereby increasing H_2 permeability. Contrarily, the CO permeability is not affected by the increase in the stage cut for 200 and 400 cm^3/min while the CO permeability increased after 25% stage cut for 600 cm^3/min. This is due to the volume of the membrane which after certain point forces the air to pass through the permeate as the driving pressure was highest for 600 cm^3/min. H_2 selectivity was lowest for 600 cm^3/min while highest for 200 cm^3/min as this is directly related to the permeability of H_2 and CO as shown in Equation (3). Highest permeability of CO at 600 cm^3/min resulted in the lowest selectivity of H_2 at this point.

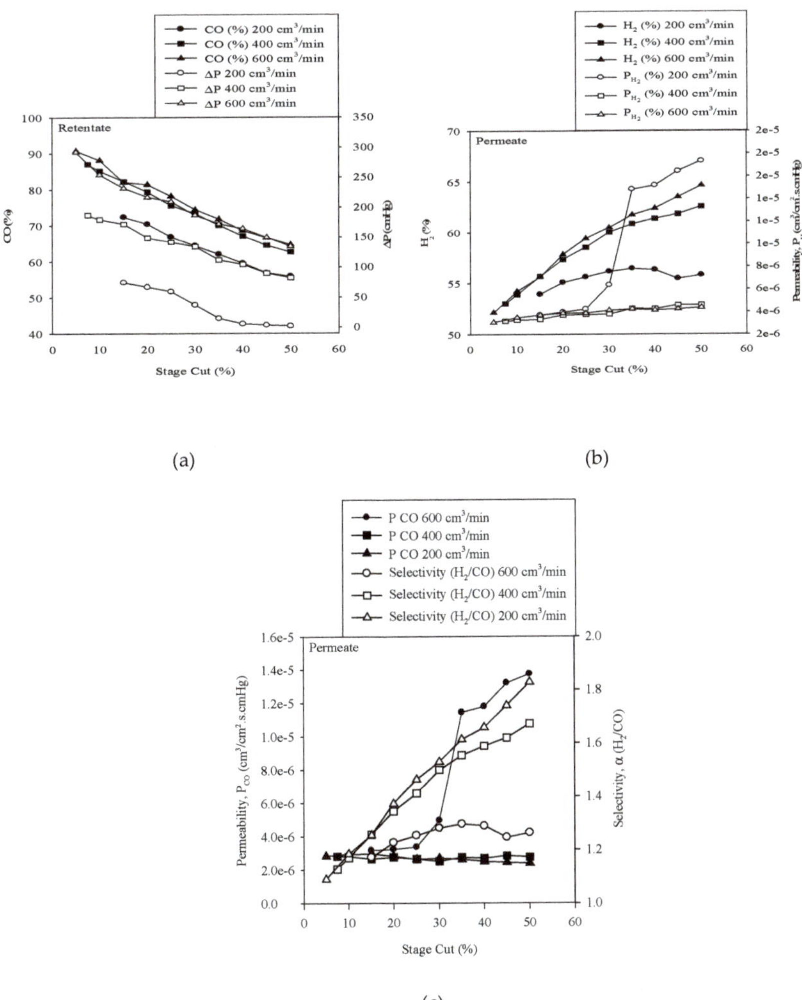

(a)

(b)

(c)

Figure 5. Characteristics of CO/H_2 separation with increasing stage cut at 6 bars inlet pressure for various inlet flow rates (**a**) CO% & ΔP, (**b**) H_2% & P_{H2}, and (**c**) P_{CO} & $\alpha(H_2/CO)$.

3.4. Effect of Varying Membrane Module Temperature

In dense polymeric materials, solution–diffusion is widely accepted to be the main mechanism of transport [27]. This mechanism is generally considered to be a three-step process. In the first step, the gas molecules are absorbed by the membrane surface in the upstream end. This is followed by the diffusion of the gas molecules through the polymer matrix. In the final step, the gas molecules are desorbed to the downstream end [28]. Thus, the permeability of gas in the polymer membrane system is affected by temperature. Furthermore, the small size of the hydrogen molecules gives H_2 its high diffusivity [29] while the low temperature suggests that the solubility of hydrogen is very low [30,31]. Therefore, the goal of the process design is to exploit the high diffusivity of the hydrogen molecule and limit the effect of the lower solubility. Usually, with an increase in module temperature, diffusivity, depending on the free volume, increases which in turn decreases the solubility. However, correlations showing the separate effects of diffusion and solubility coefficients on selectivity and permeability are more limited [32].

The effect of varying membrane module temperature on characteristics of CO/H_2 separation was determined from the experiment and is demonstrated in Figure 6a–c. A stage cut of 50% for a gas flow rate of 400 cm^3/min was considered with a gas inlet pressure of 6 bars. An increase in membrane temperature in this study increased the working pressure which however did not vary the CO and H_2 purity. This is because with an increase in membrane module temperature solubility decreases while diffusivity increases thereby developing tradeoff relation for CO and H_2 purity. The permeability of H_2 and CO gradually decreases with increase in membrane module temperature due to the increase in kinetic diameter of CO and H_2 which is increased on increasing the temperature. The kinetic diameter of H_2 and CO was 2.9 and 3.6 Å, respectively [25]. The lower flow rate with increased driving force for a constant area, resulted in decreased permeability for CO and H_2 as shown in Equations (1) and (2), a similar trend is reported in other studies [26,33,34]. Although the difference is not high, the selectivity of the membrane increased with an increase in membrane module temperature which again fell to its initial level on further increasing the membrane module temperature. Oana et al. [26] also obtained a gradual reduction trend in selectivity of H_2 on increasing the membrane module temperature in a similar temperature range of this study. The increase in membrane module temperature demonstrated the negative effect on CO/H_2 separation while using polyamide composite membrane. Contrarily, the study by Maryam et al. [35] obtained an increasing H_2 permeability and selectivity on increasing the membrane module temperature while using hollow fiber polyimide membrane. The permeabilities of the gases at the temperature studied are similar to the temperature range studied by Markovic et al. [36]. Although for different gases like He, N_2, Ar, CO_2, and C_3H_8, reduction in permeabilities of various gases were obtained on increasing the membrane module temperature.

3.5. Estimation of Multistage Membrane Required for CO Purity

Figure 7 shows a graph demonstrating the CO purity for different stages in a multistage membrane system for the input gas flow rate of 200 and 400 cm^3/min. It shows that CO purity can be obtained at every stage of the multi-stage membrane system for different stage cut and flow rates. The result obtained in this study reduces the effort to test for multistage gas separation, thus, reducing the preliminary cost. It can be inferred from the graph that at $CO:H_2$ ratio of 50:50%, 97% pure CO can be obtained in the third stage of the multi-stage membrane system for 400 cm^3/min. This, when passed through the 4th stage membrane, can generate CO purity percent of 99%. However, practically, along with the purity of CO, the recovery amount is very important for the economic feasibility of the membrane system. So, optimization of the CO purity percent and the recovery amount are indispensable.

It becomes evident that the use of single-stage membrane processes compromises the product purity and recovery as a trade-off that reduces separation performance compared to multistage membrane process [37]. In a multistage membrane system, there is a trade-off between permeate composition and permeate pressure, and, therefore, recompression costs incur. However, they are very useful in obtaining gas recovery above 95% [38]. Drioli et al. [39] suggested the multistage membrane system with highly permeable and less selective membranes in the first stage for enriching the stream of more permeable species and highly selective but less permeable membrane in the successive stages for achieving the high concentration of the desired species in retentate and permeate as a viable solution in the retentate and increase of CO as an impurity in the permeate. CO permeability initially decreased which increases on further increasing the flow rate. Initially, on increasing the gas flow rate, the separation efficiency of the membrane increased with an increase in working pressure. On further increasing the flow rate, CO was forced to pass through permeate thereby increasing the CO permeability. However, the increase in CO permeability reduces the CO/H_2 separation efficiency. The selectivity of the membrane for H_2 as it is the target product of permeate slightly increases with the increase in flow rate and decreases with a decrease in the stage cut.

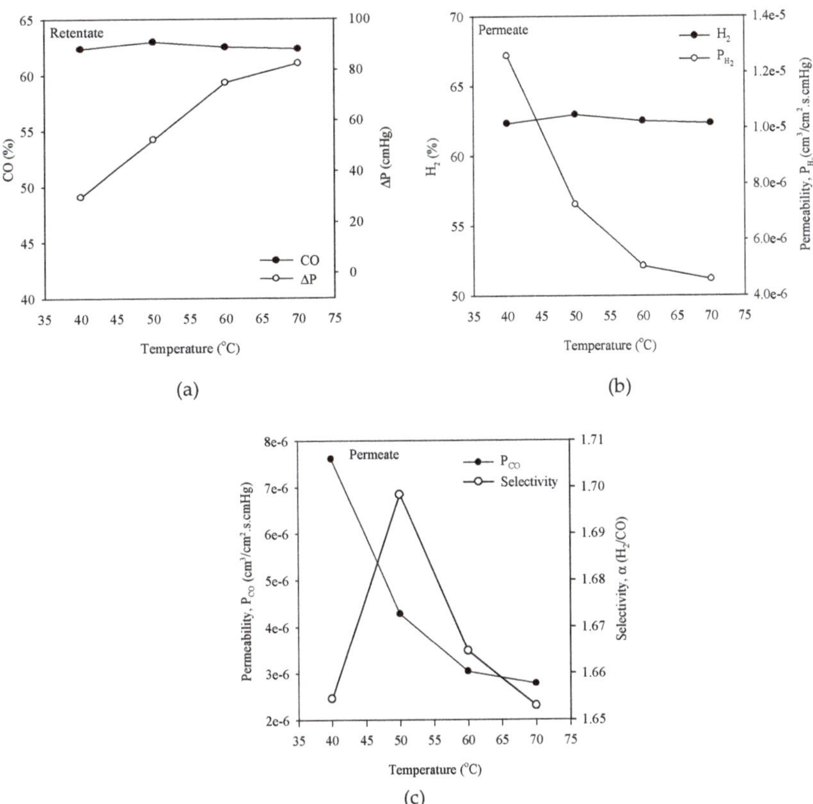

Figure 6. Characteristics of CO/H_2 separation with increasing membrane cut at 6 bars inlet pressure for 400 cm^3/min flow rate and a stage cut of 50% (**a**) CO% & ΔP, (**b**) H_2% & P_{H2}, and (**c**) P_{CO} & $\alpha(H_2/CO)$.

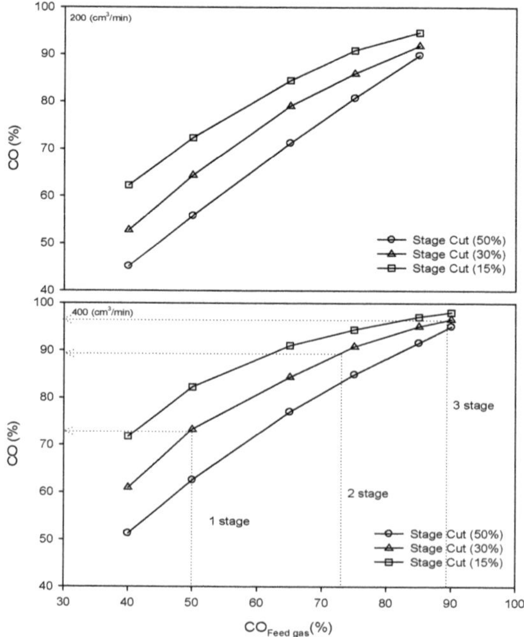

Figure 7. Figure showing the cumulative data of CO separation characteristics at various CO: H_2 concentrations for input pressure of 6 bars to estimate the multistage number required.

4. Conclusions and Future Outlooks

CO and H_2 purity, permeability, and its selectivity were studied as the process design characteristics of composite membrane based on various flow rates, stage cut, inlet pressures and membrane module temperatures. The study focused on the development of effective process design in implementing commercial polymeric membrane for syngas separation thereby reducing cost in CO/H_2 separation which otherwise needs expensive inorganic membranes. The reduction in the stage cut enhances the CO purity while H_2 purity is reduced while the permeability of CO and H_2 decreases with an increase in the stage cut. However, the variation of inlet pressure did not demonstrate significant changes in CO/H_2 separation. An increase in CO and H_2 purity was observed while increasing the gas flow rates. A notable difference in CO/H_2 separation characteristics was obtained for 200 and 400 cm^3/min flow rate while the difference was minor for 400 and 600 cm^3/min flow rate.

Increase in membrane module temperature had no effect on enhancing CO and H_2 purity. However, the reduction in CO and H_2 permeability was observed while increasing the membrane module temperature. This was due to the increase in kinetic diameter of CO and H_2 with an increase in temperature. Multistage membrane process was found to be crucial in obtaining high purity CO. Ninety-seven percent pure CO was obtained for third stage arrangement of the composite membrane. The increase in the number of membranes will result in comparatively high purity of CO.

Although the results obtained show the possibility of its application in CO/H_2 separation, further study is required to understand its performance in separating CO and H_2 with consideration of recovery of individual components. Moreover, comparative analysis of the membrane should be done with the commercially available inorganic membranes to build a base for utilizing it in large scale gas separation applications. The increase in separation efficiency with lower stage cut will increase the recovery amount, and future research should be focused on enhancing recovery with reduced stage cuts which ultimately will increase its economic feasibility.

Author Contributions: J.P. was responsible for the overall experiment, data analysis and arrangement along with preparation of the manuscript; J.H.C. was responsible for the data analysis and editing of the manuscript. S.C.O. was the supervisor of the project and was mainly responsible for the data and manuscript confirmation. All the authors were equally responsible for finalizing the manuscript and submission.

Acknowledgments: This research was supported by the National Strategic Project-Carbon Upcycling of the National Research Foundation of Korea (NRF) funded by the Ministry of Science and ICT (MSIT), the Ministry of Environment (ME) and the Ministry of Trade, Industry and Energy (MOTIE) (No. 2017M3D8A2089593) and Waste to Energy Recycling Human Resource Development Project of Korea Ministry of Environment (ME).

Conflicts of Interest: The authors declare no conflicts of interest.

References

1. Kwak, T.; Maken, S.; Lee, S.; Park, J.; Min, B.; Yoo, Y.D. Environmental Aspects of Gasification of Korean Municipal Solid Waste in a Pilot Plant. *Fuel* **2006**, *85*, 2012–2017. [CrossRef]

2. Sakai, S.; Sawell, S.; Chandler, A.; Eighmy, T.; Kosson, D.; Vehlow, J.; Van der Sloot, H.; Hartlen, J.; Hjelmar, O. World Trends in Municipal Solid Waste Management. *Waste Manag.* **1996**, *16*. [CrossRef]

3. Tchobanoglous, G.; Theisen, H.; Vigil, S.A.; Alaniz, V.M. *Integrated Solid Waste Management: Engineering Principles and Management Issues*; McGraw-Hill: New York, NY, USA, 1993.

4. Murphy, J.D.; McKeogh, E. Technical, Economic and Environmental Analysis of Energy Production from Municipal Solid Waste. *Renew. Energy* **2004**, *29*, 1043–1057. [CrossRef]

5. Alzate-Gaviria, L.M.; Sebastian, P.; Pérez-Hernández, A.; Eapen, D. Comparison of Two Anaerobic Systems for Hydrogen Production from the Organic Fraction of Municipal Solid Waste and Synthetic Wastewater. *Int. J. Hydrogen Energy* **2007**, *32*, 3141–3146. [CrossRef]

6. Luo, S.; Xiao, B.; Hu, Z.; Liu, S.; Guan, Y.; Cai, L. Influence of Particle Size on Pyrolysis and Gasification Performance of Municipal Solid Waste in a Fixed Bed Reactor. *Bioresour. Technol.* **2010**, *101*, 6517–6520. [CrossRef] [PubMed]

7. Desideri, U. *Developments and Innovation in Carbon Dioxide (CO_2) Capture and Storage Technology*; Woodhead Publishing: Cambridge, UK, 2010.

8. Nikolaidis, G.N.; Kikkinides, E.S.; Georgiadis, M.C. A Model-Based Approach for the Evaluation of New Zeolite 13X-Based Adsorbents for the Efficient Post-Combustion CO_2 Capture using P/VSA Processes. *Chem. Eng. Res. Des.* **2018**, *131*, 362–374. [CrossRef]

9. Cavenati, S.; Grande, C.A.; Rodrigues, A.E. Separation of $CH_4/CO_2/N_2$ Mixtures by Layered Pressure Swing Adsorption for Upgrade of Natural Gas. *Chem. Eng. Sci.* **2006**, *61*, 3893–3906. [CrossRef]

10. DiMartino, S.; Glazer, J.; Houston, C.; Schott, M. Hydrogen/Carbon Monoxide Separation with Cellulose Acetate Membranes. *Gas Sep. Purif.* **1988**, *2*, 120–125. [CrossRef]

11. Khanipour, M.; Mirvakili, A.; Bakhtyari, A.; Farniaei, M.; Rahimpour, M.R. Enhancement of Synthesis Gas and Methanol Production by Flare Gas Recovery Utilizing a Membrane Based Separation Process. *Fuel Process. Technol.* **2017**, *166*, 186–201. [CrossRef]

12. Dutta, N.; Patil, G. Developments in CO Separation. *Gas Sep. Purif.* **1995**, *4*, 277–283. [CrossRef]

13. Hsieh, H. *Inorganic Membranes for Separation and Reaction*; Elsevier: Amsterdam, The Netherlands, 1996.

14. Koros, W.J.; Mahajan, R. Pushing the Limits on Possibilities for Large Scale Gas Separation: Which Strategies? *J. Membr. Sci.* **2000**, *175*, 181–196. [CrossRef]

15. Ho, W.; Sirkar, K. *Membrane Handbook*; Springer Science & Business Media: Berlin/Heidelberg, Germany, 2012.

16. Bhadra, S.; Farooq, S. Separation of Methane–nitrogen Mixture by Pressure Swing Adsorption for Natural Gas Upgrading. *Ind. Eng. Chem. Res.* **2011**, *50*, 14030–14045. [CrossRef]

17. Corbett, L. *Industrial Engineering Chemistry Process Design and Development*; American Chemical Society: Washington, DC, USA, 1975; Volume 13, p. 181.

18. Lu, G.Q.; Da Costa, J.D.; Duke, M.; Giessler, S.; Socolow, R.; Williams, R.H.; Kreutz, T. Inorganic membranes for hydrogen production and purification: A critical review. *J. Colloid Interface Sci.* **2007**, *314*, 589–603. [CrossRef] [PubMed]

19. Basile, A.; Gallucci, F.; Tosti, S. Synthesis, characterization and applications of palladium membranes. *Membr. Sci. Technol.* **2008**, *13*, 255–323.

20. Sanchez, J.M.; Barreiro, MM.; Marono, M. Hydrogen enrichment and separation from synthesis gas by the use of a membrane reactor. *Biomass Bioenergy* **2011**, *35*, S132–S144. [CrossRef]

21. Shimekit, B.; Mukhtar, H. *Natural Gas Purification Technologies-Major Advances for CO_2 Separation and Future Directions*; INTECH Open Access Publisher: London, UK, 2012; pp. 235–270.

22. Robeson, L.M. The upper bound revisited. *J. Membr. Sci.* **2008**, *320*, 390–400. [CrossRef]

23. Noble, R.D. Perspectives on mixed matrix membranes. *J. Membr. Sci.* **2011**, *15*, 393–397. [CrossRef]

24. Scholes, C.A. Membrane gas separation applications in natural gas processing. *Fuel* **2012**, *96*, 15–28. [CrossRef]

25. Meulenberg, W.A.; Hansch, R.; Buchkremer, H.P.; Stöver, D. Device for Gas Separation and Method for Producing such a System. U.S. Patent 8,016,924, 13 September 2011.

26. David, O.C.; Gorri, D.; Urtiaga, A.; Ortiz, I. Mixed Gas Separation Study for the Hydrogen Recovery from $H_2/CO/N_2/CO_2$ Post Combustion Mixtures using a Matrimid Membrane. *J. Membr. Sci.* **2011**, *378*, 359–368. [CrossRef]

27. Koros, W.J.; Fleming, G. Membrane-Based Gas Separation. *J. Membr. Sci.* **1993**, *83*, 1–80. [CrossRef]

28. Javaid, A. Membranes for Solubility-Based Gas Separation Applications. *Chem. Eng. J.* **2005**, *112*, 219–226. [CrossRef]

29. Hines, A.L.; Maddox, R.N. *Mass Transfer: Fundamentals and Applications*; Prentice-Hall: Englewood Cliffs, NJ, USA, 1985.

30. Lin, H.; Freeman, B.D. Gas Solubility, Diffusivity and Permeability in Poly (Ethylene Oxide). *J. Membr. Sci.* **2004**, *239*, 105–117. [CrossRef]

31. Reid, R.C.; Prausnitz, J.M.; Poling, B.E. *The Properties of Gases and Liquids*; Mcgraw-Hill: New York, NY, USA, 1987.

32. Robeson, L.M.; Smith, Z.P.; Freeman, B.D.; Paul, D.R. Contributions of Diffusion and Solubility Selectivity to the Upper Bound Analysis for Glassy Gas Separation Membranes. *J. Membr. Sci.* **2014**, *453*, 71–83. [CrossRef]

33. El-Azzami, L.A.; Grulke, E.A. Dual Mode Model for Mixed Gas Permeation of CO_2, H_2, and N_2 through a Dry Chitosan Membrane. *J. Polym. Sci. Part B Polym. Phys.* **2007**, *45*, 2620–2631. [CrossRef]

34. Tanaka, K.; Kita, H.; Okamoto, K.; Nakamura, A.; Kusuki, Y. Gas Permeability and Permselectivity in Polyimides Based on 3,3′,4,4′-Biphenyltetracarboxylic Dianhydride. *J. Membr. Sci.* **1989**, *47*, 203–215. [CrossRef]

35. Peer, M.; Mehdi Kamali, S.; Mahdeyarfar, M.; Mohammadi, T. Separation of Hydrogen from Carbon Monoxide using a Hollow Fiber Polyimide Membrane: Experimental and Simulation. *Chem. Eng. Technol.* **2007**, *30*, 1418–1425. [CrossRef]

36. Marković, A.; Stoltenberg, D.; Enke, D.; Schlünder, E.; Seidel-Morgenstern, A. Gas Permeation through Porous Glass Membranes: Part I. Mesoporous Glasses—Effect of Pore Diameter and Surface Properties. *J. Membr. Sci.* **2009**, *336*, 17–31.

37. Zarca, G.; Urtiaga, A.; Biegler, L.T.; Ortiz, I. An Optimization Model for Assessment of Membrane-Based Post-Combustion Gas Upcycling into Hydrogen or Syngas. *J. Membr. Sci.* **2018**, *563*, 83–92. [CrossRef]

38. Bernardo, P.; Drioli, E.; Golemme, G. Membrane Gas Separation: A Review/State of the Art. *Ind. Eng. Chem. Res.* **2009**, *48*, 4638–4663. [CrossRef]

39. Drioli, M.S.B.; Barbieri, G. CO_2-CH_4 Membrane Separation. Available online: http://dx.medra.org/10.17374/CI.2015.97.5.14 (accessed on 15 September 2017).

Review

Recent Progress of Electrochemical Energy Devices: Metal Oxide–Carbon Nanocomposites as Materials for Next-Generation Chemical Storage for Renewable Energy

Dohyeong Seok †, Yohan Jeong †, Kyoungho Han, Do Young Yoon and Hiesang Sohn *

Department of Chemical Engineering, Kwangwoon University, Seoul 01897, Korea
* Correspondence: hsohn@kw.ac.kr or sonisang@gmail.com
† These authors contributed equally to this work.

Received: 30 April 2019; Accepted: 26 June 2019; Published: 5 July 2019

Abstract: With the importance of sustainable energy, resources, and environmental issues, interest in metal oxides increased significantly during the past several years owing to their high theoretical capacity and promising use as electrode materials for electrochemical energy devices. However, the low electrical conductivity of metal oxides and their structural instability during cycling can degrade the battery performance. To solve this problem, studies on carbon/metal-oxide composites were carried out. In this review, we comprehensively discuss the characteristics (chemical, physical, electrical, and structural properties) of such composites by categorizing the structure of carbon in different dimensions and discuss their application toward electrochemical energy devices. In particular, one-, two-, and three-dimensional (1D, 2D, and 3D) carbon bring about numerous advantages to a carbon/metal-oxide composite owing to the unique characteristics of each dimension.

Keywords: metal oxide; carbon; composite; energy storage; electrochemical device

---

1. Introduction

Studies on renewable energy storage devices and use are an emerging issue owing to increases in energy consumption and a reduced supply of fossil fuels [1–3]. Electrochemical energy devices including lithium (or sodium) ion batteries and supercapacitors received attention owing to their vast applications, from electronic devices to electric vehicles or energy storage systems (ESSs).

Intensive studies were conducted on the development of high-performance lithium/sodium ion-based electrochemical energy devices because of their high theoretical capacity (~3862 mAh/g for lithium ion-based batteries, and 1165 mAh/g for sodium ion-based batteries). However, despite the tremendous amount of progress made regarding electrochemical energy devices during the last 20 years, they are not suitable for application in a high-performance ESSs or electric vehicles requiring an elongated stability or large battery capacity [4–7]. More specifically, lithium-ion batteries use graphite as an anode material, which is currently available commercially [8–13]. Lithium ions are intercalated/deintercalated during the charge/discharge process to achieve a stable electrochemical reaction, but they have a relatively lower theoretical capacity of 372 mAh/g [14–17]. In sodium-ion batteries, however, it is difficult to apply the same graphite ($d_{002} = 0.334$ nm) used in lithium-ion batteries as a negative electrode material owing to the large size of sodium ions [18]. Also, commercialized supercapacitors using carbon-based materials have problems with low energy density [19,20].

It is logical to pursue novel electrode materials demonstrating their application in electrochemical devices with an improved performance. Among the many capacitive candidate materials, metal oxides received significant attention owing to their high electron storage capability and emissivity through a chemical redox reaction in an energy storage device [21–23]. However, a metal-oxide-based electrode material undergoes severe volume changes during the charge/discharge process (e.g., lithiation/delithiation and sodiation/desodiation) owing to the conversion reactions that occur with the guest ions (Li and Na) [24,25].

In this context, there were tremendous efforts made toward improving the performance of energy devices using metal oxides and addressing the low-capacity problems of carbon-based materials. If metal oxides are applied, the theoretical capacity of the carbon-based materials used in lithium and sodium ion batteries will be increased [26–31].

However, owing to the large volume change in the active metal-oxide material generated during the battery charging/discharging process, such material is separated from the electrode, thereby reducing the electrical contact, hindering the long-term cycle stability. In addition, owing to the low electrical conductivity of metal oxides, the charge transfer resistance of the electrode is high, which is unfortunate in terms of the output density and rate characteristics of the battery [32,33].

To solve the problems of low electric conductivity and large volume change of the active material during the charging/discharging process of the energy device, which are disadvantages of metal oxides, studies were carried out on the formation of a composite with carbon material. As summarized in Figure 1, synergistic effects can be achieved based on the high theoretical capacity of metal oxides, as well as the high electrical conductivity and capacity for accommodating a good volume change of the carbon-based material, thereby improving the performance of the energy device by forming a carbonaceous material-based metal-oxide composite [34–40].

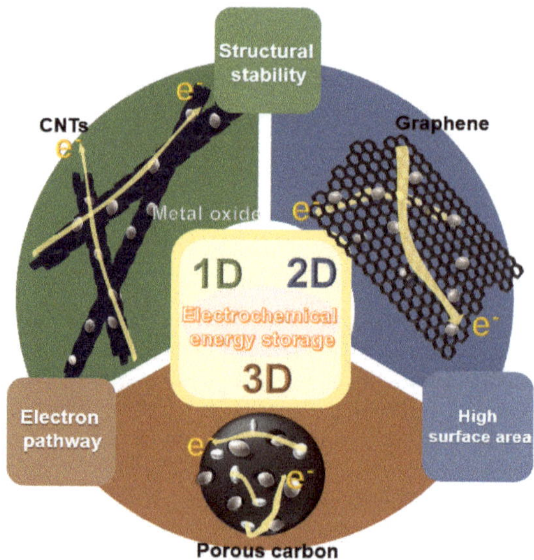

Figure 1. Schematic illustration of carbon/metal-oxide composites of various dimensions for electrochemical energy devices.

In this review, we introduce studies on improving the performance of energy devices through the combination of metal oxides and one-, two-, or three-dimensional (1D, 2D, or 3D) carbon materials. The properties of such materials and composites with metal oxides formed on them are also introduced. In addition, we discuss research applied to energy devices based on the properties of these composites.

2. Carbon/Metal-Oxide Composite Materials

As mentioned, metal oxides have a high electron storage capacity as electrode materials for electrochemical energy devices, but have a problem of low electrical conductivity. To solve these issues, numerous studies were conducted on improving the electrochemical performance of metal oxides through the formation of a composite material.

There are many different composites made of carbon and metal oxide that were applied to electrochemical energy devices. In terms of their structure, they can be divided into 1D–3D structured composites.

2.1. 1D Structured Carbon/Metal-Oxide Composite

One-dimensional structured carbon/metal-oxide composites were extensively studied owing to their unique composite nanotube/fiber structure combined with metal oxide. As a typical architecture, 1D materials of carbon nanotubes (CNTs) or carbon nanofibers (CNFs) typically form a composite structure through a homogeneous decoration of metal oxides around them. As-formed composites exhibit enhanced characteristics coming from their unique structure. Specifically, CNTs with a high aspect ratio in a continuous conductive network can facilitate the charge transport because of the reduced contact resistance with adjacent nanoparticles superior to that of a pristine metal-oxide nanostructure. That is, CNTs can serve as a "facilitated electron transport path or electron highway", which allows a charge transport along the longitude direction [41].

Considering that the conductivity of the electrode material in an energy storage device plays an important role in the power density, CNTs clearly attracted significant attention as co-electrode materials by improving the rate capability of electrochemical energy devices through their excellent charge transport properties [42].

As summarized in Figure 2, the formation of a CNT/metal-oxide composite has several advantages. Firstly, the low electrical conductivity of the metal oxide can be compensated for by the CNTs, which provide a continuous network for electron transport as an "electronic highway" [43]. CNTs can also achieve high electrical conductivity through less mass loading compared to round-shaped nanoparticles by forming a percolation network, which enables a high energy density to be achieved in energy storage devices. Secondly, the high specific surface area of the CNTs increases the contact area between the electrolyte and the electrode. Thirdly, CNTs with an enhanced mechanical toughness compared with pure metal oxide can act as a carbon scaffold through chemical and physical bonding with metal oxides [44].

Zhao et al. compared the electrical conductivities of Co_3O_4 with $Co@Co_3O_4$/CNTs at 18 MPa to investigate the effect of CNT compositing on the material properties. A $Co@Co_3O_4$/CNT nanocomposite was synthesized through an arc discharge and low-temperature oxidation (Figure 3a) [45]. Despite the low original conductivity of Co_3O_4 (7.1×10^{-4} S/m), the electrical conductivity of the $Co@Co_3O_4$/CNTs was improved by 10^4 to 7.6 S/m, suggesting the important role of CNTs as an electronic conductive highway (Figure 3b). TEM images (Figure 3c,d) clearly show the consistent formation of a conduction path for electron transport in $Co@Co_3O_4$/CNT nanocomposites through the formation of CNT conductive networks throughout the $Co@Co_3O_4$ nanoparticles (Figure 3e). It is noteworthy that a $Co@Co_3O_4$/CNT nanocomposite exhibits approximately 10,000-fold higher electrical conductivity than that of pristine Co_3O_4 without CNTs. Such an enhanced electrical conductivity of a metal-oxide/CNT composite enables a facilitated charge transport, leading to an improved electrochemical performance of energy storage devices based on a composite electrode [46].

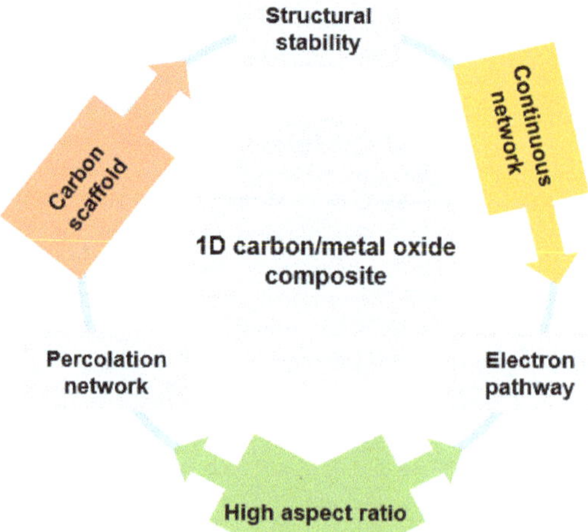

Figure 2. Overview of one-dimensional (1D) carbon/metal-oxide composite for energy storage device.

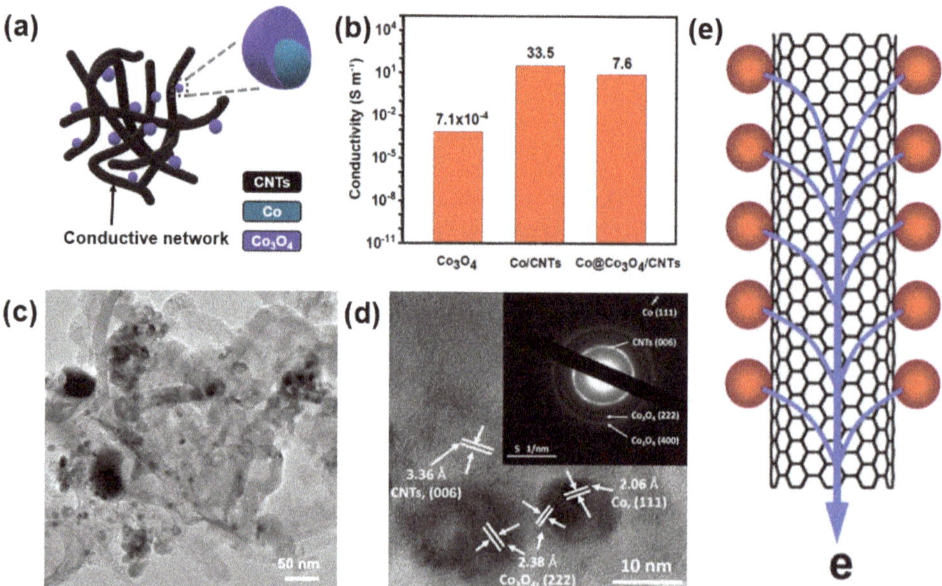

Figure 3. (**a**) Scheme of Co@Co$_3$O$_4$/carbon-nanotube (CNT) composite made up of Co$_3$O$_4$ nanoparticles with CNT networks. (**b**) Enhanced conductivity of Co@Co$_3$O$_4$/CNT composite through the formation of a composite with CNT networks. (**c**) TEM and (**d**) selected area electron diffraction (SAED) images of Co@Co$_3$O$_4$@CNTs. Reprinted with permission from Reference [45]; copyright 2018 American Chemical Society. (**e**) Electronic transport characteristics of a CNT/metal-oxide composite. Reprinted with permission from Reference [43]; copyright 2010 Royal Society of Chemistry.

Reportedly, the volume changes of the electrode material used in electrochemical energy devices (i.e., lithium-ion battery) during the charge/discharge process induce cracks or voids through the applied mechanical strain [47]. It is, therefore, logical to alleviate the mechanical stress on an electrode material to improve the electrochemical performances (i.e., cycle stability). As mentioned previously, intensive studies were conducted on reducing the mechanical strain caused by a volume change of metal-oxide-based electrode materials (Figure 4a,b).

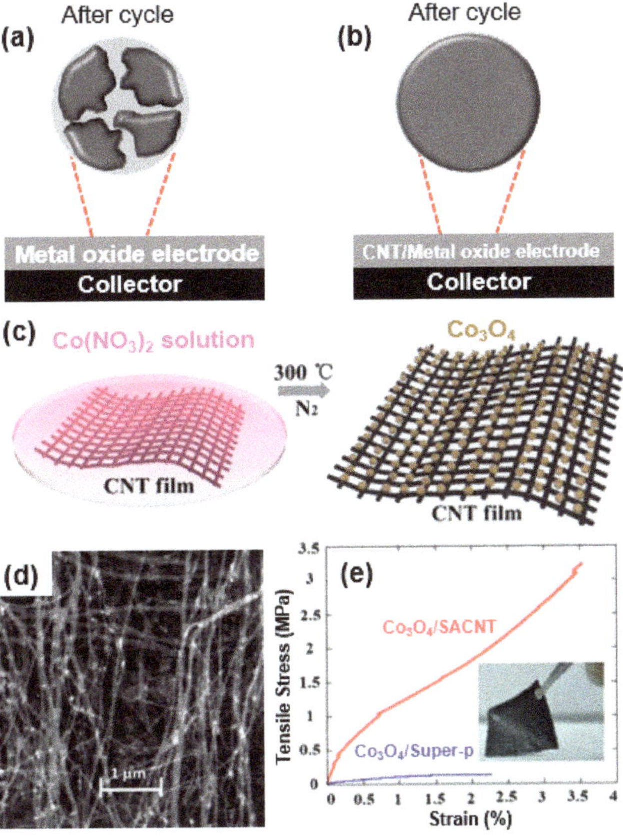

Figure 4. Scheme of morphological change of metal oxide based electrode: (**a**) Cracks and voids formed on metal oxide by volume change-induced mechanical stress during cycling. (**b**) Retained morphology of CNT/metal-oxide-based electrode after cycling. (**c**) Scheme of Co_3O_4/super-aligned CNT (SACNT) composite in nano-sized Co_3O_4 particles grown on SACNT film. (**d**) SEM images of Co_3O_4/SACNT. (**e**) Stress–strain curves of Co_3O_4/SACNT and Co_3O_4/Super-P composites; the inset is a digital photograph of a Co_3O_4/SACNT composite. Reprinted with permission from Reference [48]; copyright 2013 Royal Society of Chemistry.

For instance, He et al. prepared Co_3O_4/super-aligned CNT (SACNT) composites (Co_3O_4/SACNT) in which Co_3O_4 nanoparticles were grown homogeneously on super-aligned CNTs through a pyrolysis method (Figure 4c) to reduce the effect of the mechanical strain occurring during a volume change. As-formed Co_3O_4/SACNT composites exhibited a high Young's modulus (160 Mpa) and tensile strength (3.5 MPa) and strain (3.2%) at break (Figure 4c,d) [48]. The mechanical stability of a Co_3O_4/SACNT electrode is 6.4-fold higher than that of Co_3O_4/Super-P, and the tensile strength and strain at break are 27.2- and 1.6-fold better. Overall, a Co_3O_4/SACNT composite exhibits superior mechanical strength (higher strength and flexibility) to that of a simple mixture composite (Co_3O_4/Super-P) (Figure 4e). Such high flexibility and strength of a Co_3O_4/SACNT composite can be attributed to the high mechanical properties of the SACNT, allowing it to withstand the volumetric changes during the charge/discharge process of the energy storage device, leading to stable cycle performance.

By contrast, in a CNT/metal-oxide composite, the ratio of CNT to metal oxide plays an important role in determining the performance of an electrochemical energy device. As mentioned, there was tremendous effort to enhance the electrical conductivity of metal oxides to improve the performance (energy density) of such devices [49]. Such an enhancement was mainly carried out through the compositing of metal oxide with various nanostructured carbon-based conductors including carbon black, acetylene black, carbon nanotube, and graphene [50]. Because conductive nanostructured carbon facilitates charge transport from the redox sites present in the metal oxide, several studies showed an improved electrochemical performance when applied to an energy device [44].

To make full use of this approach, it is necessary to have the appropriate fraction of metal oxide in the capacitive composite material (CNT/metal oxide). In this respect, as shown in Figure 5a, 1D nanomaterials with high aspect ratios should be used as a conductive filler in the percolation network of the composite, enabling facilitated electron motion in the shortest conductive pathway.

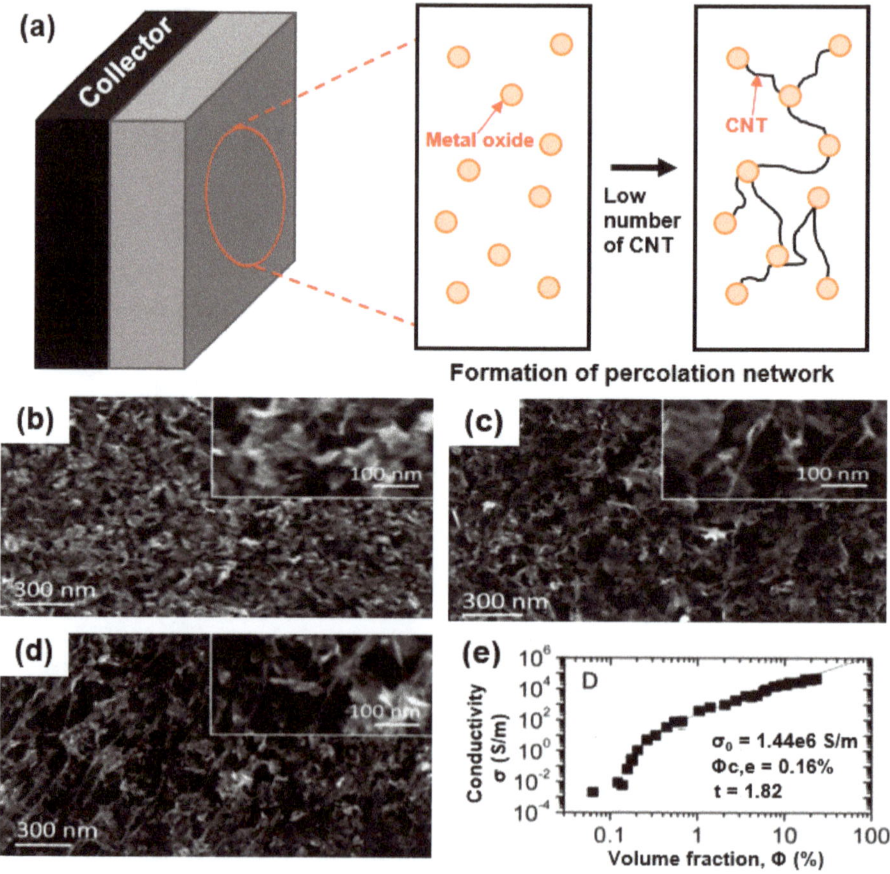

Figure 5. (**a**) Formation of percolation network in electrode material with 1D nanomaterial-based composite. (**b**) SEM images of MnO_2, (**c**) MnO_2/single-walled CNT (SWNT) composite with 1 wt.% SWNT, and (**d**) MnO_2/SWNT composite with 25 wt.% SWNT. (**e**) Percolation characteristics of 1D nanostructured CNTs: in-plane electrical conductivity of MnO_2/SWNT plotted versus SWNT volume fraction. Reprinted with permission from Reference [50]; copyright 2014 American Chemical Society.

As revealed by Higgins et al., the electrochemical performance of a supercapacitor can be significantly improved using a composite electrode of CNTs and a MnO_2 nanoplate [50]. As shown in Figure 5b–d, 2D structured MnO_2 (Figure 5b) hybridized with 1D structured CNTs (1 wt.% CNT (Figure 5c), 2.5 wt.% CNT (Figure 5d)) exhibits multi-dimensional structures. To maximize the electrochemical performance of the capacitive material (MnO_2), it is necessary to optimize the CNT amount through the formation of a smooth conductive pathway in the CNT/MnO_2 composite. Based on the percolation theory, the CNT content is represented by the volume fraction (φ) ($\varphi = V_{NT}/(V_{NT} + V_{MnO2})$). As shown in Figure 5e, the electrical conductivity increased tremendously to 10^5 S/m at φ = 25 vol.% of the CNTs, whereas an electrical conductivity of ~100 S/m occurred with an extremely small number of CNTs (1 vol.%) in the composite. Because the electrical percolation of CNTs can be satisfied beyond the threshold value of the nanoconductor (CNTs) in the volume fraction (φ), the critical volume fraction ($\varphi_{c,e}$) should be achieved in the framework, such as a nanoconductor/insulator composite. As shown in Equation (1), an extremely low electrical conductivity can be exhibited through the percolation theory [50].

$$\sigma = \sigma_0(\varphi - \varphi_{c,e}) \tag{1}$$

where σ_0 indicates the conductivity of the nanoconductor, and n is the percolation exponent.

As shown by Higgins et al., an extremely small value of $\varphi_{c,e} = 0.16$ vol.% can be obtained as a result of the fitting, where σ_0 is 1.4×10^6 S/m and the percolation exponent is 1.82. This low percolation threshold can be attributed to the unique composite structure of 1D structured CNTs with large aspect ratios hybridized with metal oxide. Based on this percolation theory, a substantial conductive pathway can be introduced without significantly reducing the amount of capacitive material (metal oxide) in the composite owing to the high electrical conductivity of a CNT network [44].

In a similar way, when used as an electrode in a storage system, the electrochemical properties of nanoconductor/insulator composites such as CNT/MnO_2 can be effectively improved using the percolation theory to overcome the low electrical conductivity of MnO_2. Specifically, the formation of a composite of metal oxide and CNTs results in an excellent theoretical capacity through an electrochemical redox reaction. This result suggests an effective improvement in the electrochemical performance of MnO_2 through the formation of a composite with a conductive agent. In addition, 1D structured CNTs can further improve the performance because of the excellent mechanical properties and charge transfer characteristics of a CNT network. Such formation of a continuous conductive percolation network can lead to an improved performance of the energy storage device without a noticeable debilitation in the capacity during the charge/discharge process [51].

2.2. 2D Structured Carbon/Metal-Oxide Composite

A 2D structured carbon/metal-oxide composite requires the use of a carbon nanosheet. As a representative carbonaceous 2D nanosheet, graphene has a honeycomb structure of sp^2-hybridized carbon atoms. Owing to the unique 2D structure, graphene has numerous advantages when used to enhance the performance of an energy device (Figure 6) [52]. Firstly, graphene facilitates the electrochemical kinetics owing to its high electrical conductivity and intrinsic carrier mobility. Secondly, because graphene has an extremely large surface area (theoretical surface area of 2620 m^2/g), as well as a very large surface-to-volume ratio, it can provide many active reaction sites and effectively shorten the ion diffusion distance. Thirdly, graphene has a high Young's modulus (1.0 TPa) and tensile strength, which improves the device stability and cycle stability. Despite the many advantages of graphene, as an electrode material, it does not achieve a good performance (or it exhibits rapid performance deterioration during the cycling) owing to its re-stacking during the electrochemical reactions [53].

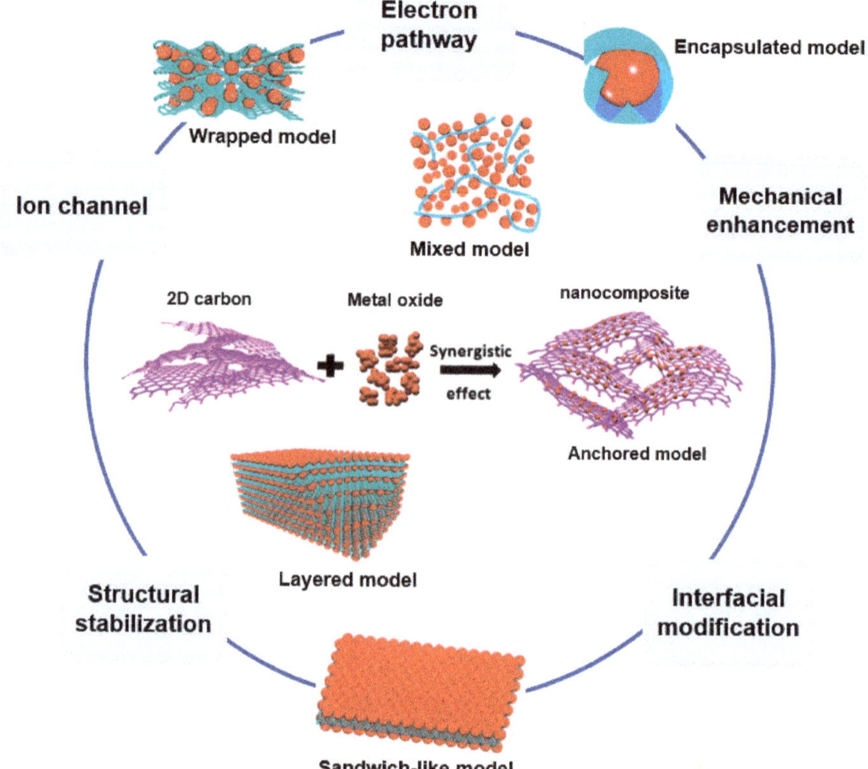

Figure 6. Advantages of graphene/metal-oxide composite in various structures. Adapted with permission from Reference [53]; copyright 2011 Elsevier.

In this context, there were intensive studies on composites of graphene and metal oxide because the advantages of graphene and the benefits of metal oxide as a capacitive material can be achieved simultaneously when used as an electrode in an electrochemical energy device [54].

As described in the previous section, metal oxide used as an electrode material of an electrochemical energy device has limited applications in commercial energy devices owing to performance deterioration from low electrical and ionic conductivity and severe volume changes, inducing a mechanical deformation of the metal oxide during the charge/discharge process [55]. There were many efforts to alleviate the above problems through the compositing of a capacitive material (metal oxide) with graphene because it can enhance the conductivity of capacitive materials, as well as mechanically accommodate the volume change-induced strain occurring during the charge/discharge. As described above, such an enhanced performance of a composite can be attributed to the unique material characteristics of graphene. Specifically, graphene with a modified surface can enhance the structural stability of a metal-oxide-based composite owing to its chemical stability and compatibility with metal oxide, thereby creating physical and chemical connection sites and achieving a high electrochemical performance (high capacity and long cycle stability).

As a common form of a graphene/metal-oxide composite, nanostructured metal oxide dispersed or anchored to a graphene nanosheet is highlighted in Figure 7a [56]. For instance, Su et al. synthesized a graphene/Fe_3O_4 (GN-Fe_3O_4) nanocomposite of Fe_3O_4 nanoparticles (7 nm) uniformly grown on a graphene sheet (Figure 7b) [57]. As shown in the TEM images, no Fe_3O_4 nanoparticles can be observed in an aggregated form or found outside a GN sheet (Figure 7b,c) owing to the strong interaction between the metal oxide and graphene.

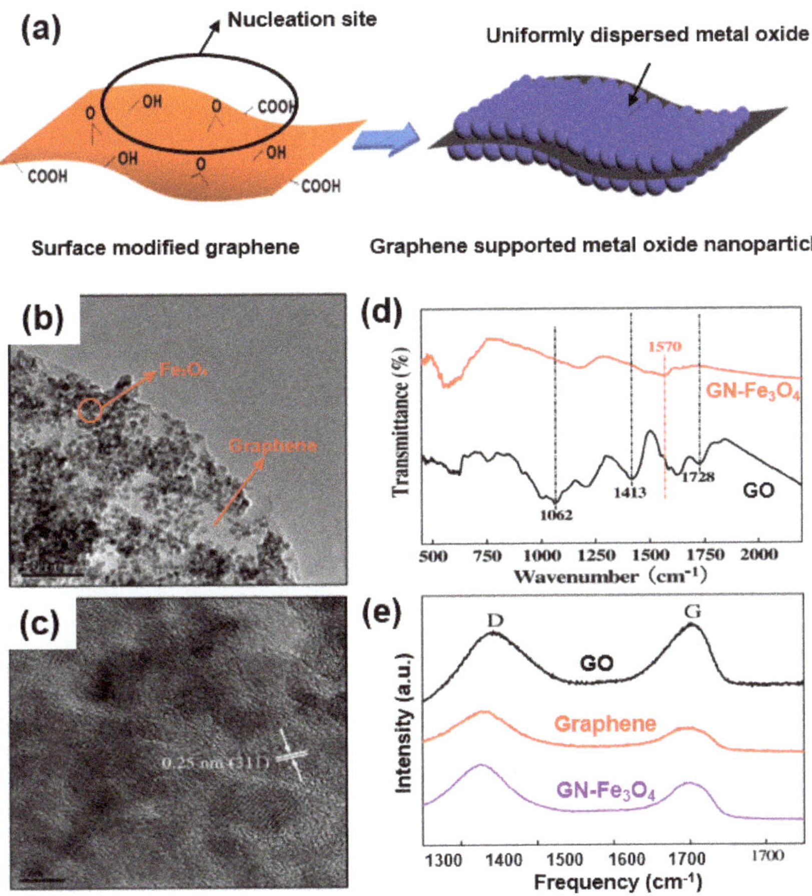

Figure 7. (**a**) Metal-oxide formation on surface-modified graphene. Reprinted with permission from Reference [56]; copyright 2012 Wiley-VCH. (**b**) High-magnification TEM images of graphene/Fe_3O_4 (GN-Fe_3O_4). (**c**) High-resolution (HR)-TEM image of Fe_3O_4 on graphene. (**d**) Fourier-transform infrared (FT-IR) spectra of graphene oxide (GO) and GN-Fe_3O_4. (**e**) Raman spectra of GO (black), GN (red), and GN-Fe_3O_4 nanocomposites (purple). Reprinted with permission from Reference [57]; copyright 2011 American Chemical Society.

The graphene oxide (GO) used in the composite was further analyzed through Fourier-transform infrared (FT-IR) spectroscopy (Figure 7d), the results of which suggest the formation of many functional groups CO (1413 cm^{-1}) and tertiary C–OH groups (1728 cm^{-1}). However, differing from the observation of oxygen-containing groups in GO, no peaks corresponding to all oxygen-containing groups were observed after the synthesis of GN-Fe_3O_4. Raman spectra (Figure 7e) applied to analyze the carbon characteristics of the composite indicate that the G-band of GO shifts to a lower wavenumber in GN-Fe_3O_4. In comparative spectra of GO, graphene, and a composite, the characteristic carbon spectra (G- and D-bands) of GO exhibit a partial shift toward a lower wavenumber in the GN-Fe_3O_4 composite, suggesting the structural retention of the graphene/Fe_3O_4 composite after long cycles of the charge/discharge process in electrochemical energy devices. Such structural stability of a composite is attributed to the strong interactions of Fe_3O_4 and graphene as a conductive network. Chemical analyses (FT-IR spectra (Figure 7d) and Raman spectra (Figure 7e)) indicate that the GN-Fe_3O_4 composite maintains structural and cyclic safety owing to the strong interactions of Fe_3O_4 and graphene. This indicates that the well-reduced graphene can also act as a conductive network.

In a similar vein, metal oxide wrapped with graphene, such as in a wrapped model, encapsulated model, or layered model, received considerable attention owing to its excellent device performance [53]. For instance, Zhou et al. synthesized a graphene-wrapped Fe_3O_4 (GNS/Fe_3O_4) composite as an anode for a lithium-ion battery (Figure 8) [58]. The SEM images and schemes of the synthesized composites are shown in Figure 8a,b. This type of graphene/metal oxide allows the graphene to have an open porous system, and the porous texture of the graphene makes it possible to achieve a flexible electrode because there is no tight connection to other adjacent nanosheets. The metal oxide confined in this flexible porous structure is controlled not only by the porous graphene matrix but also by the metal oxide present between the graphene layers, thereby suppressing the re-stacking of the graphene. Taking full advantage of this, Zhou et al. further compared the particle size of Fe_3O_4 with that of commercial Fe_3O_4 in a GNS/Fe_3O_4 composite before and after 30 cycles to confirm the structural advantages described above. Commercial Fe_3O_4 showed an average particle size of 735 nm at the initial stage and 428 nm after the cycling. In contrast, the GNS/Fe_3O_4 composite showed an initial average particle size of 196 nm (Figure 8c), whereas the sizes of composite particles showed limited growth after the cycling owing to the structural stability of the composite achieved by the wrapped graphene surrounding the Fe_3O_4 (Figure 8d). That is, the average size of the GNS/Fe_3O_4 composite was found to be similar to the average size after cycling owing to the unique graphene-wrapped structure of the GNS/Fe_3O_4 composite [58].

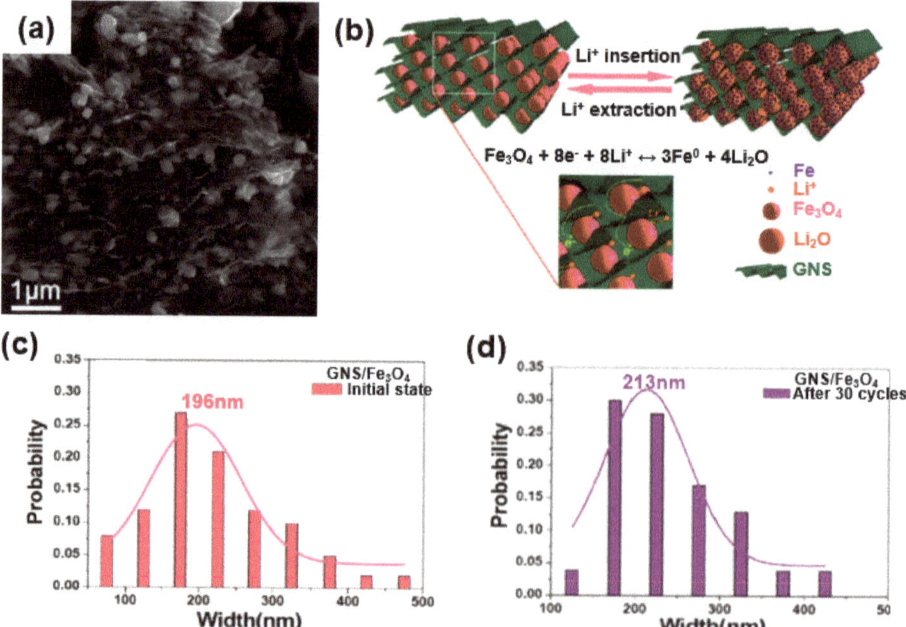

Figure 8. Wrapped model: (**a**) SEM image of graphene-wrapped Fe_3O_4 (GNS/Fe_3O_4) composite. (**b**) Mechanical stress accommodation and ion and electron transport mechanism owing to the structural advantages of the graphene-wrapped model during an electrochemical reaction. (**c**) Size distributions of the GNS/Fe_3O_4 composite during the initial state and (**d**) after 30 cycles. Reprinted with permission from Reference [58]; copyright 2010 American Chemical Society.

By contrast, a novel type of graphene (holey graphene) as a 2D structured porous material received increased attention owing to its many advantages when used as a material in an electrochemical energy device [59].

Firstly, as described in Figure 9a, holey graphene has a high surface area, which can provide more active sites and enable the effective wetting and penetration of the electrolyte through the pore channel owing to the porous structure [60]. In addition, porous graphene with interconnected structures can provide a continuous charge transfer pathway and short ion transport pathway during an electrochemical reaction [61]. Note that non-porous graphene also achieves ion diffusion near the edge or nanosheet junction, although the length of the ion diffusion has a longer transport path than direct ion diffusion occurring in porous graphene. When forming a 2D porous graphene and metal-oxide composite for electrochemical energy storage, the composite has the advantages of a reduced path length, large surface area, and facilitated electrolyte penetration, and it can effectively accommodate the strains from the volume change of metal oxide during the charge/discharge process owing to the highly porous structure [59].

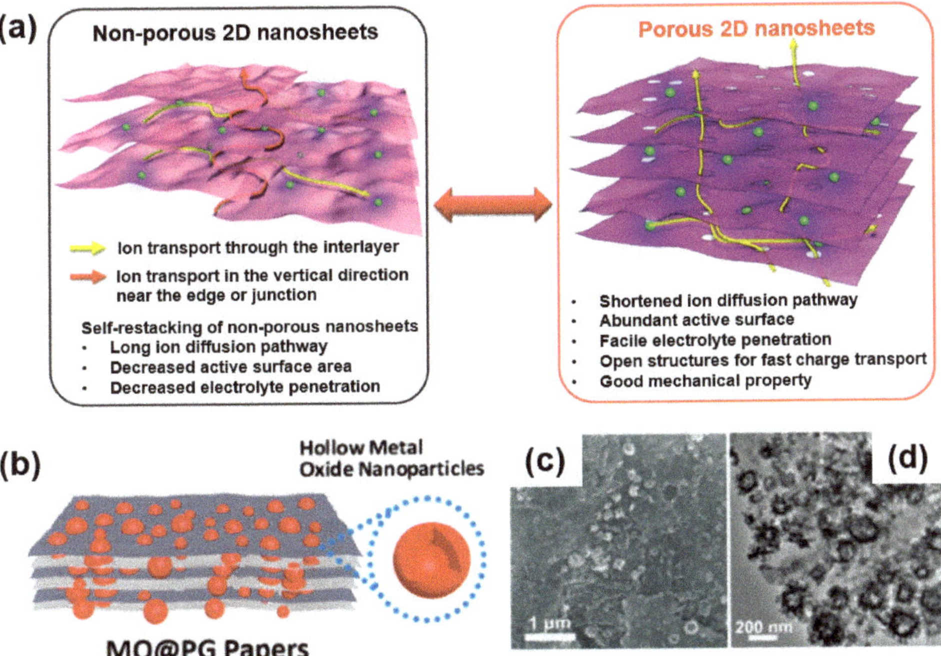

Figure 9. (**a**) Schematic of advantageous features of porous two-dimensional (2D) structured material for electrochemical energy device. Reprinted with permission from Reference [53]; copyright 2018 Wiley-VCH. (**b**) Schematic image of porous graphene and metal-oxide composite (MO@PG papers). (**c**) SEM and (**d**) TEM images of MO@PG composite after 4000 cycles at 5 A/g. Reprinted with permission from Reference [62]; copyright 2018 American Chemical Society.

For instance, Zhang's group demonstrated a composite of metal oxide and porous graphene (MO@PG) to improve the electrochemical performance as a cathode material of a lithium-ion battery (Figure 9b) [62]. As shown in SEM (Figure 9c) and TEM (Figure 9d) images, the MO@PG nanocomposite of MO is embedded in the pores of porous carbon rather than the basal plane of the graphene. In addition, as graphene forms a conductive network in the MO@PG composite, metal oxide embedded in a holey graphene composite retains structural pore stability after a lengthy electrochemical test (after 4000 cycles).

2.3. 3D Structured Carbon/Metal-Oxide Composite

The specific capacity, energy, and power density need to be maximized by improving the quality and quantity of the active materials embedded in an electrochemical energy storage device [44]. In this respect, it is important to load as much active material as possible in the electrode to enhance the capacitance, energy, and power in the device [52]. Hence, there were numerous studies conducted to load a large amount of active material into an electrode of an energy storage device without deteriorating the electrochemical performance (specific capacity, rate capability, energy density, and power density) [27]. Specifically, a significant amount of research on composites of porous metal oxide with 3D structured carbon was undertaken to address these challenges.

The characteristics of 3D porous carbon (tunable pore size, wall thickness, high pore volume and surface area, and inner connection among the pore channel) make them extremely suitable for use as a matrix of an active material [27]. As summarized in Figure 10, hierarchical porous carbon, which has macroporous, mesoporous, and microporous properties, can be used to improve the electrochemical performance of interconnected pores of different sizes in an energy storage device. In a composite of metal oxide and 3D carbon, 3D structured porous carbon plays an important role because it forms a hierarchical structure composed of different sized pores (macropores, mesopores, and micropores). Such a hierarchical structure of different sized pores can improve the electrochemical performance of an energy device through unique actions with multiple functions including ion-buffering reservoirs, short ion pathways, ion confinement, and the effective suppression of the volume change during electrochemical reactions. Because metal-oxide nanomaterials are well embedded in the pore structure of a composite, interconnected pores can effectively facilitate the electrolyte penetration and ion diffusion, leading to an enhanced structural stability [63]. In addition, the enlarged surface area of a composite through a porous structure widens the active site by creating numerous contacts with the electrolytic solution, enabling the loading of a large amount of metal oxide to realize a high energy and power density [64].

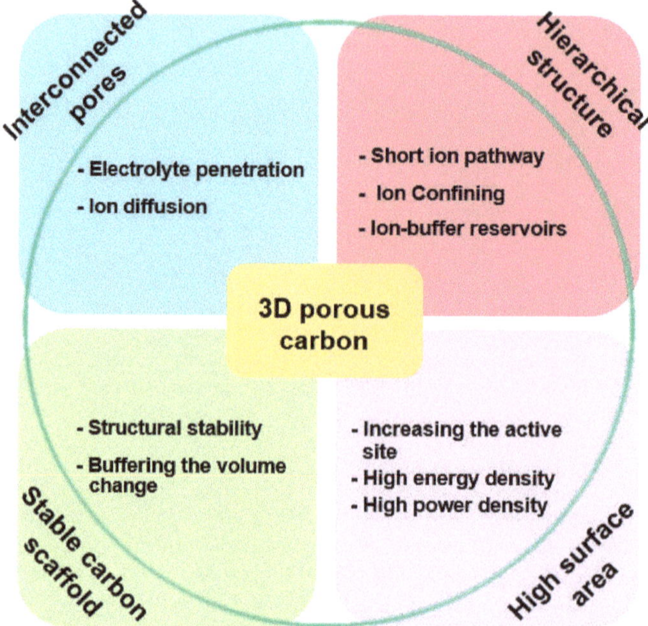

Figure 10. Structural properties of three-dimensional (3D) porous carbon and improved properties as an electrochemical energy device electrode of a 3D porous carbon/metal-oxide composite.

For instance, Wang's group reported the use of hierarchical porous graphitic carbon as an electrode material for electrochemical capacitors (Figure 11) [65]. The as-formed porous carbon exhibits a high porosity (high surface area and pore volume), suggesting its use as an appropriate electrode material for electrochemical energy devices. Specifically, the Brunauer–Emmett–Teller (BET) surface area, pore volume, and average pore diameter of porous carbon were measured to be 970 m^2/g, 0.69 cm^3/g, and 2.85 nm, respectively. In addition, the micropore volume and its ratio (micro volume in the total pore volume) were calculated as 0.3 cm^3/g and 0.43, respectively. More specifically, the nitrogen isotherm curves (Figure 11a) display a combined formation of type I in microporous materials and type II in macroporous materials, whereas hysteresis loops are well characterized as mesoporous materials. Within the pore size distribution (Figure 11b), three types of pores, namely micropores (1–2 nm), mesopores (5–50 nm), and macropores (60–100 nm), were observed, suggesting the hierarchical structure of porous carbon. As shown in Figure 11c, this hierarchical structure of 3D porous carbon can enhance the performance of electrochemical energy storage via the synergic functions of various types of pores (micro-, meso-, macropores) in different sizes. Macropores, which have a diameter of 1 μm in the composite, can extend into the particle to form ion-buffering reservoirs. Furthermore, the wall thickness of the porous graphitic carbon around the core is estimated to be less than 100 nm, indicating reduction of the diffusion distance of the electrolyte by this thin wall. The mesopore structures provide a short ion transport pathway through the wall. The micropore structures suggest facile confinement of the ions.

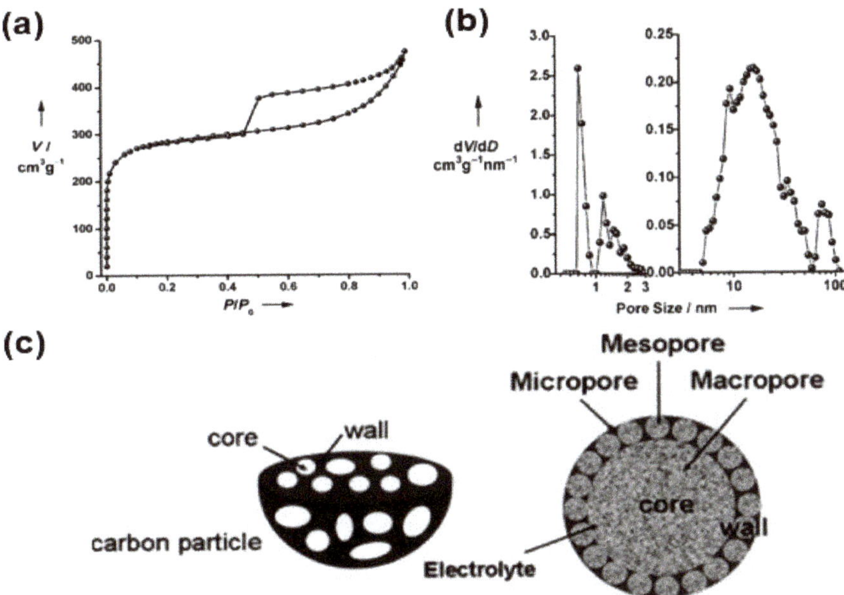

Figure 11. (a) Nitrogen adsorption–desorption isotherm of hierarchical porous graphitic carbon. (b) Pore size distribution of hierarchical porous graphitic carbon. (c) Hierarchical porous carbon textures with interconnected pores of macropores, mesopores, and micropores. Reprinted with permission from Reference [65]; copyright 2008 Wiley-VCH.

Three-dimensional porous carbon can act as an extremely good matrix for a composite with metal oxide for use as an electrode material [29]. For instance, Cao et al. synthesized an MnO/C composite with nanoparticles uniformly embedded in a porous carbon matrix to be used as an anode of a lithium-ion battery [66]. As shown in the scheme (Figure 12a) of the energy storage characteristic of the MnO/C composite, this structure facilitates the penetration of an electrolyte through an open space between internally connected pores and the aligned layers, as well as the effect of shortening the diffusion path of the Li ions during an electrochemical reaction. A high-magnification SEM image (Figure 12b) reveals a layered structure of the composite with a lotus-root-like hollow interior structure based on the distinct contrast between the outer wall and the middle cavity region. In addition, the MnO_2 metal oxide in the carbon matrix effectively reduces the mechanical stress caused by a volume change occurring from the lithiation/delithiation during the charging/discharging process. Such an advantageous structural feature leads to a reduced resistance, as shown in the electrochemical impedance spectroscopy (EIS) results (Figure 12c). Specifically, the EIS results of the MnO/C–1.0% PVA composite after 250 cycles at 0.75 A/g show that the charge transfer (R_{ct}) and Warbug impedance (Z_w) have lower values compared to those of pristine MnO. These results indicate that the 3D porous structure of a MnO/C composite improves the diffusion of lithium ions and electrons compared to pure MnO, suggesting an improved electrochemical performance in energy storage devices. Similarly, Lu et al. synthesized MnO nanoparticles with N-doped carbon coatings to improve the cycling performance and rate capability of such a device [67].

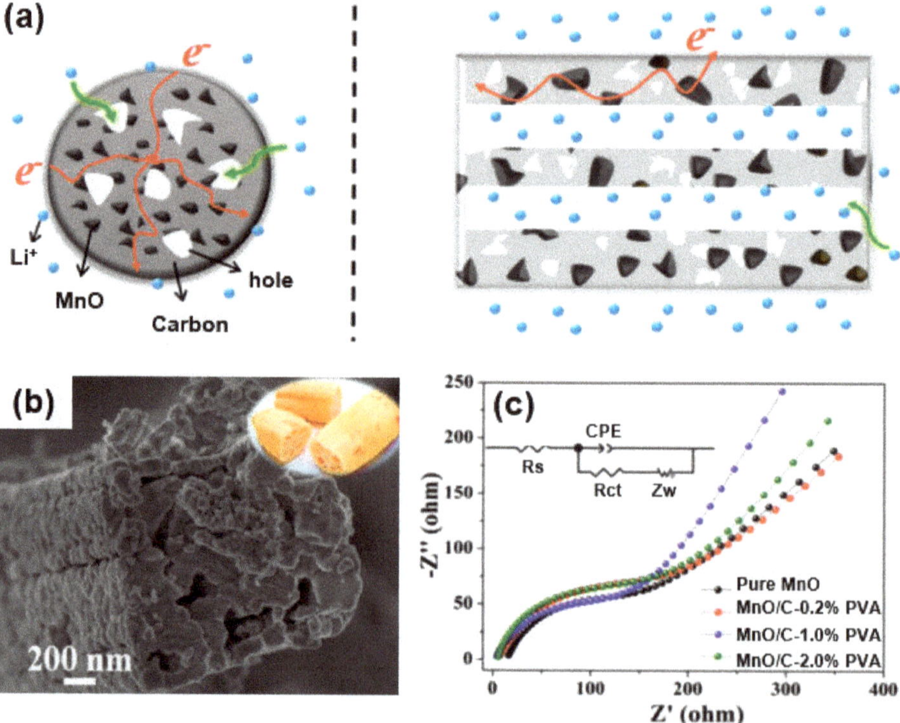

Figure 12. (a) Energy storage characteristics of MnO/C hybrids: (left) cross-sectional view and (right) axial side cutaway view. (b) High-magnification SEM image of MnO/C composite. (c) Nyquist plot of MnO/C composite after 250 cycles at 0.75 A/g. Reprinted with permission from Reference [66]; copyright 2017 American Chemical Society.

Another advantage of a composite of 3D structured porous carbon and metal oxide as an electrode of an energy storage device is the effective alleviation of the mechanical expansion and contraction of the metal oxide during the electrochemical reactions because of the hierarchical pore structure of 3D structured carbon. Such a hierarchical structure can enhance the performance of the electrode (cycle stability) [68].

For instance, Zhang et al. showed a composite (Fe_3O_4@C) in which Fe_3O_4 was confined in a nanoporous carbon framework as an anode of a lithium-ion battery through a self-assembly of $Fe(NO_3)_3$/resol/F127 [69]. SEM and TEM images (Figure 13a–f) show different morphologies of Fe_3O_4@C-1 (Figure 13a,b), Fe_3O_4@C-2 (Figure 13c,d), and Fe_3O_4@C-3 (Figure 13e,f) depending on the amount of Fe_3O_4. As shown in the nitrogen sorption isotherm curve (Figure 13g) and pore structure analyses (Figure 13h), it was found that the pore structure of Fe_3O_4@C depends on the iron precursor ($Fe(NO_3)_3$/resol/F127) content in the self-assembly process. Specifically, in the case of Fe_3O_4@C-3, a large amount of Fe_3O_4 loaded into the pores results in a decreased pore volume and surface area. In the case of Fe_3O_4@C-2, a proper amount of Fe_3O_4 was confined to the porous carbon and showed the largest surface area and pore volume. These composites enable facile contact of the active material with the electrolyte and an effective inhibition of the volume change generated during the lithiation/delithiation process. They also prevent the Fe_3O_4 from aggregating or separating from the collector, enabling a large amount of Fe_3O_4 loading into the carbon framework with an increased pore volume and surface area owing to the porous structure [70].

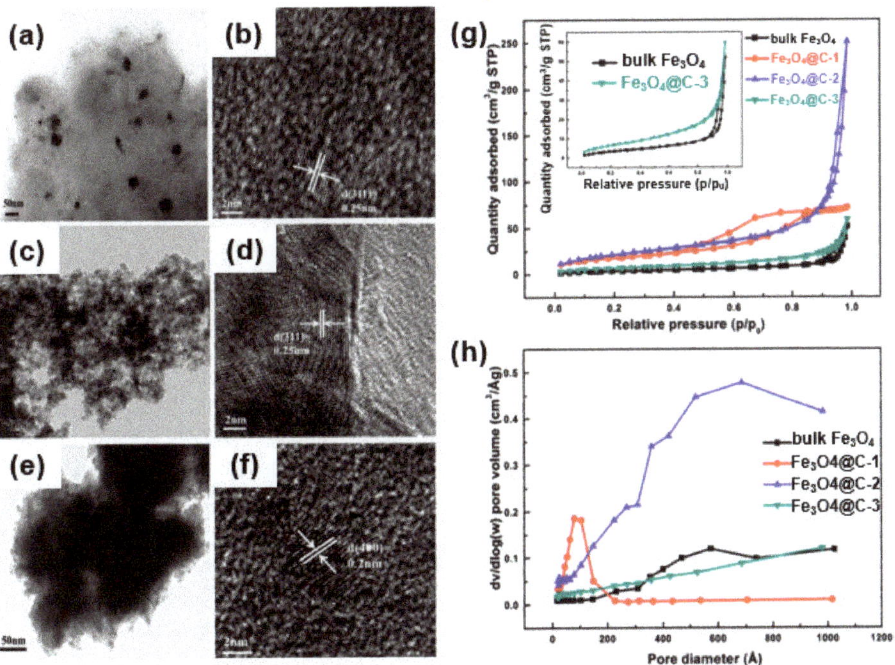

Figure 13. TEM images of Fe_3O_4 confined in various nanoporous carbon frameworks: (**a,b**) Fe_3O_4@C-1, (**c,d**) Fe_3O_4@C-2, and (**e,f**) Fe_3O_4@C-3. (**g**) Nitrogen adsorption–desorption isotherms for Fe_3O_4 and Fe_3O_4@C. (**h**) Pore size distribution. Reprinted with permission from Reference [69]; copyright 2015 Royal Society of Chemistry.

In a similar vein, Han et al. synthesized SnO_2@CMK-3 composites in the form of ultrafine SnO_2 particles encapsulated in tubular mesoporous carbon as electrodes for lithium-ion batteries (Figure 14) [71]. Figure 14a,b show TEM images of a SnO_2@CMK-3 composite before and after 100 cycles. As revealed in the TEM figures, the ordered mesostructure was observed to be in a well-maintained form in both samples, whereas the nanoparticles (SnO_2) were properly and uniformly encapsulated in the carbon matrix without aggregation. Owing to the high porosity of a porous carbon matrix, it is possible to load large amounts of SnO_2 nanoparticles into the pores, which effectively accommodates the volumetric change of SnO_2 during lithiation (Figure 14c).

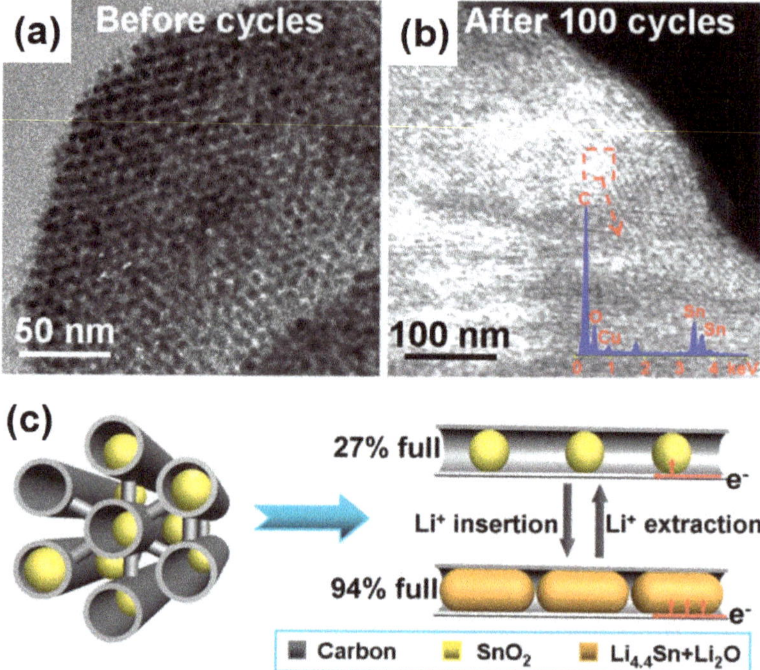

Figure 14. (**a**) TEM images of ultrafine SnO_2 particles encapsulated in tubular mesoporous carbon (SnO_2@CMK-3) before cycling. (**b**) Scanning TEM (STEM) images of SnO_2@CMK-3 after 100 cycles at 200 mA/g. (**c**) Adequate space of the mesopore channels acting as a "buffer zone" for accommodating the volume expansion and preventing the pulverization of SnO_2 nanoparticles during the charging/discharging process. Reprinted with permission from Reference [71]; copyright 2012 Royal Society of Chemistry.

Thus, 3D porous carbon with a high surface area, pore volume, and an internally well-connected hierarchical structure allows a large amount of loading of capacitive materials through the formation of a composite with metal oxide in an energy storage device. In addition, enabling an efficient contact with the electrolytes and structural stability during the cycling effectively facilitates the diffusion of lithium ions and electrons, resulting in an improved electrochemical performance.

3. Applications of Carbon/Metal-Oxide Composite Materials

3.1. Batteries

3.1.1. Lithium-Ion Batteries

Since the introduction of the first commercial lithium-ion batteries by Sony, numerous studies were conducted to improve the performance of electrochemical energy devices. Currently, the use of lithium-ion batteries is becoming increasingly widespread owing to their many advantages as energy storage devices, including a high capacity, high energy density, rate-response characteristics, and long cycle life.

Lithium-ion batteries (LIB) consist of four parts: anodes, cathodes, separators, and electrolytes. Li-ion batteries, which are currently in commercial use, are composed of $LiCoO_2$, $LiMn_2O_4$, and the like, in which a positive electrode is inserted with lithium, and graphite is used as a negative electrode. Among these options, graphite (anode) forms LiC_6 and has a theoretical capacity of 372 mAh/g during the charge/discharge of a lithium-ion battery [72]. However, because of its low theoretical capacity as an energy storage device, there were many studies conducted on metal oxides (Fe_2O_3, SnO_2, Fe_3O_4, Co_2O_3, etc.) with a high theoretical capacity. The conversion reaction of metal oxides through the lithiation/delithiation of lithium ions undergoes the following reaction (Equation (2)) [23]:

$$MO_x + 2xe^- + 2xLi^+ \leftrightarrow M^0 + xLi_2O \tag{2}$$

Despite the high theoretical capacity of metal oxide originating from the conversion reactions during the lithiation/delithiation, the metal-oxide-based LIB suffers from a low electric conductivity and low Coulombic efficiency, as well as a rapid capacity reduction caused by a large volume expansion/reduction of the active material. As mentioned in the previous section, new approaches addressed the above problems. As a representative method, carbon-based composites are currently prevailing in which a carbonaceous material is used as a matrix to form a composite with metal oxide. Such an approach can result in a high electrical conductivity of the carbon material, enabling a synergistic effect of the high theoretical capacity of the metal oxide.

As described in the previous section, metal-oxide/carbon composites, including 1D, 2D, and 3D structured composites, were intensively studied through various synthetic approaches.

Firstly, among the various types of carbon/metal-oxide composites, 1D structured CNT/metal-oxide-based composites [73] or carbon-fiber/metal-oxide-based composites [74,75] formed by bonding with a metal oxide were intensively introduced for use as anodes for energy devices (LIBs). Typically, CNTs and networks serve as a buffer layer to accommodate the mechanical strain caused by large volume changes of metal oxides, and they can act as a conductive path (electronic highway) to improve the electrical conductivity of the composite [76]. CNTs also prevent the aggregation of nano-sized metal oxides dispersed on the composites.

For instance, Luo et al. presented the performance of LIB cathodes using a composite of Mn_3O_4 mingled with well-aligned carbon nanotubes (SACNTs) [77]. Figure 15a shows that the SACNT film serves as a conduction path for electron migration and acts as a matrix to ensure the adequate dispersion of Mn_3O_4 nanoparticles. As displayed in the TEM image (Figure 15b), as-formed composites contain 4 nm of Mn_3O_4 nanoparticles, whereas the Mn_3O_4/SACNT composite exhibits a capacity of ~342 mAh/g at a high current density of 10 C (Figure 15c). Note that, as an advantageous feature, the 1D structured composite as a free-standing film does not require a conductive agent (e.g., super-P), a binder, or a cathode substrate. In a similar approach, Zhou et al. synthesized a composite of single-walled carbon nanotubes (SWCNTs) with Fe_2O_3. The composite exhibits a high capacity (1243 mAh/g) at 50 mA/g, owing to a mitigated lithium-ion path toward the dispersed Fe_2O_3 nanoparticles in the SWCNT network. In addition, the alleviation of the volume change of the composite occurring during the delithiation process results in an elongated capacity retention at an increased energy density for long cycles. Similarly, Li et al. used multi-walled CNTs (MWCNTs) as a matrix for anchoring Fe_3O_4 nanoparticles to improve the electrical conductivity and structural stability [78].

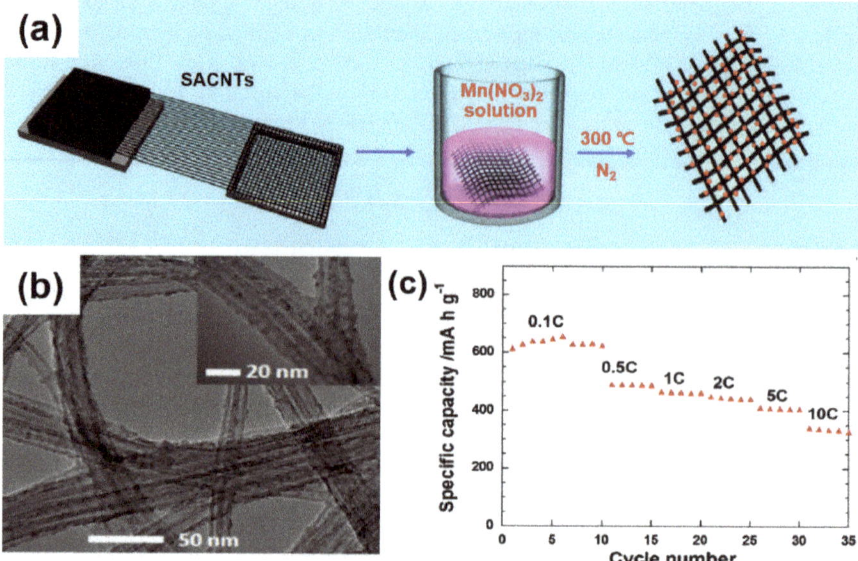

Figure 15. (**a**) Schematic illustration of the fabrication of the Mn$_3$O$_4$/SACNT composite. (**b**) TEM image of the SACNT/Mn$_3$O$_4$ composites containing 29 wt.% Mn$_3$O$_4$. (**c**) Rate capability of the Mn$_3$O$_4$/SACNT composite at various current densities. Reprinted with permission from Reference [77]; copyright 2013 Elsevier.

Metal oxides composited with a 2D structured carbon (graphene, graphene oxide, etc.) were intensively studied as energy storage materials owing to their many structural advantages, leading to an improved LIB performance. More specifically, it is logical to expect an improved LIB performance for a 2D structured carbon composite because of the unique material characteristics of graphene, including high electrical conductivity, excellent mechanical properties, and wide specific surface area [53].

In this regard, Li et al. synthesized a composite of Co$_3$O$_4$ and graphene, exhibiting a theoretical capacity of 890 mAh/g (approximately twice the theoretical capacity of graphite) through a hydrothermal synthesis followed by an annealing process [79]. The as-formed composite shows a multifunctional structure of 1D Co$_3$O$_4$ nanowires grown on a 2D graphene membrane. Note that a unique composite structure was prepared through the sequential construction processes where the graphene membrane serves as a conductive substrate and a Co$_3$O$_4$ nanowall is grown on a graphene nanosheet. Such a composite-based electrode eliminates the need for a copper current collector and a binder to increase the energy density of the battery, leading to an enhanced energy density of the active materials (Figure 16a). As confirmed through SEM images (Figure 16b), Co$_3$O$_4$ nanowires are grown on a graphene membrane. In the elongated cycle test, the composite showed good stability (Coulomb efficiency of ~100% during 500 charge/discharge cycles and a capacity of 600 mAh/g or higher (Figure 16c).

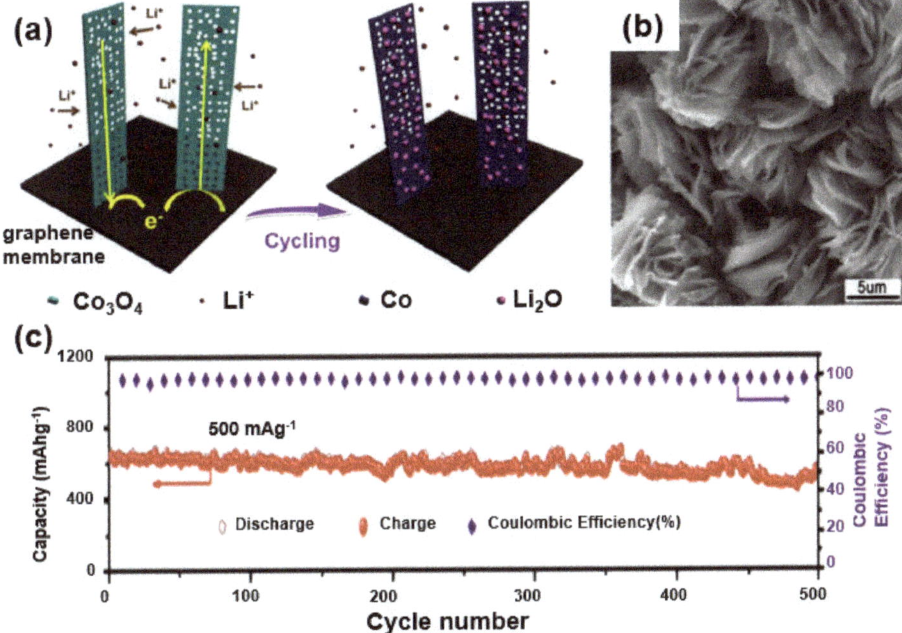

Figure 16. (**a**) A schematic of the electrochemical reaction of a Co_3O_4 nanowall/graphene membrane composite. (**b**) SEM image of Co_3O_4 nanowall/graphene membrane composite. (**c**) Long-term cycling performance of Co_3O_4 nanowall/graphene membrane composite. Reprinted with permission from Reference [79]; copyright 2014 Elsevier.

In another approach, Hu et al. published a study in which reduced graphene oxide (rGO) was surrounded by Co_3O_4 porous nanofibers to form a composite [80]. The porous nanofiber encapsulates the Co_3O_4 nanoparticles and the interconnected graphene oxide sheet on the interconnected formation, thereby suppressing the large volume change of the Co_3O_4 particles generated during the battery charging/discharging process, as well as accelerating the movement of electrons and lithium ions. Based on the structure of this composite, a dose of ~900 mAh/g was reported at a current density of 1 A/g and a capacity of ~600 mAh/g at a high current density of 5 A/g.

In addition, some studies demonstrated an improve battery performance by forming a composite in which metal-oxide nanoparticles are placed between the reduced graphene oxide sheets [81] or through a formation on a graphene sheet in the form of metal-oxide nanoparticles and nanorods [82,83]. In addition to the use of graphene-like carbon materials, a study was conducted on the application of carbon sheets as a matrix of a metal-oxide composite. Fu et al. used carbon sheets with glycine as a carbon precursor to make a composite with metal oxide [84]. Chromium (II) oxide nanoparticles are anchored onto the carbon sheets, enhancing the electrochemical performance.

Among these carbon-based composites formed using metal oxides, a 3D carbon-based composite with a porous structure is formed to enlarge the specific surface area, thereby improving the contact area with the electrolyte and facilitating the electron transfer. To improve the lithium-ion battery performance, the 3D carbon network serves as a connection path through which electrons move, and it acts as a buffer for the large volume change of the metal oxide generated during the battery charging/discharging process. Wang et al. improved the mechanical properties and ionic and electrical conductivities of the composites by synthesizing the composites in the form of MnO dispersed in a 3D porous carbon network, and obtained a capacity of 560.2 mA/g at a high current density of 4 A/g [85].

Han et al. reported a composite of SnO_2 nanoparticles embedded in ordered mesoporous carbon (CMK-3 or CMK-5) as a carbon matrix [71]. A large amount of SnO_2 nanoparticles were added to CMK-5, which has a tubular mesoporous channel with a large pore volume and a thin carbon wall, forming a composite that improves the battery performance. Compared to CMK-5, the relatively thick carbon-walled CMK-3 is unable to effectively trap SnO_2 in the mesoporous channel, leading to a relatively lower battery performance (Figure 17a). As confirmed through battery tests, SnO_2-80@CMK-5 with an increased loading of SnO_2 showed a capacity of 1039 mAh/g after 100 cycles at 200 mA/g (Figure 17b). As shown in the TEM image (Figure 17c), an even distribution of SnO_2 nanoparticles with a hexagonal structure was observed among the carbon matrices. Such a unique composite structure of SnO_2 nanoparticles uniformly dispersed in the mesoporous channel of CMK-5 can effectively alleviate the large volume changes in the mesoporous carbon wall of CMK-5 during the charge/discharge processes, leading to an enhanced electrochemical performance (Figure 17c).

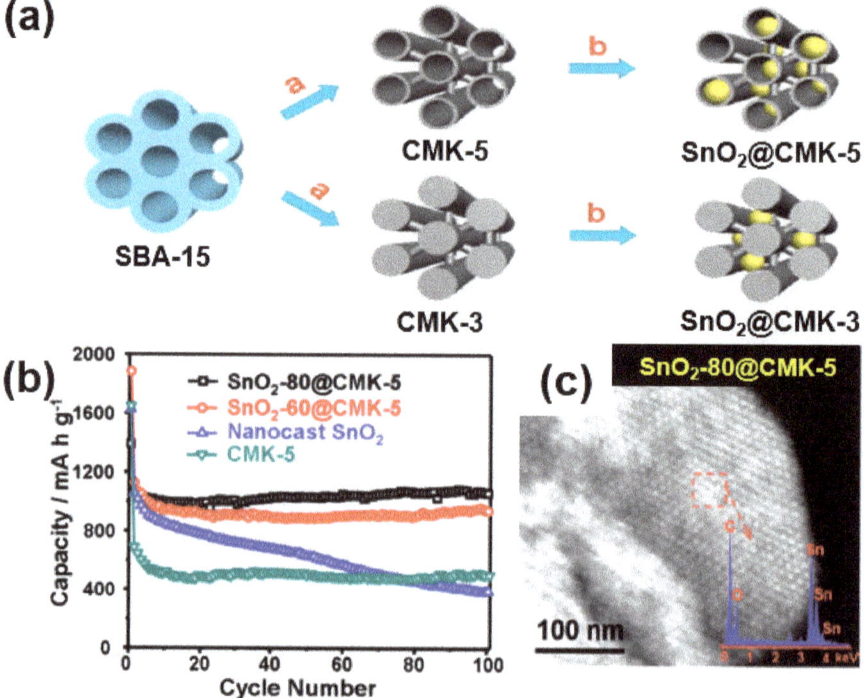

Figure 17. (**a**) Schematic illustration of fabrication of composites of SnO_2 nanoparticles embedded in ordered mesoporous carbon (SnO_2@CMK-5 and CMK-3). (**b**) Cycling performance of SnO_2-80@CMK-5, SnO_2-60@CMK-5, nano-casted SnO_2, and CMK-5. (**c**) STEM image of SnO_2-80@CMK-5 composite; the inset shows STEM energy-dispersive X-ray spectroscopy (EDS) analysis. Reprinted with permission from Reference [71]; copyright 2012 Royal Society of Chemistry.

3.1.2. Sodium-Ion Batteries

As discussed in the previous section, although LIBs can achieve a high energy density, lithium, an essential element of an LIB, is quite expensive [86]. In this regard, sodium (Na)-ion batteries (SIB) recently gained attention owing to the abundance and lower cost of sodium compared to lithium. However, sodium-based batteries have several disadvantages compared to lithium-based batteries, including a lower standard electrochemical potential (sodium, 2.71 V; lithium, 3.04 V), a relatively larger ionic radius (Na^+, 1.02 Å; Li^+, 0.59 Å), a relatively slow dynamic response, and a smaller energy density [86,87]. In particular, the almost double ionic radius of sodium results in a much larger volume change of the active material than that of lithium during the charge/discharge process. In this context, there were intensive studies addressing the above issues of SIBs through the formation of composites of metal oxide and carbonaceous materials with a 1D, 2D, or 3D structure.

Firstly, CNTs as 1D structured carbon materials were used as a matrix in which metal-oxide nanoparticles are well adhered/distributed on the surface of the carbon nanotubes. Using CNTs as a fast-moving path for the electrons, the dispersed metal-oxide nanoparticles can quickly cause an electrochemical reaction of sodium ions and electrons. In addition, CNTs can retard the decrease in battery capacity caused by a large volume change of the metal-oxide particles during the charge/discharge processes.

For instance, Wang et al. reported an improved performance of SIBs by adding SnO_2 nanoparticles to CNTs into a composite formation [88]. Rahman et al. also studied the application of Co_3O_4, which was extensively studied in LIBs, in sodium-ion batteries [89]. A Co_3O_4/CNTs composite was formed using a liquid plasma (Figure 18a). SEM and TEM images show that the CNTs are wrapped in a composite of Co_3O_4 particles (Figure 18b). The CNTs are wrapped around the Co_3O_4 nanoparticles to improve the contact properties of the CNTs by reducing the volumetric change in the active materials during the charge/discharge processes and by increasing the electrical conductivity. It can be seen from the EIS measurement results that the charge transfer resistance of the Co_3O_4/CNT composite is relatively smaller than that of Co_3O_4, which indicates that the conductivity of the composite battery is improved (Figure 18c). As a result of testing the charge/discharge cycles, it was confirmed that the capacity of the Co_3O_4/CNT composite is maintained at 403 mAh/g even after 100 cycles (Figure 18d). The results indicate that the performance of an SIB can be improved through the composite formation of CNTs with metal oxides.

To improve the performance of the SIBs, it is necessary to form a composite with a carbon material that acts as a small nanoparticle-sized metal oxide, creating a passage through which the electrons can properly move. Graphene not only increases the electrical conductivity but also acts as a matrix in which the metal-oxide particles can be uniformly dispersed and bonded. Fe_2O_3, one of the metal oxides applicable to sodium-ion batteries, has the following reaction (Equation (3)) [90]:

$$Fe_2O_3 + 6Na^+ + 6e^- \leftrightarrow 2Fe^0 + 3Na_2O \tag{3}$$

Through such a reaction, the Fe nanoparticles are dispersed in the Na_2O matrix and cause a large volume change.

Liu et al. demonstrated an Fe_2O_3/rGO composite-based SIB with an improved device performance [91]. Specifically, they prepared a composite of Fe_2O_3 nanoparticles homogeneously dispersed on the surface of reduced graphene through a microwave-assisted method, and showed that the capacity of 289 mAh/g was maintained at 50 mA/g after 50 cycles. Such an enhanced performance can be attributed to the reduced graphene, which alleviates the volume change of the Fe_2O_3 nanoparticles and compensates for the low electrical conductivity of the metal oxide. In addition, the large specific surface area of the reduced graphene further improves the contact between the sodium ions and the composite, facilitating a charge transfer reaction, resulting in an improved battery performance.

When nano-sized metal-oxide particles such as SnO_2, exhibiting an alloying/dealloying reaction, are dispersed on a graphene surface to form a composite, and a charge/discharge process is performed, the reaction is as shown in Figure 19A. The sodiation/desodiation process can be expressed as follows (Equations (4) and (5)):

$$SnO_2 + 4Na \rightarrow Sn + 2Na_2O \tag{4}$$

$$Sn + 3.75Na \leftrightarrow Na_{3.75}Sn\ (Na_{15}Sn_4) \tag{5}$$

When sodiation occurs, the metal-oxide particles and Na^+ alloy with each other, causing the volume to expand and decrease when a dissociation of alloy occurs. The SnO_2/rGO composite maintains a stable solid electrolyte interface (SEI) during this process, resulting in improved battery performance [92].

In a similar approach, Su et al. prepared SnO$_2$@graphene composites using a hydrothermal process for application toward SIBs [93]. SEM and TEM images show the homogeneously dispersed SnO$_2$ nanoparticles on graphene (Figure 19B-a). The performance of the composite was confirmed through battery charge/discharge tests. As a result, the SnO$_2$@graphene composite showed a capacity of 1942 mAh/g during the first cycle and 741 mAh/g during the second cycle (Figure 19B-b). The rapidly decreasing capacity during the second cycle can be seen as irreversible capacity reduction owing to the generation of an SEI. However, a relatively higher performance is shown compared to SnO$_2$ nanoparticles and graphene alone owing to the structural advantage of the composite formation of SnO$_2$ combined with graphene. Based on a rate characteristic test, when the current density was returned to 20 mA/g after reaching a current density of 640 mA/g, it was confirmed that the capacity was reversibly restored, indicating that the graphene composites retain their structure well even under a high current density.

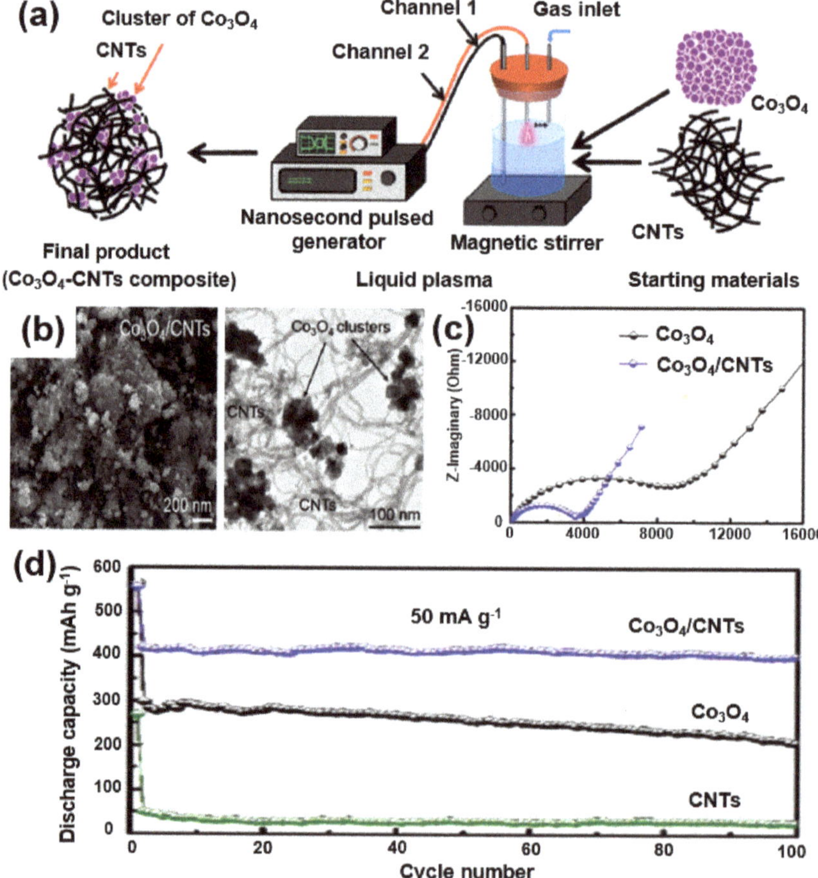

Figure 18. (**a**) Schematic illustrations of the synthesis process of Co$_3$O$_4$/CNT composites. (**b**) SEM and TEM images of Co$_3$O$_4$/CNT composites. (**c**) Nyquist plots of Co$_3$O$_4$ and Co$_3$O$_4$/CNT composites before cycling. (**d**) Cycling performance of Co$_3$O$_4$/CNTs composites, Co$_3$O$_4$, and CNTs at a current density of 50 mA/g. Reprinted with permission from Reference [89]; copyright 2015 Royal Society of Chemistry.

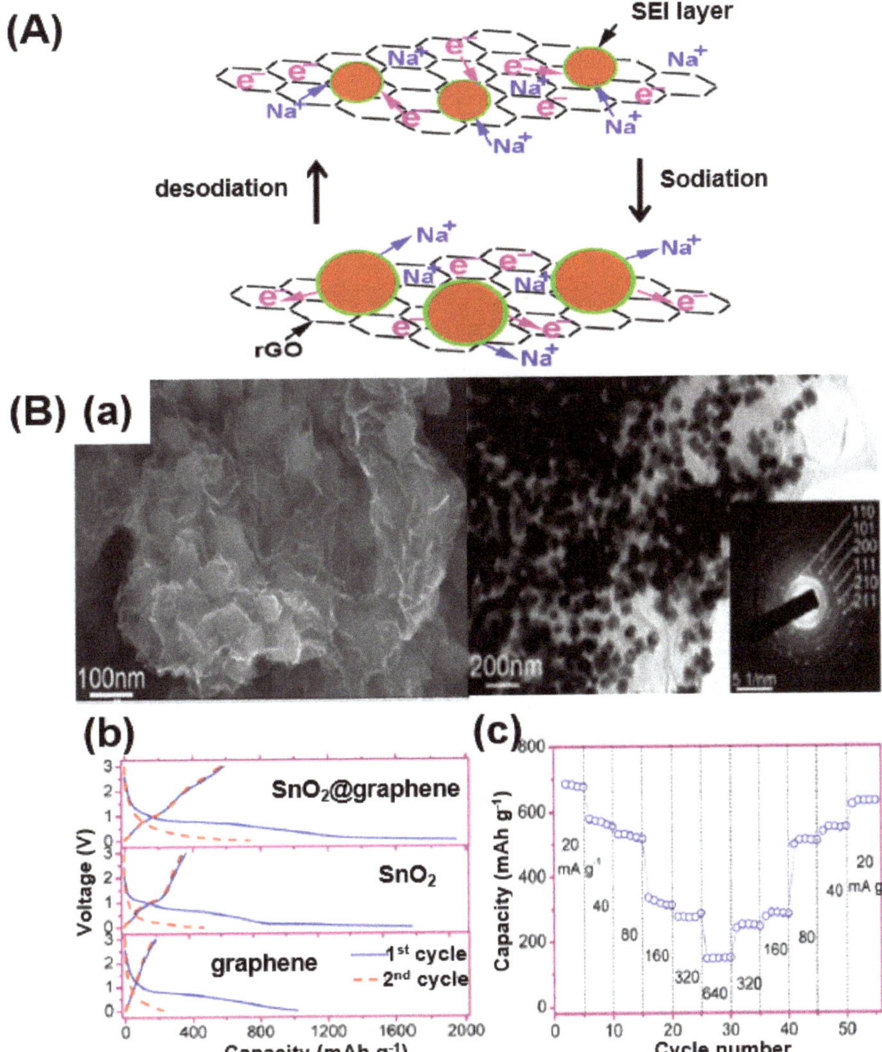

Figure 19. (**A**) A schematic illustration of the sodiation/desodiation procedure in an SnO$_2$/reduced GO (rGO) composite. Reprinted with permission from Reference [92]; copyright 2015 Elsevier. (**B**) (**a**) Field-emission (FE)-SEM and TEM images of a SnO$_2$@graphene nanocomposite. (**b**) First and second discharge/charge voltage profiles of SnO$_2$@graphene nanocomposite, SnO$_2$, and graphene. (**c**) Rate capability of SnO$_2$@graphene nanocomposite at different current densities. Reprinted with permission from Reference [93]; copyright 2013 Royal Society of Chemistry.

In the case of a CNT (1D structure) or graphene (2D structure), a metal-oxide composite with short CNTs or graphene with a small size may cause a high contact resistance. To address these issues, Zhao et al. developed SnO$_2$@CEM composites by growing SnO$_2$ nanosheets on a carbonized eggshell membrane (CEM) with a porous structure as a matrix to improve the 1D and 2D structures [94]. Firstly, these 3D hierarchical structures also lead to an enhanced electric conductivity. Using a porous CEM with a unique 3D structure, the composite can be used as a substitute for a copper substrate used in a conventional battery electrode (cathode), leading to an improved energy density of the battery

(Figure 20a). Secondly, because the porous structure facilitates the diffusion of the electrolyte by facilitating the entry and exit of the electrolyte, it can reduce the volume change of the metal oxide, which is a significant problem in SIBs (Figure 20a). An SEM image (Figure 20b) shows that the SnO_2 nanosheets grow on the CEM matrix to a size of 300 nm and are in intimate contact with each other. It is anticipated that, when making electrodes with an SnO2@CEM composite, conductive additives such as binders and carbon black may not be needed. Figure 20c shows the cycle stability of the SnO_2@CEM composite at 0.1 A/g. Except for the initial irreversible capacity owing to the formation of a solid electrolyte interface, the Coulombic efficiency of the composite was maintained at 92% after the second cycle and reached up to 99% after 100 cycles.

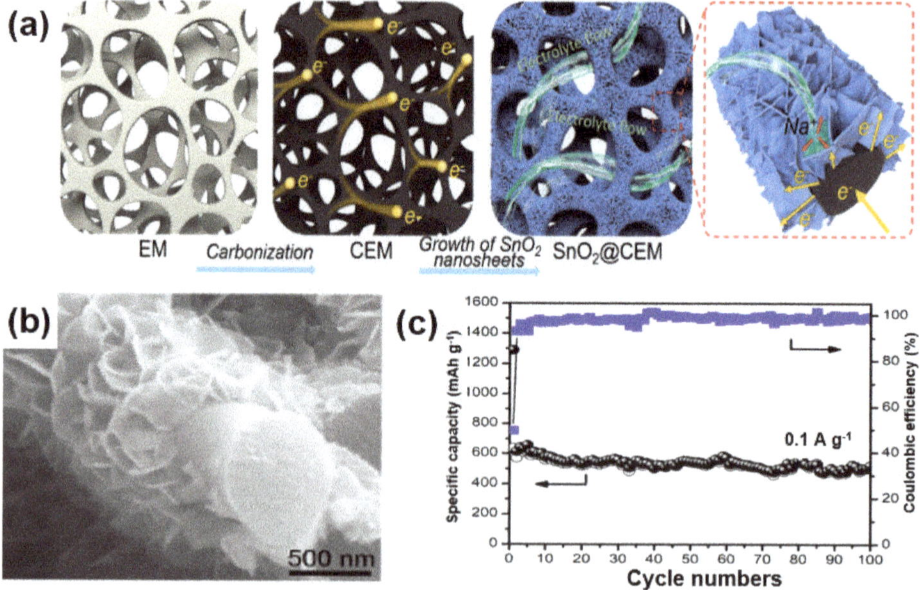

Figure 20. (**a**) Schematic illustrations of the synthesis procedure of SnO_2 nanosheets on a carbonized eggshell membrane (SnO_2@CEM) composite and its structural advantage as an electrode for a sodium-ion battery (SIB). (**b**) SEM image and (**c**) cycle performance of SnO_2@CEM composite. Reprinted with permission from Reference [94]; copyright 2018 American Chemical Society.

3.2. Supercapacitors

A supercapacitor is used as an energy storage device similar to lithium- or sodium-ion batteries. The power density of a supercapacitor is higher than a battery and, thus, it can be rapidly charged and discharged [95]. However, because the energy density is relatively low, a performance improvement is required. The supercapacitor stores energy through two mechanisms: electrochemical double-layer capacitance (EDLC) and pseudo-capacitance. Currently, commercialized electrodes are made of carbon, and the cyclic stability of a supercapacitor is sufficiently improved owing to the excellent chemical stability of carbon. However, low-capacitance carbon-based supercapacitors generally have an energy density of 3 to 5 Wh/kg, which is a very low energy density (10 to 250 Wh/kg is achieved for a lithium-ion battery) compared to batteries [44]. To increase the capacitance and energy density, a supercapacitor using metal oxide was studied.

When oxidized metal is used alone, the charge transfer and sheet resistances of the supercapacitor electrode increase owing to the low electrical conductivity of the metal oxide (e.g., 10^{-5} to 10^{-6} S/cm in MnO_2) [96]. In addition, a large reduction in capacitance occurs at a high current density. Moreover, like lithium- and sodium-ion batteries, during the charging/discharging process of the supercapacitor, the volume of an electrode made of metal oxide is changed, thereby decreasing both the contact between the electrodes and the long cycle stability. It is, therefore, necessary to improve the performance of the supercapacitor through synergy by forming a composite of metal oxide and a carbon material. In addition to mitigating the volume change of metal oxide, which is an advantage of carbon, through the formation of a carbon/metal-oxide composite, as well as an improvement in the low electrical conductivity of the electrode, a supercapacitor with a high energy density can be realized.

The high charge/discharge rate of a supercapacitor can be obtained through a short ion movement path, which is an advantage of a 1D structure. For instance, Sankapal et al. developed a ZnO/MWNT composite by growing ZnO on the surfaces of carbon nanotubes (Figure 21a) [97]. Using continuous ion-layer adsorption, ZnO was deposited onto the carbon nanotubes (Figure 21b). A long-term stability test at a current density of 200 mV/s showed a capacity retention rate of close to 83% after 5000 cycles despite the relatively high specific capacitance at 100 F/g (Figure 21c). This is due to the synergistic effects with ZnO based on the improved electrical conductivity, excellent mechanical properties, and short ion transport pathways owing to the 1D structure of carbon nanotubes.

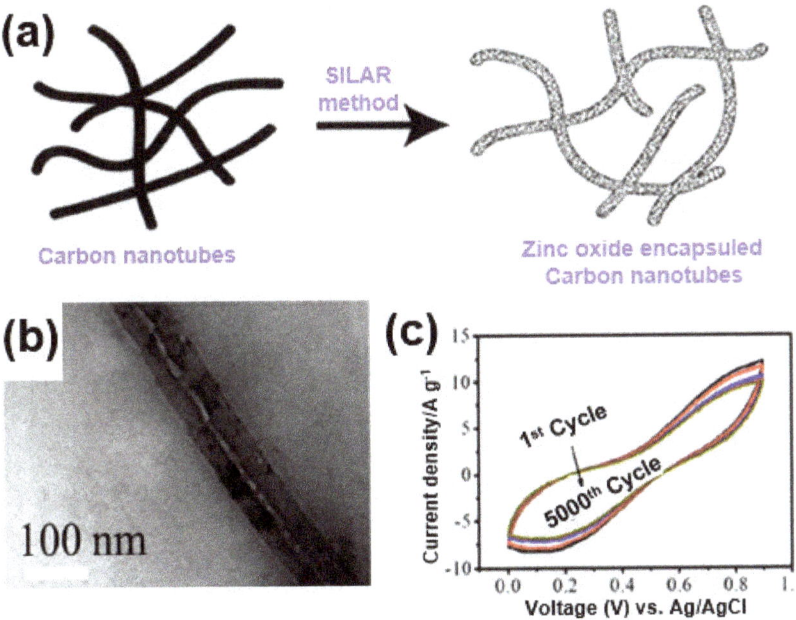

Figure 21. (a) A schematic illustration and (b) TEM image of a ZnO/multi-walled CNT (MWNT) composite. (c) Cycle stability performance of ZnO/MWNT composite. Reprinted with permission from Reference [97]; copyright 2016 Elsevier.

A similar approach was used by Yu et al. to study the formation of nanosheet-like NiO particles and composites on carbon nanotubes [98]. NiO nanosheets reduce the ion path and increase the electrochemical active sites through a large specific surface area, enabling a rapid reaction. When NiO nanosheets are present alone, NiO nanosheet particles are dispersed onto carbon nanotubes to prevent the performance degradation of a supercapacitor as the reaction progresses, thereby improving the performance. A high specific capacitance of 1177 F/g was obtained at a current density of 2 A/g.

Studies were conducted in an attempt to achieve application to wearable equipment or small devices by synthesizing a fiber-based supercapacitor with a 1D structure [99,100]. Shi et al. synthesized MnO$_2$@MWCNT composite fibers by twisting MWCNT sheets after injecting amorphous MnO$_2$ onto a sheet of aligned MWCNTs in a solid-state fiber-based supercapacitor, resulting in excellent mechanical stability, high electrical conductivity, fast ion-diffusion enhanced energy density, rate-limiting properties, and cycle stability of the supercapacitor [100].

Graphene with a 2D structure has a wide specific surface area and excellent electrical conductivity. In addition, graphene oxide and reduced graphene oxide have hydrophilic properties compared to hydrophobic graphene owing to their large number of functional groups on the surface, which is advantageous for synthesizing a composite with metal oxide. In addition, the functional groups present on the surface may act as a redox center of the redox reaction and contribute to a pseudo-capacitance.

Xiang et al. demonstrated Co_3O_4/rGO composites prepared using Co_3O_4 nanoparticles distributed on reduced graphene oxide through a hydrothermal synthesis [101]. As mentioned earlier, the main pseudo-capacitance comes from the metal oxide. The electrochemical reactions of electrochemically active Co_3O_4 are as follows (Equations (6) and (7)):

$$Co_3O_4 + OH^- + H_2O \leftrightarrow 3CoOOH + e^- \tag{6}$$

$$CoOOH + OH^- \leftrightarrow CoO_2 + H_2O + e^- \tag{7}$$

Through these reactions, a highly specific capacitance can be obtained, and the electron transfer between Co_3O_4 nanoparticles is accelerated through a reduced graphene matrix, resulting in an energy density of 39.0 Wh/kg and a power density of 8.3 kW/kg. Zhou et al. reported that the supercapacitor performance is improved through the synthesis of Co_3O_4 nanoparticles and reduced graphene oxide composites in the form of rolls [102]. Sodium dodecyl sulfate (SDS), a surfactant, and graphene oxide were bonded to each other based on the hydrophilic property. A rolled Co_3O_4/rGO composite was synthesized through the ionic bonding of functional groups on the surface of the graphene oxide and Co^{2+} cations (Figure 22a). This unique structure improves the electrical conductivity of the supercapacitor electrodes and tightens the bond between the Co_3O_4 and reduced graphene oxide to achieve a 93% specific capacitance retention after 1000 cycles at a scan rate of 20 mV/s, demonstrating a specific capacitance of more than 140 F/g at a current density of 10 A/g and 163.8 F/g at a current density of 1 A/g (Figure 22b,c).

A high energy density, power density, and specific capacitance are essential to improve the performance of a supercapacitor, allowing it to store more capacitance as the loading of the active material increases [103–105].

Li et al. improved the performance of a supercapacitor by forming a composite with Fe_3O_4 on hollow porous graphitized carbon with a double shell [103]. The hollow porous carbon increases the specific surface area and electrical conductivity of the metal oxide, and serves as a matrix in which the Fe_3O_4 nanoparticles can be uniformly dispersed in the porous carbon (Figure 23a). The composite has a porous formation mixed with micropores and mesopores, and the Fe_3O_4 nanoparticles are uniformly dispersed in the graphitized carbon shell (Figure 23b). Owing to the 3D porous structure, the electrolyte penetration of Fe_3O_4 nanoparticles can be expected to increase the rate-limiting behavior and cycle stability of the supercapacitor by facilitating a redox reaction at the active sites of the Fe_3O_4 nanoparticles. Figure 23c shows a capacitance maintenance rate of 96.7% even after 8000 cycles at a current density of 1.5 A/g, which can be seen in the results obtained through the structural advantages of the hollow porous graphitized carbon/Fe_3O_4 composite.

A 3D structure may be formed not only through the use of a porous carbon material, but also by connecting carbon materials with 1D and 2D structures. Dong et al. synthesized graphene/Co_3O_4 composites by forming Co_3O_4 nanowires through chemical vapor deposition on a 3D porous matrix using graphene foam, and then measured the performance of the supercapacitor [106]. Chen et al. introduced a study on the synthesis of a composite of MnO_2 and CNTs in the form of a 3D sponge structure [107].

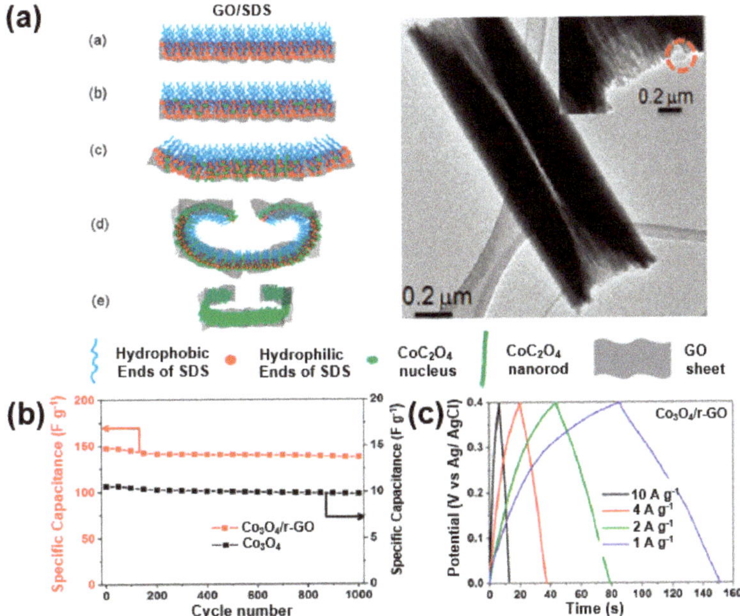

Figure 22. (**a**) Schematic illustrations of CoC$_2$O$_4$ scrolls fabricated in the presence of GO; the inset shows a TEM image of a single CoC$_2$O$_4$/GO scroll). Adapted with permission from Reference [102]; copyright 2011 Royal Society of Chemistry. (**b**) Specific capacitances of Co$_3$O$_4$ and Co$_3$O$_4$/rGO composite. (**c**) Charge/discharge profiles of Co$_3$O$_4$/rGO composite at various current densities. Reprinted with permission from Reference [102]; copyright 2011 Royal Society of Chemistry.

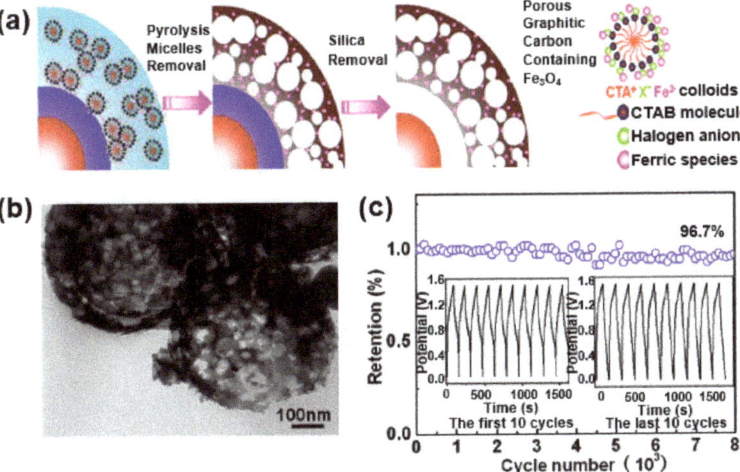

Figure 23. (**a**) Schematic illustration of the synthesis of double-shelled carbon (C-C)/Fe$_3$O$_4$ hollow spheres with a porous structural composite. (**b**) TEM image of the C-C/Fe$_3$O$_4$ composite. (**c**) Cycle stability of the C-C/Fe$_3$O$_4$ composite electrode. Reprinted with permission from Reference [103]; copyright 2016 Elsevier.

4. Conclusions

This review discussed the use of carbon/metal-oxide composites as electrodes for electrochemical energy devices. Despite the large capacity and high energy density of metal-oxide-based electrochemical energy devices, the low electrical conductivity of the metal oxide and cracks induced through a volumetric change during the electrochemical reactions lead to a deterioration in the electrochemical performances (degradation of the rate capability and cycle stability). This review discussed the solutions to these issues by means of the formation of a carbon/metal-oxide composite, which exhibits synergetic effects and new properties. As described, 1D, 2D, and 3D carbon-based composites bring about the following advantages owing to the unique properties of the composite for each dimension:

1. A 1D carbon nanostructure (e.g., CNT) provides a continuous network for metal oxides with low electrical conductivity, where 1D carbon with a high aspect ratio enables the formation of percolation networks in small quantities. In addition, the carbon in the composite makes it possible to stabilize the metal oxide during cycling owing to its high mechanical robustness.
2. A 2D carbon nanostructure (e.g., graphene) has high electrical conductivity, mechanical strength, and high surface area, being a very good complement to the disadvantages of metal oxides mentioned above. In particular, the metal-oxide/2D-carbon composites of various structures further improve the structural stability of the composite, where 2D carbon with a porous structure enables improved kinetics in electron and ion transport.
3. A 3D carbon nanostructure enables a high loading of the active materials through a high storage capacity (porosity), allowing an increase in the active site for ions during the electrochemical reactions. In addition, the hierarchical structure enables a facilitation of the electrochemical reactions depending on the pore characteristics (e.g., micro/meso/macro) and pore connectivity (or interconnectivity).

Because these different dimensions of carbon have different characteristics owing to the structural characteristics of the material, we focused on describing the effects of composites with metal oxide and carbon for each dimension based on the performance of an electrochemical energy device.

Research on improvement in the performance of electrochemical energy devices is currently underway, and a study on the carbon/metal-oxide composites reviewed in this article is expected to provide an alternative to the energy problems facing mankind.

Author Contributions: D.S. and Y.J. contributed equally to this work. H.S. conceived the idea and H.S., D.S., Y.J., performed the basic study and survey. H.S., D.S. and Y.J. wrote the manuscript with the support from K.H. and D.Y.Y. All authors discussed the results and commented on the manuscript.

Funding: This research was supported by the National Strategic Project Carbon Upcycling of the National Research Foundation of Korea (NRF) funded by the Ministry of Science and Information and Communications Technology (MSIT), the Ministry of Environment (ME), and the Ministry of Trade, Industry, and Energy (MOTIE) (2017M3D8A2086014).

Acknowledgments: We thanks to Sang Wook Kang, Jong-Min Oh, Chulhwan Park, Gun Youl Park, Young Jun Ji, Won Jik Kim and Si Hun Oh owing to their contributions of the preliminary data survey, manuscript preparation and fruitful discussions.

Conflicts of Interest: The authors declare no conflicts of interest. The funders had no role in the design of the study; in the collection, analyses, or interpretation of data; in the writing of the manuscript, or in the decision to publish the results.

References

1. Shen, X.; Liu, H.; Cheng, X.B.; Yan, C.; Huang, J.Q. Beyond Lithium Ion Batteries: Higher Energy Density Battery Systems based on Lithium Metal Anodes. *Energy Storage Mater.* **2018**, *12*, 161–175. [CrossRef]
2. Xiao, Q.; Sohn, H.; Chen, Z.; Toso, D.; Mechlenburg, M.; Zhou, Z.H.; Poirier, E.; Dailly, A.; Wang, H.; Wu, Z.; et al. Mesoporous Metal and Metal Alloy Particles Synthesized by Aerosol-Assisted Confined Growth of Nanocrystals. *Angew. Chem. Int. Ed.* **2012**, *51*, 10546–10550. [CrossRef]

3.   Dai, F.; Zai, J.; Yi, R.; Gordin, M.L.; Sohn, H.; Chen, S.; Wang, D. Bottom-up synthesis of high surface area mesoporous crystalline silicon and evaluation of its hydrogen evolution performance. *Nat. Commun.* **2014**, *5*, 3605. [CrossRef]

4.   Chen, L.; Shaw, L.L. Recent Advances in Lithium–Sulfur Batteries. *J. Power Sources* **2014**, *267*, 770–783. [CrossRef]

5.   Bruce, P.G.; Freunberger, S.A.; Hardwick, L.J.; Tarascon, J.M. Li–O$_2$ and Li–S Batteries with High Energy Storage. *Nat. Mater.* **2012**, *11*, 19–29. [CrossRef]

6.   Balogun, M.S.; Luo, Y.; Qiu, W.; Liu, P.; Tong, Y. A Review of Carbon Materials and Their Composites with Alloy Metals for Sodium Ion Battery Anodes. *Carbon* **2016**, *98*, 162–178. [CrossRef]

7.   Sohn, H. Deposition of Functional Organic and Inorganic Layer on the Cathode for the Improved Electrochemical Performance of Li-S Battery. *Korean Chem. Eng. Res.* **2017**, *55*, 483–489.

8.   Fukuda, K.; Kikuya, K.; Isono, K.; Yoshio, M. Foliated Natural Graphite as the Anode Material for Rechargeable Lithium-Ion Cells. *J. Power Sources* **1997**, *69*, 165–168. [CrossRef]

9.   Sohn, H.; Chen, Z.; Jung, Y.S.; Xiao, Q.; Cai, M.; Wang, H.; Lu, Y. Robust lithium-ion anodes based on nanocomposites of iron oxide–carbon–silicate. *J. Mater. Chem. A* **2013**, *1*, 4539–4545. [CrossRef]

10.   Sohn, H.; Kim, D.H.; Yi, R.; Tang, D.; Lee, S.E.; Jung, Y.S.; Wang, D. Semimicro-size agglomerate structured silicon-carbon composite as an anode material for high performance lithium-ion batteries. *J. Power Sources* **2016**, *334*, 128–136. [CrossRef]

11.   Sohn, H.; Gordin, M.L.; Xu, T.; Chen, S.; Lv, D.; Song, J.; Manivannan, A.; Wang, D. Porous spherical carbon/sulfur nanocomposites by aerosol-assisted synthesis: The effect of pore structure and morphology on their electrochemical performance as lithium/sulfur battery cathodes. *ACS Appl. Mater. Interfaces* **2014**, *6*, 7596–7606. [CrossRef]

12.   Aurbach, D.; Zinigrad, E.; Cohen, Y.; Teller, H. A Short Review of Failure Mechanisms of Lithium Metal and Lithiated Graphite Anodes in Liquid Electrolyte Solutions. *Solid State Ion.* **2002**, *148*, 405–416. [CrossRef]

13.   Jiang, J.; Dahn, J.R. Effects of Solvents and Salts on the Thermal Stability of LiC$_6$. *Electrochim. Acta* **2004**, *49*, 4599–4604. [CrossRef]

14.   Sohn, H.; Gordin, M.L.; Regula, M.; Kim, D.H.; Jung, Y.S.; Song, J.; Wang, D. Porous spherical polyacrylonitrile-carbon nanocomposite with high loading of sulfur for lithium–sulfur batteries. *J. Power Sources* **2016**, *302*, 70–78. [CrossRef]

15.   Hwang, K.; Sohn, H.; Yoon, S. Mesostructured niobium-doped titanium oxide-carbon (Nb-TiO$_2$-C) composite as an anode for high-performance lithium-ion batteries. *J. Power Sources* **2018**, *378*, 225–234. [CrossRef]

16.   Sohn, H.; Kim, D.; Lee, J.; Yoon, S. Facile synthesis of a mesostructured TiO$_2$–graphitized carbon (TiO$_2$–gC) composite through the hydrothermal process and its application as the anode of lithium ion batteries. *RSC Adv.* **2016**, *6*, 39484–39491. [CrossRef]

17.   Gu, Y.; Wu, A.; Sohn, H.; Nicoletti, C.; Iqbal, Z.; Federici, J.F. Fabrication of rechargeable lithium ion batteries using water-based inkjet printed cathodes. *J. Manuf. Process.* **2015**, *20*, 198–205. [CrossRef]

18.   Ge, P.; Fouletier, M. Electrochemical Intercalation of Sodium in Graphite. *Solid State Ion.* **1988**, *28*, 1172–1175. [CrossRef]

19.   Frackowiak, E.; Beguin, F. Carbon Materials for the Electrochemical Storage of Energy in Capacitors. *Carbon* **2001**, *39*, 937–950. [CrossRef]

20.   Stoller, M.D.; Magnuson, C.W.; Zhu, Y.; Murali, S.; Suk, J.W.; Piner, R.; Ruoff, R.S. Interfacial Capacitance of Single Layer Graphene. *Energy Environ. Sci.* **2011**, *4*, 4685–4689. [CrossRef]

21.   Kim, H.; Seo, D.H.; Kim, H.; Park, I.; Hong, J.; Park, K.Y.; Kang, K. Multicomponent Effects on the Crystal Structures and Electrochemical Properties of Spinel-Structured M$_3$O$_4$ (M = Fe, Mn, Co) Anodes in Lithium rechargeable batteries. *Chem. Mater.* **2012**, *24*, 720–725. [CrossRef]

22.   Laruelle, S.; Grugeon, S.; Poizot, P.; Dolle, M.; Dupont, L.; Tarascon, J.M. On the Origin of the Extra Electrochemical Capacity Displayed by MO/Li cells at Low Potential. *J. Electrochem. Soc.* **2002**, *149*, A627–A634. [CrossRef]

23.   Poizot, P.; Laruelle, S.; Grugeon, S.; Dupont, L.; Tarascon, J.M. Nano-sized Transition-Metal Oxides as Negative-Electrode Materials for Lithium-Ion Batteries. *Nature* **2000**, *407*, 496–499. [CrossRef] [PubMed]

24.   Zhang, W.M.; Wu, X.L.; Hu, J.S.; Guo, Y.G.; Wan, L.J. Carbon coated Fe$_3$O$_4$ Nanospindles as a Superior Anode Material for Lithium-Ion Batteries. *Adv. Funct. Mater.* **2008**, *18*, 3941–3946. [CrossRef]

25. Jiang, J.; Li, Y.; Liu, J.; Huang, X.; Yuan, C.; Lou, X.W. Recent Advances in Metal Oxide-based Electrode Architecture Design for Electrochemical Energy Storage. *Adv. Mater.* **2012**, *24*, 5166–5180. [CrossRef]

26. Jayaprakash, N.; Jones, W.D.; Moganty, S.S.; Archer, L.A. Composite Lithium Battery Anodes based on Carbon@Co_3O_4 Nanostructures: Synthesis and Characterization. *J. Power Sources* **2012**, *200*, 53–58. [CrossRef]

27. Chen, Y.; Song, B.; Li, M.; Lu, L.; Xue, J. Fe_3O_4 Nanoparticles Embedded in Uniform Mesoporous Carbon Spheres for Superior High-Rate Battery Applications. *Adv. Funct. Mater.* **2014**, *24*, 319–326. [CrossRef]

28. Xia, Y.; Xiao, Z.; Dou, X.; Huang, H.; Lu, X.; Yan, R.; Gan, Y.; Zhu, W.; Tu, J.; Zhang, W.; et al. Green and Facile Fabrication of Hollow Porous MnO/C Microspheres from Microalgaes for Lithium-Ion Batteries. *ACS Nano* **2013**, *7*, 7083–7092. [CrossRef]

29. Wang, X.; Li, Z.; Yin, L. Nanocomposites of SnO_2@Ordered Mesoporous Carbon (OMC) as Anode Materials for Lithium-Ion Batteries with Improved Electrochemical Performance. *CrystEngComm* **2013**, *15*, 7589–7597. [CrossRef]

30. Li, Z.; Liu, N.; Wang, X.; Wang, C.; Qi, Y.; Yin, L. Three-Dimensional Nanohybrids of Mn_3O_4/Ordered Mesoporous Carbons for High Performance Anode Materials for Lithium-Ion Batteries. *J. Mater. Chem.* **2012**, *22*, 16640–16648. [CrossRef]

31. Gao, G.; Jin, Y.; Zeng, Q.; Wang, D.; Shen, C. Carbon Nanotube-Wrapped Fe_2O_3 Anode with Improved Performance for Lithium-Ion Batteries. *Beilstein J. Nanotechnol.* **2017**, *8*, 649–656. [CrossRef] [PubMed]

32. Ma, J.; Wang, J.; He, Y.S.; Liao, X.Z.; Chen, J.; Wang, J.Z.; Yuan, T.; Ma, Z.F. A Solvothermal Strategy: One-Step in situ Synthesis of Self-Assembled 3D Graphene-based Composites with Enhanced Lithium Storage Capacity. *J. Mater. Chem. A* **2014**, *2*, 9200–9207. [CrossRef]

33. Zhu, J.; Zhu, T.; Zhou, X.; Zhang, Y.; Lou, X.W.; Chen, X.; Zhang, H.; Hng, H.H.; Yan, Q. Facile Synthesis of Metal Oxide/Reduced Graphene Oxide Hybrids with High Lithium Storage Capacity and Stable Cyclability. *Nanoscale* **2011**, *3*, 1084–1089. [CrossRef] [PubMed]

34. Noerochim, L.; Wang, J.Z.; Chou, S.L.; Li, H.J.; Liu, H.K. SnO_2-Coated Multiwall Carbon Nanotube Composite Anode Materials for Rechargeable Lithium-Ion Batteries. *Electrochim. Acta* **2010**, *56*, 314–320. [CrossRef]

35. Qiu, S.; Gu, H.; Lu, G.; Liu, J.; Li, X.; Fu, Y.; Yan, X.; Hu, C.; Guo, Z. Rechargeable Co_3O_4 Porous Nanoflake Carbon Nanotube Nanocomposite Lithium-Ion Battery Anodes with Enhanced Energy Performances. *RSC Adv.* **2015**, *5*, 46509–46516. [CrossRef]

36. Xia, H.; Lai, M.; Lu, L. Nanoflaky MnO_2/Carbon Nanotube Nanocomposites as Anode Materials for Lithium-Ion Batteries. *J. Mater. Chem.* **2010**, *20*, 6896–6902. [CrossRef]

37. Nam, I.; Kim, N.D.; Kim, G.P.; Park, J.; Yi, J. One step preparation of Mn_3O_4/Graphene Composites for Use as an Anode in Li Ion Batteries. *J. Power Sources* **2013**, *244*, 56–62. [CrossRef]

38. Zou, Y.; Kan, J.; Wang, Y. Fe_2O_3-Graphene Rice-on-Sheet Nanocomposite for High and Fast Lithium Ion Storage. *J. Phys. Chem. C* **2011**, *115*, 20747–20753. [CrossRef]

39. Sun, H.; Liu, Y.; Yu, Y.; Ahmad, M.; Nan, D.; Zhu, J. Mesoporous Co_3O_4 Nanosheets-3D Graphene Networks Hybrid Materials for High-Performance Lithium Ion Batteries. *Electrochim. Acta* **2014**, *118*, 1–9. [CrossRef]

40. Sassin, M.B.; Mansour, A.N.; Pettigrew, K.A.; Rolison, D.R.; Long, J.W. Electroless Deposition of Conformal Nanoscale Iron Oxide on Carbon Nanoarchitectures for Electrochemical Charge Storage. *ACS Nano* **2010**, *4*, 4505–4514. [CrossRef]

41. De las Cases, C.; Li, W. A Review of Application of Carbon Nanotubes for Lithium Ion Battery Anode materials. *J. Power Sources* **2012**, *208*, 74–85. [CrossRef]

42. Goriparti, S.; Miele, E.; De Angelis, F.; Di Fabrizio, E.; Zaccaria, R.P.; Capiglia, C. Review on Recent Progress of Nanostructured Anode Materials for Li-ion battery. *J. Power Sources* **2014**, *257*, 421–443. [CrossRef]

43. Zhang, D.W.; Xu, B.; Jiang, L.C. Functional Hybrid Materials based on Carbon Nanotubes and Metal Oxide. *J. Mater. Chem.* **2010**, *20*, 6383–6391. [CrossRef]

44. Zhi, M.; Xiang, C.; Li, J.; Li, M.; Wu, N. Nanostructured Carbon-Metal Oxide Composite Electrode for Supercapacitor: A Review. *Nanoscale* **2013**, *5*, 72–88. [CrossRef] [PubMed]

45. Zhao, Y.; Dong, W.; Riaz, M.S.; Ge, H.; Wang, X.; Liu, Z.; Huang, F. "Electron-sharing" Mechanism Promote Co@Co_3O_4/CNTs Composite as the High Capacity Anode Materials of Lithium-Ion Battery. *ACS Appl. Mater. Interfaces* **2018**, *10*, 43641–43649. [CrossRef] [PubMed]

46. Park, J.; Moon, W.G.; Kim, G.P.; Nam, I.; Park, S.; Kim, Y.; Yi, J. Three-Dimensional Aligned Mesoporous Carbon Nanotubes Filled with Co_3O_4 Nanoparticles for Li-Ion Battery Anode Applications. *Electrochim. Acta* **2013**, *105*, 110–114. [CrossRef]

47. Abbas, S.M.; Hussain, S.T.; Ali, S.; Ahmad, N.; Munawar, K.S. Synthesis of Carbon Nanotubes Anchored with Mesoporous Co_3O_4 Nanoparticles as Anode Material for Lithium-Ion Batteries. *Electrochim. Acta* **2013**, *105*, 481–488. [CrossRef]

48. He, X.; Wu, Y.; Zhao, F.; Wang, J.; Jiang, K.; Fan, S. Enhanced Rate Capabilities of Co_3O_4/Carbon Nanotube Anodes for Lithium Ion Battery Applications. *J. Mater. Chem. A* **2013**, *1*, 11121–11125. [CrossRef]

49. He, Y.; Huang, L.; Cai, J.S.; Zheng, X.M.; Sun, S.G. Structure and Electrochemical Performance of Nanostructured Fe_3O_4/Carbon Nanotube Composites as Anodes for Lithium Ion Batteries. *Electrochim. Acta* **2010**, *55*, 1140–1144. [CrossRef]

50. Higgins, T.M.; McAteer, D.; Coelho, J.C.M.; Sanchez, B.M.; Gholamvand, Z.; Moriarty, G.; McEvoy, N.; Berner, N.C.; Duesberg, G.S.; Nicolosi, V.; et al. Effect of Percolation on the Capacitance of Supercapacitor Electrodes Prepared from Composites of Manganese Dioxide Nanoplatelets and Carbon Nanotubes. *ACS Nano* **2014**, *8*, 9567–9579. [CrossRef]

51. Zhang, H.; Song, H.; Chen, X.; Zhou, J.; Zhang, H. Preparation and Electrochemical Performance of SnO_2@ Carbon Nanotube Core–Shell Structure Composites as Anode Material for Lithium-Ion Batteries. *Electrochim. Acta* **2012**, *59*, 160–167. [CrossRef]

52. Candelaria, S.L.; Shao, Y.; Zhou, W.; Li, X.; Xiao, J.; Zhang, J.G.; Wang, Y.; Liu, J.; Li, J.; Cao, G. Nanostructured Carbon for Energy Storage and Conversion. *Nano Energy* **2012**, *1*, 195–220. [CrossRef]

53. Wu, Z.S.; Zhou, G.; Yin, L.C.; Ren, W.; Li, F.; Cheng, H.M. Graphene/Metal Oxide Composite Electrode Materials for Energy Storage. *Nano Energy* **2012**, *1*, 107–131. [CrossRef]

54. Lian, P.; Zhu, X.; Xiang, H.; Li, Z.; Yang, W.; Wang, H. Enhanced Cycling Performance of Fe_3O_4–Graphene Nanocomposite as an Anode Material for Lithium-Ion Batteries. *Electrochim. Acta* **2010**, *56*, 834–840. [CrossRef]

55. Zhong, C.; Wang, J.; Chen, Z.; Liu, H. SnO_2–Graphene Composite Synthesized via an Ultrafast and Environmentally Friendly Microwave Autoclave Method and Its Use as a Superior Anode for Lithium-Ion Batteries. *J. Phys. Chem. C* **2011**, *115*, 25115–25120. [CrossRef]

56. Luo, B.; Wang, B.; Li, X.; Jia, Y.; Liang, M.; Zhi, L. Graphene-confined Sn nanosheets with enhanced lithium storage capability. *Adv. Mater.* **2012**, *24*, 3538–3543. [CrossRef] [PubMed]

57. Su, J.; Cao, M.; Ren, L.; Hu, C. Fe_3O_4–Graphene Nanocomposites with Improved Lithium Storage and Magnetism Properties. *J. Phys. Chem. C* **2011**, *115*, 14469–14477. [CrossRef]

58. Zhou, G.; Wang, D.W.; Li, F.; Zhang, L.; Li, N.; Wu, Z.S.; Wen, L.; Lu, G.Q.; Cheng, H.M. Graphene-Wrapped Fe_3O_4 Anode Material with Improved Reversible Capacity and Cyclic Stability for Lithium Ion Batteries. *Chem. Mater.* **2010**, *22*, 5306–5313. [CrossRef]

59. Peng, L.; Fang, Z.; Zhu, Y.; Yan, C.; Yu, G. Holey 2D Nanomaterials for Electrochemical Energy Storage. *Adv. Energy. Mater.* **2018**, *8*, 1702179. [CrossRef]

60. Xu, Y.; Lin, Z.; Zhong, X.; Huang, X.; Weiss, N.O.; Huang, Y.; Duan, X. Holey Graphene Frameworks for Highly Efficient Capacitive Energy Storage. *Nat. Commun.* **2014**, *5*, 4554. [CrossRef]

61. Zhao, X.; Hayner, C.M.; Kung, M.C.; Kung, H.H. Flexible Holey Graphene Paper Electrodes with Enhanced Rate Capability for Energy Storage Applications. *ACS Nano* **2011**, *5*, 8739–8749. [CrossRef] [PubMed]

62. Zhang, X.; Zhou, J.; Chen, X.; Song, H. Pliable Embedded-Type Paper Electrode of Hollow Metal Oxide@ Porous Graphene with Abnormal but Superior Rate Capability for Lithium-Ion Storage. *ACS Appl. Energy. Mater.* **2017**, *1*, 48–55. [CrossRef]

63. Guo, S.; Lu, G.; Qiu, S.; Liu, J.; Wang, X.; He, C.; Wei, H.; Yan, X.; Guo, Z. Carbon-coated MnO microparticulate porous nanocomposites serving as anode materials with enhanced electrochemical performances. *Nano Energy* **2014**, *9*, 41–49. [CrossRef]

64. Chang, P.Y.; Huang, C.H.; Doong, R.A. Ordered Mesoporous Carbon–TiO_2 Materials for Improved Electrochemical Performance of Lithium Ion Battery. *Carbon* **2012**, *50*, 4259–4268. [CrossRef]

65. Wang, D.W.; Li, F.; Liu, M.; Lu, G.Q.; Cheng, H.M. 3D Aperiodic Hierarchical Porous Graphitic Carbon Material for High-Rate Electrochemical Capacitive Energy Storage. *Angew. Chem. Int. Ed.* **2008**, *48*, 373–376. [CrossRef] [PubMed]

66. Cao, Z.; Shi, M.; Ding, Y.; Zhang, J.; Wang, Z.; Dong, H.; Yin, Y.; Yang, S. Lotus-root-like MnO/C Hybrids as Anode Materials for High-Performance Lithium-Ion Batteries. *J. Phys. Chem. C* **2017**, *121*, 2546–2555. [CrossRef]

67. Lu, G.; Qiu, S.; Lv, H.; Fu, Y.; Liu, J.; Li, X.; Bai, Y.J. Li-ion storage performance of MnO nanoparticles coated with nitrogen-doped carbon derived from different carbon sources. *Electrochim. Acta* **2014**, *146*, 249–256. [CrossRef]
68. Kang, E.; Jung, Y.S.; Cavanagh, A.S.; Kim, G.H.; George, S.M.; Dillon, A.C.; Kim, J.K.; Lee, J. Fe_3O_4 Nanoparticles Confined in Mesocellular Carbon Foam for High Performance Anode Materials for Lithium-Ion Batteries. *Adv. Funct. Mater.* **2011**, *21*, 2430–2438. [CrossRef]
69. Zhang, X.; Hu, Z.; Xiao, X.; Sun, L.; Han, S.; Chen, D.; Liu, X. Fe_3O_4@Porous Carbon Hybrid as the Anode Material for a Lithium-Ion Battery: Performance Optimization by Composition and Microstructure Tailoring. *New. J. Chem.* **2015**, *39*, 3435–3443. [CrossRef]
70. Yoon, T.; Chae, C.; Sun, Y.K.; Zhao, X.; Kung, H.H.; Lee, J.K. Bottom-up in situ formation of Fe_3O_4 Nanocrystals in a Porous Carbon Foam for Lithium-Ion Battery Anodes. *J. Mater. Chem.* **2011**, *21*, 17325–17330. [CrossRef]
71. Han, F.; Li, W.C.; Li, M.R.; Lu, A.H. Fabrication of Superior-Performance SnO_2@C Composites for Lithium-Ion Anodes Using Tubular Mesoporous Carbon with Thin Carbon Walls and High Pore Volume. *J. Mater. Chem.* **2012**, *22*, 9645–9651. [CrossRef]
72. Yun, Q.; Lu, Q.; Zhang, X.; Tan, C.; Zhang, H. Three-Dimensional Architectures Constructed from Transition-Metal Dichalcogenide Nanomaterials for Electrochemical Energy Storage and Conversion. *Angewan. Chem. Int. Ed.* **2018**, *57*, 626–646. [CrossRef] [PubMed]
73. Zhou, G.; Wang, D.W.; Hou, P.X.; Li, W.; Li, N.; Liu, C.; Li, F.; Cheng, H.M. A Nanosized Fe_2O_3 Decorated Single-Walled Carbon Nanotube Membrane as a High-Performance Flexible Anode for Lithium Ion Batteries. *J. Mater. Chem.* **2012**, *22*, 17942–17946. [CrossRef]
74. Lv, H.; Qiu, S.; Lu, G.; Fu, Y.; Li, X.; Hu, C.; Liu, J. Nanostructured antimony/carbon composite fibers as anode material for lithium-ion battery. *Electrochim. Acta* **2015**, *151*, 214–221. [CrossRef]
75. Lyu, H.; Liu, J.; Qiu, S.; Cao, Y.; Hu, C.; Guo, S.; Guo, Z. Carbon composite spun fibers with in situ formed multicomponent nanoparticles for a lithium-ion battery anode with enhanced performance. *J. Mater. Chem. A* **2016**, *4*, 9881–9889. [CrossRef]
76. Qiu, S.; Lu, G.; Liu, J.; Lyu, H.; Hu, C.; Li, B.; Yan, X.; Guo, J.; Guo, Z. Enhanced electrochemical performances of MoO_2 nanoparticles composited with carbon nanotubes for lithium-ion battery anodes. *RSC Adv.* **2015**, *5*, 87286–87294. [CrossRef]
77. Luo, S.; Wu, H.; Wu, Y.; Jiang, K.; Wang, J.; Fan, S. Mn_3O_4 Nanoparticles Anchored on Continuous Carbon Nanotube Network as Superior Anodes for Lithium Ion Batteries. *J. Power Sources* **2014**, *249*, 463–469. [CrossRef]
78. Li, X.; Gu, H.; Liu, J.; Wei, H.; Qiu, S.; Fu, Y.; Lv, H.; Lu, G.; Wang, Y.; Guo, Z. Multi-walled carbon nanotubes composited with nanomagnetite for anodes in lithium ion batteries. *RSC Adv.* **2015**, *5*, 7237–7244. [CrossRef]
79. Li, L.; Zhou, G.; Shan, X.Y.; Pei, S.; Li, F.; Cheng, H.M. Co_3O_4 Mesoporous Nanostructures@Graphene Membrane as an Integrated Anode for Long-Life Lithium-Ion Batteries. *J. Power Sources* **2014**, *255*, 52–58. [CrossRef]
80. Hu, R.; Zhang, H.; Bu, Y.; Zhang, H.; Zhao, B.; Yang, C. Porous Co_3O_4 Nanofibers Surface-Modified by Reduced Graphene Oxide as a Durable, High-Rate Anode for Lithium Ion Battery. *Electrochim. Acta* **2017**, *228*, 241–250. [CrossRef]
81. Park, S.K.; Seong, C.Y.; Yoo, S.; Piao, Y. Porous Mn_3O_4 Nanorod/Reduced Graphene Oxide Hybrid Paper as a Flexible and Binder-Free Anode Material for Lithium Ion Battery. *Energy* **2016**, *99*, 266–273. [CrossRef]
82. Lin, J.; Raji, A.R.O.; Nan, K.; Peng, Z.; Yan, Z.; Samuel, E.L.; Natelson, D.; Tour, J.M. Iron Oxide Nanoparticle and Graphene Nanoribbon Composite as an Anode Material for High-Performance Li-Ion Batteries. *Adv. Funct. Mater.* **2014**, *24*, 2044–2048. [CrossRef]
83. Tao, L.; Zai, J.; Wang, K.; Zhang, H.; Xu, M.; Shen, J.; Su, Y.; Qian, X. Co_3O_4 Nanorods/Graphene Nanosheets Nanocomposites for Lithium Ion Batteries with Improved Reversible Capacity and Cycle Stability. *J. Power Sources* **2012**, *202*, 230–235. [CrossRef]
84. Fu, Y.; Gu, H.; Yan, X.; Liu, J.; Wang, Y.; Huang, J.; Li, X.; Lv, H.; Wang, X.; Guo, J.; et al. Chromium(III) oxide carbon nanocomposites lithium-ion battery anodes with enhanced energy conversion performance. *Chem. Eng. J.* **2015**, *277*, 186–193. [CrossRef]
85. Wang, S.; Xing, Y.; Xu, H.; Zhang, S. MnO Nanoparticles Interdispersed in 3D Porous Carbon Framework for High Performance Lithium-Ion Batteries. *ACS Appl. Mater. Interfaces* **2014**, *6*, 12713–12718. [CrossRef] [PubMed]

86. Wang, S.; Wang, N.; Zhang, Q. Ni-and/or Mn-Based Layered Transition Metal Oxides as Cathode Materials for Sodium Ion Batteries: Status, Challenges and Counter Measurements. *J. Mater. Chem. A* **2019**, *7*, 10138–10158. [CrossRef]

87. Ma, C.; Li, X.; Deng, C.; Hu, Y.Y.; Lee, S.; Liao, X.Z.; He, Y.S.; Ma, Z.F.; Xiong, H. Coaxial Carbon Nanotube Supported TiO_2@ MoO_2@Carbon Core–Shell Anode for Ultrafast and High-Capacity Sodium Ion Storage. *ACS Nano* **2018**, *13*, 671–680. [CrossRef]

88. Wang, Y.; Su, D.; Wang, C.; Wang, G. SnO_2@MWCNT Nanocomposite as a High Capacity Anode Material for Sodium-Ion Batteries. *Electrochem. Commun.* **2013**, *29*, 8–11. [CrossRef]

89. Rahman, M.M.; Sultana, I.; Chen, Z.; Srikanth, M.; Li, L.H.; Dai, X.J.; Chen, Y. Ex-situ Electrochemical Sodiation/desodiation Observation of Co_3O_4 Anchored Carbon Nanotubes: A High Performance Sodium-Ion Battery Anode Produced by Pulsed Plasma in a Liquid. *Nanoscale* **2015**, *7*, 13088–13095. [CrossRef]

90. Jian, Z.; Zhao, B.; Liu, P.; Li, F.; Zheng, M.; Chen, M.; Shi, Y.; Zhou, H. Fe_2O_3 Nanocrystals Anchored onto Graphene Nanosheets as the Anode Material for Low-Cost Sodium-Ion Batteries. *Chem. Commun.* **2014**, *50*, 1215–1217. [CrossRef]

91. Liu, X.; Chen, T.; Chu, H.; Niu, L.; Sun, Z.; Pan, L.; Sun, C.Q. Fe_2O_3-Reduced Graphene Oxide Composites Synthesized via Microwave-Assisted Method for Sodium Ion Batteries. *Electrochim. Acta* **2015**, *166*, 12–16. [CrossRef]

92. Zhang, Y.; Xie, J.; Zhang, S.; Zhu, P.; Cao, G.; Zhao, X. Ultrafine Tin Oxide on Reduced Graphene Oxide as High-Performance Anode for Sodium-Ion Batteries. *Electrochim. Acta* **2015**, *151*, 8–15. [CrossRef]

93. Su, D.; Ahn, H.J.; Wang, G. SnO_2@Graphene Nanocomposites as Anode Materials for Na-Ion Batteries with Superior Electrochemical Performance. *Chem. Commun.* **2013**, *49*, 3131–3133. [CrossRef]

94. Zhao, X.; Luo, M.; Zhao, W.; Xu, R.; Liu, Y.; Shen, H. SnO_2 Nanosheets Anchored on a 3D, Bicontinuous Electron and Ion Transport Carbon Network for High-Performance Sodium-Ion Batteries. *ACS Appl. Mater. Interfaces* **2018**, *10*, 38006–38014. [CrossRef] [PubMed]

95. Borenstein, A.; Hanna, O.; Attias, R.; Luski, S.; Brousse, T.; Aurbach, D. Carbon-based Composite Materials for Supercapacitor Electrodes: A Review. *J. Mater. Chem. A* **2017**, *5*, 12653–12672. [CrossRef]

96. Bélanger, D.; Brousse, L.; Long, J.W. Manganese Oxides: Battery Materials Make the Leap to Electrochemical Capacitors. *Electrochem. Soc. Interface* **2008**, *17*, 49.

97. Sankapal, B.R.; Gajare, H.B.; Karade, S.S.; Salunkhe, R.R.; Dubal, D.P. Zinc Oxide Encapsulated Carbon Nanotube Thin Films for Energy Storage Applications. *Electrochim. Acta* **2016**, *192*, 377–384. [CrossRef]

98. Yu, W.; Li, B.Q.; Ding, S.J. Electroless Fabrication and Supercapacitor Performance of CNT@ NiO-Nanosheet Composite Nanotubes. *Nanotechnology* **2016**, *27*, 075605. [CrossRef]

99. Choi, C.; Sim, H.J.; Spinks, G.M.; Lepró, X.; Baughman, R.H.; Kim, S.J. Elastomeric and Dynamic MnO_2/CNT Core–Shell Structure Coiled Yarn Supercapacitor. *Adv. Energy Mater.* **2016**, *6*, 1502119. [CrossRef]

100. Shi, P.; Li, L.; Hua, L.; Qian, Q.; Wang, P.; Zhou, J.; Sun, G.; Huang, W. Design of Amorphous Manganese Oxide@Multiwalled Carbon Nanotube Fiber for Robust Solid-State Supercapacitor. *ACS Nano* **2016**, *11*, 444–452. [CrossRef]

101. Xiang, C.; Li, M.; Zhi, M.; Manivannan, A.; Wu, N. A Reduced Graphene Oxide/Co_3O_4 Composite for Supercapacitor Electrode. *J. Power Sources* **2013**, *226*, 65–70. [CrossRef]

102. Zhou, W.; Liu, J.; Chen, T.; Tan, K.S.; Jia, X.; Luo, Z.; Cong, C.; Yang, H.; Li, C.H.; Yu, T. Fabrication of Co_3O_4-Reduced Graphene Oxide Scrolls for High-Performance Supercapacitor Electrodes. *Phys. Chem. Chem. Phys.* **2011**, *13*, 14462–14465. [CrossRef] [PubMed]

103. Li, X.; Zhang, L.; He, G. Fe_3O_4 Doped Double-Shelled Hollow Carbon Spheres with Hierarchical Pore Network for Durable High-Performance Supercapacitor. *Carbon* **2016**, *99*, 514–522. [CrossRef]

104. Wang, R.; Ma, Y.; Wang, H.; Key, J.; Brett, D.; Ji, S.; Yin, S.; Shen, P.K. A Cost Effective, Highly Porous, Manganese Oxide/Carbon Supercapacitor Material with High Rate Capability. *J. Mater. Chem. A* **2016**, *4*, 5390–5394. [CrossRef]

105. Liu, T.; Zhang, L.; You, W.; Yu, J. Core–Shell Nitrogen-Doped Carbon Hollow Spheres/Co_3O_4 Nanosheets as Advanced Electrode for High-Performance Supercapacitor. *Small* **2018**, *14*, 1702407. [CrossRef] [PubMed]

106. Dong, X.C.; Xu, H.; Wang, X.W.; Huang, Y.X.; Chan-Park, M.B.; Zhang, H.; Wang, L.H.; Huang, W.; Chen, P. 3D Graphene–Cobalt Oxide Electrode for High-Performance Supercapacitor and Enzymeless Glucose Detection. *ACS Nano* **2012**, *6*, 3206–3213. [CrossRef]
107. Chen, W.; Rakhi, R.B.; Hu, L.; Xie, X.; Cui, Y.; Alshareef, H.N. High-Performance Nanostructured Supercapacitors on a Sponge. *Nano Lett.* **2011**, *11*, 5165–5172. [CrossRef] [PubMed]

Communication

Sustainable Treatment for Sulfate and Lead Removal from Battery Wastewater

Hong Ha Thi Vu [1,2], Shuai Gu [1], Thenepalli Thriveni [1], Mohd Danish Khan [3], Lai Quang Tuan [3,4] and Ji Whan Ahn [1,*]

1   Center for Carbon Mineralization, Mineral Resources Division, Korea Institute of Geoscience and Mineral Resources, 124 Gwahak-ro, Gajeong-dong, Yuseong-gu, Daejeon 34132, Korea
2   Faculty of Biotechnology, Chemistry and Environmental Engineering, Phenikaa University, Hanoi 100000, Vietnam
3   Resources Recycling Department, University of Science and Technology, 217, Gajeong-ro, Yuseong-gu, Daejeon 34113, Korea
4   Tectonic and Geomorphology Department, Vietnam Institute of Geoscience and Mineral Resources (VIGMR), 67 Chienthang Street, Hadong district, Hanoi 151170, Vietnam
*   Correspondence: ahnjw@kigam.re.kr

Received: 30 April 2019; Accepted: 18 June 2019; Published: 26 June 2019

Abstract: In this study, we present a low-cost and simple method to treat spent lead–acid battery wastewater using quicklime and slaked lime. The sulfate and lead were successfully removed using the precipitation method. The structure of quicklime, slaked lime, and resultant residues were measured by X-ray diffraction. The obtained results show that the sulfate removal efficiencies were more than 97% for both quicklime and slaked lime and the lead removal efficiencies were 49% for quicklime and 53% for slaked lime in a non-carbonation process. After the carbonation step, the sulfate removal efficiencies were slightly decreased but the lead removal efficiencies were 68.4% for quicklime and 69.3% for slaked lime which were significantly increased compared with the non-carbonation process. This result suggested that quicklime, slaked lime, and carbon dioxide can be a potential candidate for the removal of sulfate and lead from industrial wastewater treatment.

Keywords: sustainable; battery wastewater; sulfate removal; lead removal

---

1. Introduction

Currently, lead–acid battery is an important industry in the world and has been commonly employed as secondary sources of energy due to its low cost, high energy density, high specific energy, high-rate discharge capability, and safety [1,2]. The lead–acid battery is generally used in vehicles as an energy storage device, backup power supply, and stationary applications [3–7]. However, lead–acid batteries have a fundamental disadvantage of low life expectancy. The life expectancy of the lead batteries is no more than five years. Thus, a huge amount of batteries become expired and are discarded every year. In a spent lead–acid battery recycling plant, the acid electrolyte is regularly gathered and assign to further purification [8]. However, the spent electrolyte is discharged and collected into a tank in a normal disassembling lead–acid battery recycling company thus the used acidic electrolyte cannot recycle by purification because the spent electrolyte was contaminated with pollutants in the tank.

The high lead concentration and low pH, due to the presence of excessive sulfuric acid in the solution perform, comprise the main contaminations of battery wastewater (BWW). The average concentration of lead in wastewater is about 3–15 mg/L and the pH of wastewater falls in the range of 1.6-2.9 [9]. If the battery wastewater is not treated well before discharge to environment, lead can contaminate food and water, and be present in nature. This can cause extreme harm to human health such as damage to organs, kidney, and nervous systems [10,11]. Moreover, high sulfate concentration in drinking water can be responsible for serious cases of diarrhea and dehydration to human and animals [12]. The industrial wastewater is limited to under 500 mg/L sulfate in the legislation of many countries [13]. Several research have been reported for efficient removal of lead and sulfate from wastewater such as ion exchange [14–16], electrochemical reduction [17], activated carbon [18,19], biosorption [20], adsorption [21–23], and membrane separation [24]. The efficiency of these methods depends on the lead and sulfate concentration, pH, the economics involved, and social factor set in the standard of government agencies. Nevertheless, these methods are expensive and problems related to the disposal of concentrate or regenerating solution. Besides, the chemical precipitation method has low cost, simplicity, and high removal efficiency of pollutants. Of the precipitation methods, the hydroxide precipitation is the most widely used method. This process involves increased pH to convert heavy metal ions to their respective hydroxides which can be easily removed by filtration [25–28].

This research proposes a low-cost method for treating spent lead–acid battery wastewater by quicklime and slaked lime which are generally cheap due to their abundance in nature. The precipitation method is used to efficiently remove sulfate and lead from the wastewater. In addition, carbon dioxide gas was bubbled into the reaction to increase lead removal efficiencies as well as reduce the pH value to about 7 to meet relevant standards of environmental regulations. Furthermore, the utilization of CO_2 in the carbonation method also contributes to the international targets on greenhouse gas reduction.

2. Materials and Method

2.1. Materials

The quicklime CaO and slaked lime $Ca(OH)_2$ were collected from a company in Republic of Korea. The composition of quicklime and slaked lime are shown in Table 1. In both samples, the dominant component is CaO with 91.25 wt % for quicklime and 66.63 wt % for slaked lime. The raw lead–acid battery wastewater sample was generated from a lead–acid battery company and kept in plastic bottles. The battery company had no recycling system; therefore, the sulfuric acid from the used lead–acid battery was directly poured into a storage tank. The main contaminated compositions in the wastewater were sulfate and lead (Table 2). Carbon dioxide (CO_2) used was of 99.9% purity and obtained from Chungang Gas Company in Republic of Korea.

Table 1. Chemical composition (by mass) of quicklime and slaked lime (wt %).

Compound Name	SiO_2	Al_2O_3	Fe_2O_3	CaO	MgO	K_2O	Na_2O	TiO_2	MnO	P_2O_5	Ig.Loss
Quicklime (CaO)	1.98	0.31	0.38	91.25	1.27	0.05	<0.02	0.02	0.02	0.01	4.75
Slaked lime $Ca(OH)_2$	1.84	0.33	0.40	66.63	3.36	0.08	<0.02	0.02	0.02	0.02	26.96

Ig.Loss: Loss on ignition.

Table 2. Spent lead–acid wastewater compositions.

Parameter	Value
Sulfate (mg/L)	147,000
Lead (mg/L)	3.01
pH	−0.52

2.2. Experiment

The experiments were carried out in a 500 mL reactor with two-wings-oar-type impeller rotating at 400 rpm at room temperature. A glass electrode of pH meter was dipped into solution in the reactor for measuring pH of the mixture during a reaction. Then, 200 mL raw spent lead–acid battery wastewater was slowly added into the 500 mL reactor containing 200 mL of calcium oxide CaO with concentration 2.6 M under vigorous stirring. The reaction was kept stirred until a homogenous mixture was obtained at pH 12 and 90 minutes for precipitation of the SO_4^{2-}. In order to separate residual sludge from residual solution, the suspension was filtered under vacuum using a paper filter, and then the solid residue was washed several times by distilled water. The residual solution was further filtered through 0.45 μm Hyundai Micro syringe filter and was diluted to 500 mL with distilled water for further ion chromatography and inductively coupled plasma-optical emission spectrometry measurements' purpose. The solid sludge was dried in an oven at 100 °C for overnight. The slaked lime ($Ca(OH)_2$) sample was also mixed with battery wastewater using a similar methodology to compare results with quicklime.

Unlike the non-carbonation process, with carbonation, when reaction reached stable pH (about pH 12), carbon dioxide (CO_2) was injected into the reaction system via porous bubbler with 1 L/min flow rate to further stabilize lead. When the pH was reduced to a range of 6–7 the reaction was stopped. The solid and solution residues were collected to analysis as same way with the non-carbonation process.

The residual sulfate and lead were analyzed by ion chromatography and inductively coupled plasma-optical emission spectrometry. The contamination removal efficiency E (%) was determined using the following equation:

$$E\% = \frac{C_0 - C}{C_0} \times 100$$

where C_0 is the primary concentration of total sulfate or lead (mg L^{-1}) and C is the concentration of total sulfate or lead after treatment (mg L^{-1}).

2.3. Physical Characterization

The chemical composition of quicklime and slaked lime samples were performed using X-ray fluorescence spectrometer (XRF, Shimadzu, Japan). X-ray diffraction (XRD, D/MAX 2200, Rigaku, Japan) with X-ray source of Cu Kα ($\lambda = 0.15406$ nm) in the scan range of diffraction angle 2θ from 10° to 90° was used to identify the crystallization of the lime (CaO), quick lime, and solid residues samples. The pH was measured by Orion Versa Star Pro (Thermo Scientific, USA) pH meter with a glass electrode. Before measuring, the pH meter was calibrated by different buffer solutions for an accurate pH. The concentration of sulfate before and after treatment was measured by ion chromatography (ICS-3000, Dionex). Lead concentration analysis was determined using an inductively coupled plasma-optical emission spectrometry (ICP-OES, Optima 8300, PerkinElmer).

3. Results and Discussion

The pH value of the suspensions indicates the progress of the reaction. In the non-carbonation process, the reaction reaches chemical equilibrium when the pH is stable in the solution. Initially, the pH was increased rapidly within 5 minutes and then increased slowly until equilibrium was attained. The pH surpasses 12 after 90 minutes reaction therefore the sulfate ions were removed and OH^- ion from excessive calcium hydroxide ($Ca(OH)_2$) in the mixture (Figure 1a,c).

In the carbonation process, the mixtures reached equilibrium after 90 minutes of reaction time. Then CO_2 gas was injected into the suspension through porous bubbler. The pH value of the suspension decreased because hydroxide OH^- ion was reduced due to the reaction between carbon dioxide with calcium hydroxide and calcium carbonate was formed (Figure 1b,d). When the pH of the mixture was about 7, which was suitable for a number of applications, carbon dioxide gas was stopped being injected into the solution.

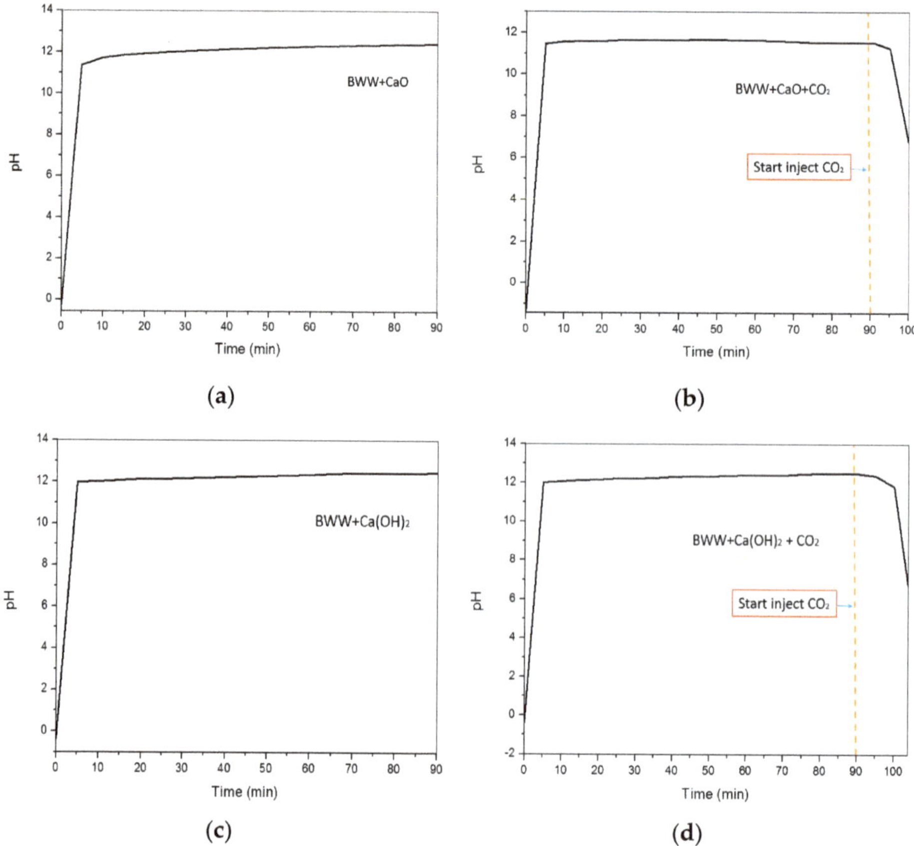

Figure 1. Changes in pH values at different time for (**a**) battery wastewater (BWW) + quicklime, (**b**) battery wastewater + quicklime + CO_2, (**c**) battery wastewater + slaked lime, and (**d**) battery wastewater + slaked lime + CO_2.

Figure 2 shows XRD patterns of quicklime and slaked lime raw samples. The raw quicklime is highly crystalline and all the observed diffraction peaks are good agreement with cubic lime phase CaO of space group Fm-3m (PDF#99-0070) (Figure 2a). The XRD peaks at $2\theta = 32.2°$, $37.3°$, $53.8°$, $64.1°$, $67.3°$, $79.6°$, and $88.5°$ corresponded to (111), (200), (220), (311), (222), (400), and (331) plane of calcium oxide phase, respectively. The dominant peak is observed at (200) plane. For the slaked lime case, the major XRD peaks resembled very well the characteristic peaks of portlandite phase with space group P-3m1 (PDF#87-0673) [29]. The peaks at 18.0°, 28.7°, 34.1°, 47.1°, 50.8°, 54.3°, 56.1°, 62.6°, 64.2°, and 77.7° 2θ were assigned to the (001), (100), (101), (102), (110), (111), (003), (021), (013), and (004) plane of hexagonal portlandite $Ca(OH)_2$ phase, respectively. Besides, a minor peak was detected at $2\theta = 29.3$ which corresponds to (104) plane of rhombohedral calcite phase with space group R-3c since the portlandite $Ca(OH)_2$ reacted with CO_2 gas from surrounding air and calcite formed [30].

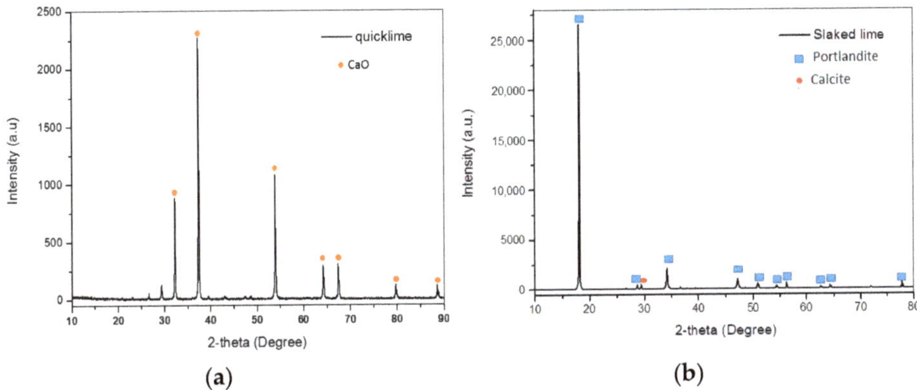

Figure 2. X-ray diffraction (XRD) patterns of (**a**) quicklime and (**b**) slaked lime.

Regarding the identification of crystal structure of solid residues after reaction, the samples was investigated by X-ray diffraction, as shown in Figure 3. Figure 3a,c show the XRD patterns of white precipitation of the non-carbonation process. Four substances can be observed which are gypsum, calcium hemihydrate ($CaSO_4(H_2O)_{0.5}$), calcium hydroxide ($Ca(OH)_2$), and calcium sulfate ($CaSO_4$), respectively. The precipitates were presented portlandite $Ca(OH)_2$ owing to the excessive calcium hydroxide from quicklime or slaked lime in the mixtures.

For the carbonation process of quicklime and slaked lime, gypsum, calcite and calcium sulfate substances were formed during the precipitation process and they can be found in the XRD peaks (Figure 3b,d). The calcium hydroxide was not found in this case due to complete carbonation.

In addition, the XRD patterns of solid residues of both the non-carbonation process and the carbonation process could not detect any peaks related to lead precipitates, such as $Pb(OH)_2$ and $PbCO_3$, because the lead precipitate concentrations may be a very small amount in the solid residues.

The removal of sulfate was dominated by calcium sulfate ($CaSO_4$) precipitation. The results showed that more than 97% sulfate removal efficiencies within 90 minutes contact time were achieved in both quicklime and slaked lime in non-carbonation process. The sulfate removal efficiencies were 97.87 and 97.95% for quicklime and slaked lime, respectively.

Figure 4 and Table 3 show the removal efficiencies of lead in normal process and carbonation step. The lead removal efficiency of quicklime in the carbonation process was 68.4%, which was 38.7% higher than that of the case without carbonation (49.3%). The removal efficiency of lead in slaked lime was 53.5% for the normal process and it was increased to 69.3% for the carbonation process. Both quicklime and slaked lime had almost the same effects on increasing lead removal efficiency in the carbonation process. Chen et al. report that lead removal efficiency was significantly decreased since $Pb(OH)_2$ was dissolved when the pH of the solution was above 10 [25,28]. For non-carbonation, the pH of the mixture was higher than 12 after reaction due to high concentration of OH^- ion from $Ca(OH)_2$. Therefore, lead hydroxide was dissolved back into the mixture. For the carbonation process, the pH of the solution was about 7 after the reaction, thus $Pb(OH)_2$ was stable. In addition, when CO_2 gas was injected into the mixture, at this point the excessive calcium hydroxide was transformed to $CaCO_3$, then lead can be adsorbed on the surface of calcium carbonate and the cerussite $PbCO_3$ precipitate can be generated as lead ion and carbonate ion both exist in the mixture by the following Reactions (1), (2), and (3). For these reasons, the removal efficiency of lead in carbonation process was higher than that in the non-carbonation process.

$$CO_2 + H_2O = H_2CO_3 \tag{1}$$

$$H_2CO_3 = 2H^+ + CO_3^{2-} \tag{2}$$

$$Pb^{2+} + CO_3^{2-} = PbCO_3(s) \tag{3}$$

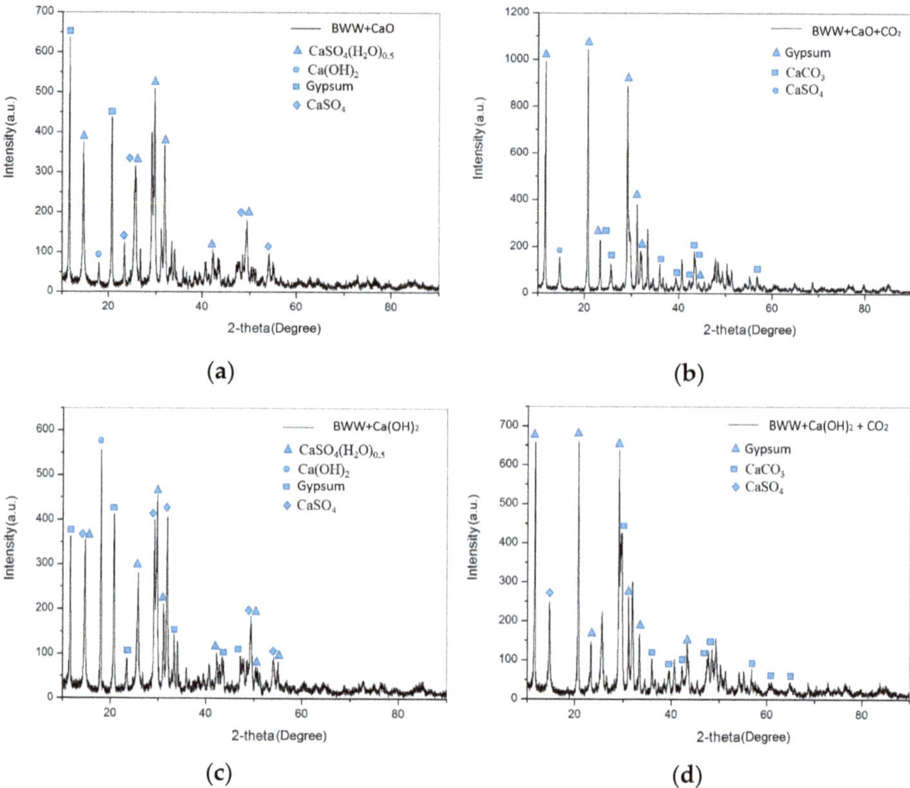

Figure 3. XRD patterns of (**a**) solid residue of battery wastewater + quicklime, (**b**) solid residue of battery wastewater + quicklime + CO₂, (**c**) solid residue of battery wastewater + slaked lime and (**d**) solid residue of battery wastewater + slaked lime + CO₂.

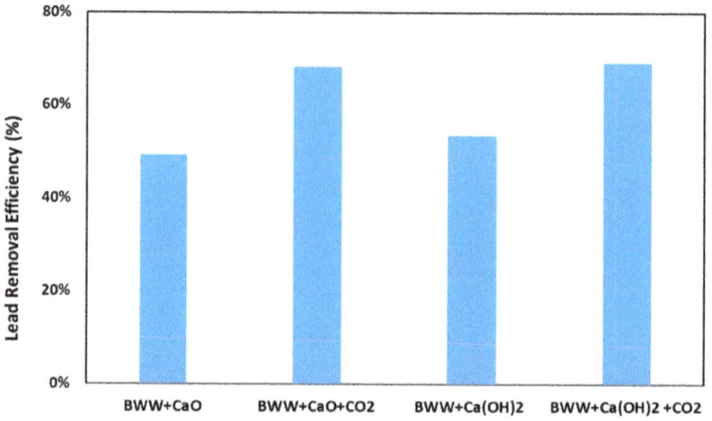

Figure 4. Lead removal efficiency with and without carbonation process.

Table 3. Sulfate and lead removal efficiencies with different experiments.

Experiment	Lead Removal Efficiency
BWW + CaO	49.3%
BWW + CaO + CO_2	68.4%
BWW + Ca(OH)$_2$	53.5%
BWW + Ca(OH)$_2$ + CO_2	69.3%

BWW: Battery wastewater.

4. Conclusions

In this work, the sulfate and lead were successfully treated by using quicklime and slaked lime through a precipitation method. For both quicklime and slaked lime samples, the sulfate SO_4^{2-} removal efficiencies were more than 97% and the lead removal efficiencies were 49% for quicklime and 53% for slaked lime. The removal efficiency of lead was increased after using a carbonation step with 68% for quicklime and 69% for slaked lime. The carbonation process not only enhanced the lead removal efficiency in the battery wastewater but also reduced pH to meet requirements of environmental regulations. In addition, the carbonation method is an environment-friendly method since the technology can utilize CO_2 generated from industry. Due to the environmental benefit, this carbonation technology is recommended as a simple and low-cost method for industrial wastewater treatment.

Author Contributions: H.H.T.V. and S.G. planned and designed the experiment; H.H.T.V. and L.Q.T. carried out the experiments; H.H.T.V. and M.D.K. analyzed the data and wrote the paper; T.T and J.W.A. reviewed and revised the manuscript.

Acknowledgments: This research was supported by the National Strategic Project-Carbon Mineralization Flagship Center of the National Research Foundation of Korea (NRF) funded by the Ministry of Science and ICT (MSIT), the Ministry of Environment (ME), and the Ministry of Trade, Industry and Energy (MOTIE).(2017M3D8A2084752).

Conflicts of Interest: The authors declare no conflicts of interest.

References

1. Murariu, T.; Morari, C. Time-dependent analysis of the state-of-health for lead-acid batteries: An EIS study. *J. Energy Storage* **2019**, *21*, 87–93. [CrossRef]
2. Iwai, T.; Murakami, M.; Takai, S.; Yabutsuka, T.; Yao, T. Chemical transformation of PbO_2 due to local cell reaction on the cathode of lead acid battery. *J. Alloy. Compd.* **2019**, *780*, 85–89. [CrossRef]
3. May, J.G.; Davidson, A.; Monahov, B. Lead batteries for utility energy storage: A review. *J. Energy Storage* **2018**, *15*, 145–157. [CrossRef]
4. Karden, E. Development trends for future automobiles and their demand on the battery. In *Lead-acid Batteries for Future Automobiles*; Garche, J., Karden, E., Moseley, P.T., Rand, D.A., Eds.; Elsevier Science: Aachen, Germany, 2017; pp. 3–25.
5. Cooper, A. Development of a high-performance lead-acid battery for new-generation vehicles. *J. Power. Sour.* **2005**, *144*, 385–394. [CrossRef]
6. Ambrose, H.; Gershenson, D.; Gershenson, A.; Kammen, D. Driving rural energy access: A second-life application for electric-vehicle batteries. *Environ. Res. Lett.* **2014**, *9*, 1–9. [CrossRef]
7. Wagner, R. Stationary application. I. Lead-acid batteries for telecommunication and UPS. In *Industrial Application of Batteries*; Broussely, M., Pistoia, G., Eds.; Elsevier Science: Amsterdam, The Netherlands, 2007; pp. 395–454.
8. Sun, Z.; Cao, H.; Zhang, X.; Lin, X.; Zheng, W.; Cao, G.; Sun, Y.; Zhang, Y. Spent lead acid battery recycling in China- A review and sustainable analyses on mass flow of lead. *Waste. Manag.* **2017**, *64*, 190–201. [CrossRef]
9. Sheng, P.X.; Ting, Y.P.; Chen, J.P.; Hong, L. Sorption of lead, copper, cadmium, zinc, and nickel by marine algal biomass: characterization of biosorptive capacity and investigation of mechanisms. *J. Coll. Interface Sci.* **2004**, *275*, 131–141. [CrossRef] [PubMed]
10. Drinking Water Contaminants—Standard and Regualtions. Available online: http://water.epa.gov/drink/contaminants/index.cfm#Inorganic (accessed on 30 January 2012).

11. Khaoya, S.; Pancharoen, U. Removal of Lead (II) from industry wastewater by HFSLM. *Int. J. Chem. Eng. Appl.* **2012**, *3*, 98–103. [CrossRef]
12. WHO. *Guidelines for Drinking-water Quality*; World Health Organization: Geneva, Switzerland, 2011.
13. Silva, A.M.; Lima, R.; Leão, V.A. Mine water treatment with limestone for sulphate removal. *J. Hazard. Mater.* **2012**, *221–222*, 45–55.
14. Petruzzelli, D.; Pagano, M.; Tiravanti, G.; Passino, R. Lead removal and recovery from battery wastewaters by natural zeolite clinoptilolite. *Solv. Extr. Ion Exc.* **1999**, *17*, 677–694. [CrossRef]
15. Dabrowski, A.; Hubicki, Z.; Podkoscielny, P.; Robens, E. Selective removal of the heavy metal ions from waters and industrial wastewaters by ion-exchange method. *Chemosphere* **2004**, *56*, 91–106. [CrossRef] [PubMed]
16. Guimarães, D.; Leão, V.A. Batch and fixed-bed assessment of sulphate removal by the weak base ion exchange resin Amberlyst A21. *J. Hazard. Mater.* **2014**, *280*, 209–215. [CrossRef]
17. Lin, S.W.; Navarro, R.M.F. An innovative method for removing Hg^{2+} and Pb^{2+} in ppm concentrations from aqueous media. *Chemosphere* **1999**, *39*, 1809–1817. [CrossRef]
18. Acharya, J.; Sahu, J.N.; Mohanty, C.R.; Meikap, B.C. Removal of lead(II) from wastewater by activated carbon developed from Tamarind wood by zinc chloride activation. *Chem. Eng. J.* **2009**, *149*, 249–262. [CrossRef]
19. Singh, C.K.; Sahu, J.N.; Mahalik, K.K.; Mohanty, C.R. Studies on the removal of Pb(II) from wastewater by activated carbon developed from Tamarind wood activated with sulphuric acid. *J. Hazard. Mater.* **2008**, *153*, 221–228. [CrossRef]
20. Saeed, A.; Iqbal, M.; Akhtar, M.W. Removal and recovery of lead (II) from single and multimetal (Cd, Cu, Ni,Zn) solutions by crop milling waste (black gram husk). *J. Hazard. Mater.* **2005**, *117*, 65–73. [CrossRef]
21. Gunay, A.; Arslankaya, E.; Tosun, I. Lead removal from aqueous solution by natural and pretreated clinoptilolite: Adsorptionequilibrium and kinetics. *J. Hazard. Mater.* **2007**, *146*, 362–371. [CrossRef] [PubMed]
22. Doyurum, S.; Celik, A. Pb(II) and Cd(II) removal from aqueous solutions by olive cake. *J. Hazard. Mater.* **2006**, *138*, 22–28. [CrossRef] [PubMed]
23. Runtti, H.; Luukkonen, T.; Niskanen, M.; Tuomikoski, S.; Kangas, T.; Tynjälä, P.; Tolonen, E.T.; Sarkkinen, M.; Kemppainen, K.; Rämö, J. Sulphate removal over barium-modified blast-furnace-slag geopolymer. *J. Hazard. Mater.* **2016**, *317*, 373–384. [CrossRef] [PubMed]
24. Bader, M.S. Separation of chloride and sulphate ions in univalent and divalent cation forms from aqueous streams. *J. Hazard. Mater.* **2000**, *73*, 269–283. [CrossRef]
25. Chen, Q.; Yao, Y.; Li, X.; Lu, J.; Zhou, J.; Huang, Z. Comparison of heavy metal removals from aqueous solution by chemical precipitation and characteristics of precipitates. *J. Water Process Eng.* **2018**, *26*, 289–300. [CrossRef]
26. Mahmood, M.B.; Balasim, A.A.; Najab, M.A. Removal of heavy metals using chemicals precipitation. *Eng. Technol. J.* **2011**, *29*, 1–18.
27. Mallampati, R.; Valiyaveettil, S. Co-precipitation with calcium carbonate—a fast and nontoxic method for removal of nanopollutants from water? *RSC Adv.* **2015**, *5*, 11023–11028. [CrossRef]
28. Azimi, A.; Azari, A.; Rezakazemi, M.; Ansarpour, M. Removal of heavy metals from industrial wastewaters: A review. *Chem. Bio. Eng. Rev.* **2017**, *4*, 1–24. [CrossRef]
29. Vu, H.H.T.; Khan, M.D.; Chilakala, R.; Lai, T.Q.; Thenepalli, T.; Ahn, J.W.; Park, D.U.; Kim, J. Utilization of lime mud waste from paper mills for efficient phosphorus removal. *Sustainability* **2019**, *11*, 1524. [CrossRef]
30. Khan, M.D.; Ahn, J.W.; Nam, G. Environmental benign synthesis, characterization and mechanism studies of green calcium hydroxide nano-plates derived from waste oyster shells. *J. Environ. Manag.* **2018**, *223*, 947–951. [CrossRef]

MDPI

St. Alban-Anlage 66

4052 Basel

Switzerland

Tel. +41 61 683 77 34

Fax +41 61 302 89 18

www.mdpi.com

Sustainability Editorial Office

E-mail: sustainability@mdpi.com

www.mdpi.com/journal/sustainability